Food and the Elements

Proceedings of the Oxford Symposium on Food and Cookery

Series Editor: Mark McWilliams

The Oxford Symposium on Food and Cookery has been held annually for more than 40 years, making it the longest-running food conference in the world. It is held each summer in Oxford, United Kingdom, in recent years at St Catherine's College over a weekend. Also, since Covid, it has been both an in-person event and a global, online conference.

Keynote speakers are invited each year to address a chosen topic and to guide delegates' thinking in new directions. Selected papers from the Symposium are published in an annual volume.

Food and the Elements

Proceedings of the Oxford Symposium on Food and Cookery 2025

Edited by Mark McWilliams

equinox

SHEFFIELD UK BRISTOL CT

Published by Equinox Publishing Ltd.

UK: Office 415, The Workstation, 15 Paternoster Row, Sheffield, South Yorkshire S1 2BX

USA: ISD, 70 Enterprise Drive, Bristol, CT 06010

www.equinoxpub.com

First published 2026

British Library Cataloguing-in-Publication Data

A catalogue record for this book is available from the British Library.

ISBN-13	978 1 80050 921 4	(paperback)
	978 1 80050 922 1	(ePDF)
	978 1 80050 923 8	(ePub)

Library of Congress Cataloging-in-Publication Data

Names: Oxford Symposium on Food & Cookery (2025 : Oxford, England) author | McWilliams, Mark editor
Title: Food and the elements : proceedings of the Oxford Symposium on Food and Cookery 2025 / edited by Mark McWilliams.
Other titles: Proceedings of the Oxford Symposium on Food and Cookery 2025
Description: Bristol, CT : Equinox Publishing Ltd, 2026. | Includes bibliographical references. | Summary: "The annual volume of papers selected from the 2025 Oxford Symposium will be devoted to the exploration of the theme "food and the elements". The papers examine not just the four elements of classical Western philosophy - fire, water, earth, and air - but also the elements of flavor, of chemistry, of culture, of care. In short, all the building blocks of the culinary world"-- Provided by publisher.
Identifiers: LCCN 2026014648 (print) | LCCN 2026014649 (ebook) | ISBN 9781800509214 paperback | ISBN 9781800509221 pdf | ISBN 9781800509238 epub
Subjects: LCSH: Food--History | Food--Social aspects | Four elements (Philosophy) | LCGFT: Conference papers and proceedings
Classification: LCC TX345 .O94 2025 (print) | LCC TX345 (ebook)
LC record available at https://lccn.loc.gov/2026014648
LC ebook record available at https://lccn.loc.gov/2026014649

Typeset by Scribe Inc.

Printed and bound by CPI Group (UK) Ltd, Croydon, CR0 4YY

Contents

Contents

Contents

Foreword

Each year the Oxford Food Symposium ends in raucous democracy, as symposiasts debate the topic for the meeting three years hence. The chaos sometimes produces eerily perfect results, as when 2017 participants voted for 'Food Away from the Table' as the theme for 2020 – when the Symposium was forced online by the pandemic. Other times selection energy coalesces around something more amorphous, like 'Food and the Elements'. I confess I was not clear on what we had just voted for in 2022, but the papers in this volume demonstrate the creativity, rigour, and urgency with which symposiasts defined the topic.

From specific chemical elements to ancient Chinese philosophies, from the elements of the food system to the ancient divisions of fire, air, water, and earth, symposiasts repeatedly returned to problems – and opportunities – caused by imbalance and change. Some of those changes are historical, and symposiasts chronicled shifts in human migration, traced adaptations to dying industries, and corrected historical records where our understanding of the past is incomplete or flat-out wrong. The biggest threats, though, come from anthropogenic climate change, which endangers lives, uproots agricultural practices, and challenges the very idea of terroir. Our ability to confront such dangers is hindered by polarized societies and lives lived online, social changes Claude Fischler fears might undo millennia of shared dining practices.[1]

These are concerns of the greatest scale. And yet symposiasts found cause for hope even in such a moment. Some pointed to efforts to revitalize ancient practices; others to new ways to grow the food we need without increasing planetary destruction. Even politics offers possibilities: Marion Nestle expressed shock at the unexpected, stunning rearrangements of the elements of the food industry and the political environment. Critiques that activists like her have long levelled at government support for ultra-processed foods now come from governments themselves, with the Secretary of the US Department of Health and Human Services using much the same language to express the same concerns.[2] In such a moment, Ole Mouritsen wondered what such radical shifts might mean for accelerating the widespread adoption of plant-based diets.[3]

Symposiasts clung to hope even facing the greatest dangers, the catastrophic impacts of a warming world amid political reversals and radical cuts to humanitarian aid. Those impacts were vividly traced by the UN Food and Agriculture Organization's Théodore Kaboré, who described seemingly unstoppable desertification as climate change accelerates the sweep of the Sahara south into huge, highly populated regions. In the same moment, however, Kaboré explained how the Indigenous peoples of the Sahel are confronting these forces with tools as simple as a hoe. Images as uplifting as any I have seen revealed that they are regreening their land – and at scales best appreciated from space.[4]

Such realistic hope seems to characterize symposiasts, and perhaps that tenacity helps explain why we keep coming back together, both in Oxford and in our virtual meetings. The heart of our gatherings – the core elements of the Symposium – remains the papers themselves: this book, alongside its predecessors, represents the real legacy of the Oxford Food Symposium. For almost ten years now, these volumes have been made possible by generous grants from Tokyo's RINRI Institute of Ethics, testifying to the Symposium's international community and its commitment to lifelong learning. With that backing, the papers in this volume, in print and online, will reach an audience many thousands larger than the capacity of St Catz's Bernard Sunley lecture hall. I thank those who helped see it to press, including Cathy Kaufman, Richard Shepro, David Matchett, Naomi Duguid, Carolyn Steel, Elisabeth Luard, and Janet Beizer, along with Equinox Publishing's Sarah Lee and Janet Joyce. Richly deserving of separate praise is Jake Tilson, whose artwork has made the volumes since 2019 the most striking in our history. The real appreciation goes, of course, to the authors of these papers.

Mark McWilliams
Editor, Oxford Symposium on Food and Cookery

About the Editor

Mark McWilliams is the author of *Food and the Novel in Nineteenth-Century America* and *The Story behind the Dish: Classic American Foods*. He has served as the Symposium's editor since 2011.

Notes

1 Claude Fischler, 'Gastronomy in the Age of Gastro-Anomy', YouTube, 13 July 2025 <https://www.youtube.com/watch?v=O7KSlcuqy6Q> [accessed 15 March 2025].

2 Marion Nestle, 'The Elements: Food, Nutrition, Health, and Politics, 2025 Style', YouTube, 11 July 2025 <https://www.youtube.com/watch?v=7UaVzofG27M> [accessed 15 March 2025].

3 Ole Mouritsen, 'The Elements of a Plant-Forward Cuisine', YouTube, 12 July 2025 <https://www.youtube.com/watch?v=8oggufb3kvI> [accessed 15 March 2025].

4 Théodore Kaboré, 'Agricultural Practices and Community Efforts to Restore the Fertility and Productive Value of Land in the Face of Climate Elements in the Sahel', YouTube, 12 July 2025 <https://www.youtube.com/watch?v=gq1aVe99EMI> [accessed 15 March 2025].

I

Mastering the Elements with Doña Petrona

Culinary Narratives of Fire, Water, Earth, and Air in Argentine and Latin American Households

Felipe Acevedo Riquelme

Introduction

As I write this essay, I have beside me my mother-in-law's copy of *El libro de Doña Petrona*. It is the fifty-eighth edition, published in 1963, and one of the wedding gifts she received in that same decade. What distinguishes this copy is a cloth pouch, which she made herself on a sewing machine when she noticed the spine loosening and the pages beginning to detach. This gesture does not strike me as unusual. It belongs to what might be described as an ethic and aesthetic of domestic care that until recently remained deeply embedded in Chilean households.

With great care, I remove the pouch and begin to skim through the book before pausing to read. It does not take long before I come upon a pencilled note written next to an illustrated guide on how to make 'little bunnies' from hard-boiled eggs. The annotation – 'Conejitos', handwritten in soft lead at the top of the page in Spanish – seems to fill a gap where a printed title perhaps once belonged or was never placed. But before I follow this speculative path, I am struck by how this handwritten note triggers another image: the faint pencil marks my mother used to make on recipes kept in a blue folder in the kitchen cabinet – recipes I now keep in a drawer since her passing. The connection between these domestic culinary writings becomes so vivid in my mind that I begin to notice a resemblance between their penmanship.

It is here that I become aware of the tricks memory can play on us – and of how the connections I propose in this essay invite a weaving of readings that do not necessarily depend on explicit or material links between texts. My mother and my mother-in-law never met. And yet, a link emerges between them through the possible readings of their shared participation in a domestic network of culinary knowledge transmission, which I attempt to trace throughout this paper. The pencil markings of both women speak of culinary knowledge as a dynamic process, and in that sense, I propose a way of reading this archive formed by what Amy L. Tigner and Allison Carruth have called 'vernacular literary practices': 'quotidian modes of writing that develop within specific historical contexts and that intermix rhetorical and aesthetic craft with the dissemination of applied knowledge that is variously empirical, sensory, instructive, interactive, and intergenerational'.[1]

This convergence of annotations, hands that both cook and write, and inherited texts is the point of departure for a broader inquiry: how do culinary texts such as *El libro de Doña Petrona* encode knowledge that is at once elemental (literally and figuratively) and transgenerational? First published in 1933 and continuously reedited throughout the twentieth century, Doña Petrona's cookbook is often regarded as the cornerstone of modern Argentine domesticity. But beyond its recipe content, it functions as a cultural artifact whereby the kitchen becomes a stage for negotiating modernity itself as women learn to manage a new world of shifting social meanings in domestic work.

As I handle the book, what stands out are the marks of its use: browned edges that have once hovered too close to a flame or water stains that trace accidents long absorbed into routine. Even the faint lift of air when a page turns feels like a reminder of movement in the kitchen in rhythm with the heat of the burners. These traces are not incidental. They record the ongoing negotiation between the body and the material world, with the kitchen as a place where touch and cognition fold into one another. In this sense, the cookbook is not only a repository of instructions but a record of living practices reshaped through years of repetition.

Thus, this essay proposes a reading of *El libro de Doña Petrona* that focuses on the symbolic and material presence of the classical elements – fire, water, earth, and air – as they shape culinary techniques, household practices, and modes of feminine authorship. Each element will be considered in both its literal application in the kitchen and its metaphorical function in the construction of cultural narratives. This book not only instructs on how to roast a cut of beef, knead dough, and beat egg whites: it stages the transmission of knowledge through flame, steam, soil, and breath.

This particular edition, published in 1963, belongs to a transitional moment when gas and log stoves still coexisted in many Latin American households. It called my attention precisely because Doña Petrona addresses this shift in her instructions, offering guidance not only for different technologies but for different temporalities of cooking. In doing so, she enacts a feminine epistemology anchored in practice and care. Doña Petrona C. de Gandulfo is not merely an author or cook – she becomes a guide of the domestic sphere, leading her readers through an initiation into mid-century culinary modernity. Her instructions mediate the transition from wood stoves to gas ranges, from rural improvisation to urban standardization, from hand-me-down wisdom to printed authority. And yet, the margins of the text – handwritten annotations, worn covers, food stains – speak to a continued layering of knowledge outside the printed page, within what may be understood as an archive of vernacular domestic literacy.

To trace this multilayered archive, I attempt a close reading of *El libro de Doña Petrona* as both a practical manual and a literary document. This approach includes an analysis of visual materials (illustrations, *mise-en-page*), specific recipes, and paratextual elements, alongside historical context and gender theory. The work of Tigner and Carruth offers a crucial conceptual tool here: culinary texts as vernacular literary practices. Following this line of thought, I read Gandulfo's work in relation to the domestic spaces that shaped it, attending to how techniques learned by doing take on cultural meaning over time. The elemental practices in the cookbook – managing heat, tending bubbling liquids, working doughs, etc. – register the broader transformations of mid-twentieth-century Argentina, where new appliances, new ideals of efficiency, and shifting expectations of women's labour reshaped the kitchen as an imaginative workspace.

Far from being solely a site of domestic confinement, the kitchen emerges here as a generative space of knowledge production and community formation. Mastery of the elements, knowing when to stir or when to wait, becomes more than just a technical skill, as it is also a way of engaging with the world that echoes the intellectual legacy of figures like Sor Juana Inés de la Cruz, who once asked: '¿Qué podemos saber las mujeres sino filosofías de cocina?' ('What can we women know but kitchen philosophies?').[2] In Doña Petrona's work, as in Sor Juana's anecdote, the domestic sphere becomes a site of experimental inquiry, where culinary practice enacts a form of embodied reasoning. Doña Petrona's pedagogical voice thus invites her readers into a shared epistemic terrain, where cooking is not merely a form of care or survival but a mode of cultural authorship and intergenerational transmission – a vernacular philosophy articulated in flour, fire, and annotated margins in a similar manner as Megan J. Elias points out when referring to these pieces of evidence of the lives of countless women.[3]

Fire: Culinary Alchemy and the Feminine Mastery of Heat

In the classical schema of the elements, fire is the force of transformation – intensely powerful, ambivalent, and foundational to the act of cooking. Within *El libro de Doña Petrona*, fire occupies a central role as a medium that must be directed through practice. It becomes the site where domestic skill is tested and where the figure of the cook asserts authority through timing and care. Doña Petrona C. de Gandulfo appears here as a caring instructor who also shaped the conditions under which women learned to work with new kitchen technologies and to understand themselves as competent within them. In this sense, her figure invites comparison with Betty Crocker – the corporate creation of General Mills in the United States – who similarly instructed generations of women in mastering the domestic sphere.[4] Yet unlike Crocker, whose authority was imagined and disembodied, Doña Petrona was both present and performative: a real woman whose public demonstrations and media presence made her a charismatic conduit of culinary instruction. As Barbara Haber and Arlene Voski Avakian note, figures like Crocker represented a cultural ideal of postwar femininity, linking domestic competence to national identity and consumer responsibility.[5] Doña Petrona, shaped by different socioeconomic dynamics, similarly naturalized the link between fire and feminine modernity, making the kitchen a harmony of tradition and transformation.[6]

Doña Petrona's instructions are saturated with a pedagogical sensibility that frames fire as a resource to be harnessed with precision and purpose. Her repeated emphasis on oven efficiency – 'Hay que aprovechar el tiempo del horno prendido' ('Take advantage of the oven time while it's already on') – functions as technical guidance expressed within daily kitchen practices. In this logic, heat must be used with care and directed to where it is most needed across multiple dishes to avoid waste. This attention to the orchestration of fire aligns with her role as *ecónoma* for the Argentine subsidiary of British Gas, a position that placed her at the intersection of corporate modernization campaigns and domestic labour. As Rebekah E. Pite notes, Gandulfo's early publications and media appearances aimed to facilitate the transition from wood-fired stoves to gas-powered ranges, positioning her as a mediator between technological change and everyday practice.[7]

This domestication of fire in Doña Petrona's kitchen echoes a much older and broader human story – one that links the control of fire with the origins of culture itself. Michael Pollan, in his chapter on fire in *Cooked*, links the taming of fire to the very emergence of human culture, quoting Richard Wrangham's hypothesis that cooking with fire contributed to the evolutionary leap of *Homo sapiens*.[8] While Pollan focuses on the pitmaster and the male figure at the grill, Doña Petrona's relation to fire takes shape in the repetition of everyday cooking. She teaches by showing how heat is brought into rhythm with the pace of the kitchen: when to wait, when to turn, when to move on. Her authority rests in this accumulated experience that she shares through the directness of her instruction.

Yet fire is also messy, visceral, and gendered in ways that extend beyond the printed page. As Begoña Alberdi argues in her study of *Mil fórmulas de cocina 'La Negra'*, the meat industry in early twentieth-century Argentina relied on women's labour to handle the most physically intense and polluting aspects of food preparation work, which involved handling animal remains and maintaining heat for long periods in enclosed, physically demanding environments. In these settings, heat shaped the pace and strain of labour. It was present in the rendering of fat or the treatment of by-products, but it rarely appeared as an element associated with skill or domestic refinement. Instead, it marked the effort exerted by women whose work remained largely unacknowledged in the public narratives of cooking.[9] While Doña Petrona's readers – largely white, urban, and middle class – were encouraged to see fire as an ally of cleanliness and control, in the slaughterhouse and canning lines, fire signified danger, excess, and the abject. Alberdi notes that *La Negra*'s cookbooks sanitized this labour through aesthetics and branding, offering white homemakers canned goods 'prepared' by an idealized Black domestic figure while occluding the actual conditions of the working-class women who handled the heat. A similar operation occurs in the Chilean cookbook by Lawe, *La Negrita Doddy* (1911), where a fictionalized Black woman is positioned as the bearer of culinary wisdom yet stripped of a historical voice or labouring body. Unlike the embodied pedagogical presence of Doña Petrona, Doddy functions as a reassuring symbol of domestic knowledge, racialized and aestheticized for a white readership, without participating in the epistemic authority granted to named, visible authors.

Against this backdrop of industrial heat and invisibilized labour, *El libro de Doña Petrona* offers a counter-stage – one in which fire is reimagined through discipline, aesthetics, and domestic authorship. The cookbook, then, becomes a stage where the symbolic meanings of fire are reworked. Doña Petrona's visual materials – colour plates, photographs, and diagrams – construct a language of control and beauty, where the element is disciplined into soufflés, roasts, and elaborate moulded puddings. Her pedagogical tone and technical detail elevate the labour of cooking from the realm of intuitive know-how to a form of formal instruction, echoing the logics of factory management even as she domesticates them. Doña Petrona draws on techniques associated with efficiency and standardization, but her authority emerges through performativity rather than managerial mechanization, marking her work as inhabiting the kitchen rather than aligned with industrial oversight.

But fire, in Doña Petrona's hands, also serves as a metaphor for empowerment – though one that is shaped by the racialized and classed hierarchies of mid-twentieth-century Argentina. Her mastery of the oven, the flame, and the timing of heat becomes a way of asserting authority within a domestic domain historically confined to the private sphere and devalued as routine. Yet this reconfiguration of culinary labour into epistemic authority is one accessible largely to her white, urban, middle-class readership. Her pedagogy does not reject the elemental power of fire; rather, it feminizes it by transforming it into a medium of collective instruction and domestic authorship. As explored earlier through Sor Juana Inés de la Cruz's reflection on women's kitchen philosophies, Doña Petrona likewise claims the hearth as a space of intellectual and cultural production. Drawing on the framework developed in my doctoral research, this feminization of fire entails its containment within a logic of care, order, and repetition – a culinary epistemology in which heat is not wild or excessive but choreographed, sequenced, and made teachable. While far from subaltern, Doña Petrona's role as a guide and technician of fire exemplifies a form of feminine modernity in which cooking becomes not merely a labour of love but a method of regulated, transmissible knowledge.

Thus, in this elemental reading, fire is not only about combustion; it is about transformation. It alters ingredients, but it also reshapes domestic hierarchies, pedagogical models, and imaginaries of feminine modernity. Through Doña Petrona, fire becomes a form of cultural authorship: the hearth as a printing press, the recipe as an act of inscription, the oven as a crucible of tradition and innovation.

Water: Broths, Bonds, and the Feminine Syntax of Sustenance

If fire enacts the visible drama of transformation, water sustains the quiet labour of continuity. In *El libro de Doña Petrona*, water manifests not only in its material role – as the basis for soups, stews, and gelatinous preparations – but also as an epistemic and relational force: it carries embodied culinary knowledge and enables the slow, attentive forms of care that shape everyday domestic relations. Through boiling, poaching, and emulsifying, water enables a culinary logic of absorption and patience. It preserves rather than destroys. As Claude Lévi-Strauss classically proposed, the contrast between 'exo-cuisine' and 'endocuisine' aligns with the opposition between dry, direct flame (public, masculine) and wet, contained cooking (private, feminine).[10] In Doña Petrona's recipes, this endocuisine becomes a narrative mode, one that mirrors the rhythms of care and continuity, as recipes unfold in measured steps, where ingredients are coaxed into union through gentle simmering rather than violent searing.

Doña Petrona's approach to water-based cooking exemplifies a kind of culinary literacy rooted in care and composition. Techniques such as steaming, boiling, and braising require attentiveness and embodied memory: knowing how much water to use, how to time the reduction of liquid, and how to test a spoonful by scent and viscosity, as she carefully describes in the process of making sugar syrup for different purposes.[11] The wet cooking pot functions like a second stomach – breaking down and softening food in ways that extend digestion outside the body. It is a vessel not of spectacle but of transformation through absorption. This also implies the potential of the container that is able to conserve more nutrients, in an image that well evokes a nourishing, protective, warm maternal and primeval womb, as Sonia Montecino describes the pot in *La olla deleitosa*.[12] In Doña Petrona's practice, soups and stews open the meal by preparing the palate and the body. The movement within the pot is gradual, guided by small adjustments in temperature and taste. Her instructions call for staying with the pot as it changes, paying attention to how the liquid thickens and how the aroma deepens, so the cook responds to the dish as it is cooked in real time.

This logic also governs the syntax – or structured sequencing – of Doña Petrona's recipes. In her water-based dishes, cooking becomes a grammar of transformation, where each ingredient must enter at the right moment, in the right form, to achieve a cohesive whole. The structure of these recipes requires attentiveness to timing, layering, and consistency – a choreography of flavours that produces more than the sum of its parts. The recipes demand both mimetic knowledge and linguistic mediation: the cook reads, tastes, watches, and remembers. In this sense, the written recipe is always haunted by the body that once enacted it. It is here that Gandulfo's cookbook most powerfully intersects with the oral and gestural transmission of knowledge described by Sonia Montecino and Luce Giard: the flick of a wrist, the way a mother teaches her daughter to test doneness by smell or consistency, the invisible gestures inscribed in the kitchen's heat and steam.[13]

Water, then, becomes the element of transmission and community. Its slow processes echo the maternal temporality of feeding others daily, often without recognition.[14] The 'female syntax' of care becomes the architecture of recipes in what Montecino highlights as a grammar of food. This is also where the cookbook as an object becomes an archive: splattered, annotated, worn at the margins. These marks – pencil notes, water stains, flour residue – bear witness to women's relational labour, to the hands that turned words into nourishment, and to the embodied knowledge that circulates beneath the written surface.

By reading *El libro de Doña Petrona* through the element of water, we are invited to reimagine the domestic as a site not of limitation but of deep epistemological work. Cooking in water – as Montecino, Pollan, and Lévi-Strauss suggest – might seem humble or even mundane, but it is precisely in the ritual of boiling, braising, and blending that another kind of power resides: the slow forging of taste, memory, and intergenerational solidarity. Water is not flashy. It does not sear or scorch. It permeates and preserves.

Earth: Ingredients, Labour, and the Material Imagination

If water sustains, earth provides. It grounds the culinary arts in substance, weight, and origin. In *El libro de Doña Petrona*, earth appears through the raw materials that come from the soil – tubers, legumes, grains, herbs, and vegetables – but also through the embodied labour and sensory work required to transform these materials into nourishment. Earth is the realm of preparation: peeling, grinding, kneading, touching. It is the element of proximity – what clings to the hands, stains the apron, powders the kitchen tiles. Through this material closeness, Doña Petrona's cookbook articulates a relationship to food that is not abstract but haptic, bodily (because, as Le Breton puts it, it emerges through the conjugation of the senses), and profoundly gendered.[15] This is illustrated as Gandulfo recalls 'the forgotten art of meat carving' that was performed once by men, but now it has become the duty of women to carry on with this important tradition (p. 29).

From the perspective of the new materialisms, earth resonates with Jane Bennett's notion of 'vibrant matter': the idea that things – food, tools, ingredients – take part in the work of shaping how cooking feels, tastes, and unfolds.[16] In Doña Petrona's text, potatoes respond to heat and pressure, and flour changes under moisture and handling, so the cook learns from what these ingredients do as they change. Recipes instruct alongside measurements; these recipes teach how to gauge with the hands – how to feel when a dough is ready, how to test a puree by its grain, how to manipulate form with the fingers (p. 37). These gestures are not trivial: they encode memory and identity, the habits passed across kitchens, and the persistence of lineage through embodied repetition.[17]

Doña Petrona's work also reflects this entanglement. Her recipes move between fresh and preserved ingredients, between handcrafted dishes and those made from prepackaged products. What seems like a celebration of domestic efficiency – the use of blenders, standardized measurements, oven temperatures – also testifies to a shift in culinary temporality: from the intuitive and seasonal to the quantified and eternal. The 'earth' in her book is not untouched nature but a curated substratum – a set of ingredients and practices already selected, processed, and presented through the lens of modern food systems – already mediated by markets, corporations, and gendered expectations.

Still, it is precisely through these tensions that the cookbook makes visible a unique feminine material imagination. Some of Doña Petrona's recipes include sculptural, decorative elements – 'pollitos', 'conejitos', and 'chinitos' made from eggs and vegetables – each a miniature composition that blends play, pedagogy, and sensory appeal, explained through photographs for better clarity. These are not frivolous embellishments; they are material acts of expression and instruction, translating earthbound ingredients into edible symbols. Their presence affirms the creative labour of women in a tradition often dismissed as unskilled or repetitive (p. 160).

Earth, then, in Doña Petrona's culinary writing, is more than a source of food. It is a surface of inscription. It shows like a dough the pressure and movement of work done on it, carrying traces of repeated contact. The cookbook becomes a sedimentary archive, layered with what has accumulated on its pages through cooking and note-taking over time. Through this elemental lens, *El libro de Doña Petrona* reveals how groundedness – far from being static – is a dynamic relation between the material surface of the kitchen and the lives that shape it.

Air: Imagination, Instruction, and the Breath of Culinary Discourse

Among the classical elements, air is the most elusive. It cannot be held or contained; it circulates, surrounds, and slips through. In *El libro de Doña Petrona*, air is not only a necessary medium for leavening, steaming, or aerating – it is also the element of language, imagination, and instruction. It is the breath of the teacher – Doña Petrona's assertive, demonstrative tone – and the rhythm of her recipes, which guide readers through a carefully paced culinary script. In this sense, air represents not just technique but inspiration in the fullest sense of the word: from the Latin *inspirare*, to breathe into. Her cookbook offers a setting where ideas are introduced in language before they are tried at the stove. The reader encounters a voice that explains, demonstrates, and sets the pace for the work to come. The connection between teacher and reader emerges through this shared tempo, carried from the page into the kitchen.

Cookbooks are rarely solitary performances of authorship. They are vernacular epistemologies woven through communal gestures and familial rhythms – texts that evolve through the repetition of cooking and the circulation of knowledge within the home. As Megan J. Elias observes, the 'little direct evidence of cooking that occurs in physical cookbooks comes in the form of such personal annotations and stains', which 'can be read for clues to women's lives'.[18] These marks, such as stains, notes, and altered instructions, testify to a collective domestic labour that extends beyond the author's voice, so the cookbook becomes a living document shaped by the hands and rhythms of its readers.

Air is also the element of imagination and, in Doña Petrona's case, of performance. Her cookbook contains over seventy-six colour plates in the fifty-eighth edition, alongside black-and-white illustrations and decorative flourishes. These images do more than instruct; they invite. They construct a world of aspiration – of proper taste, good homemaking, and domestic elegance. The image helps bridge the gap between the real and the ideal, between what is done and what might be done. It blows a breath of life into the static recipe, allowing the reader to envision a successful outcome and, more crucially, a competent self. Additionally, the hand gesture imprinted in the colour plates illustrations somehow reminds us of the handmade sensorial aspect of creativity materialized.

But air, in *El libro de Doña Petrona*, also carries traces of social positioning. Like the soufflé or the whipped cream, success depends on knowing how not to collapse. These recipes demand attention to how the mixture holds its shape and how it responds to touch – techniques that metaphorically echo the ideal of middle-class femininity circulating in Gandulfo's work: held in a state that appears effortless, always at risk of deflation if mishandled. As Rebekah E. Pite has shown, Doña Petrona's published and televised persona was instrumental in shaping a modern Argentine domestic ideal – one that required women to master the physical setting of the kitchen, the skills required to work within it, and the demeanour expected while doing so.[19] Megan J. Elias similarly observes that mid-century cookbooks placed the responsibility for both culinary performance and social harmony squarely on women's shoulders, requiring them to be at once creative, competent, and graceful under pressure.[20] In this context, the lightness of whipped air becomes a metaphor for gendered aspiration: a carefully sustained form of domestic elegance that is the result of repeated practice rather than an inherent quality.

This management of tone also has gendered implications. As Janet Theophano and Megan J. Elias suggest, the feminine culinary voice in cookbooks has historically navigated a delicate balance between intimacy and authority, between domestic warmth and technical precision.[21] Theophano's analysis of handwritten and community cookbooks highlights how women used culinary writing to share recipes while also establishing relational forms of knowledge that blend affection with instruction. Doña Petrona's tone negotiates this balance with remarkable dexterity: she is firm but not cold, precise but not aloof. In this way, air becomes a mode of relationality by linking the person who writes to the person who cooks.

Finally, air evokes breath – and by extension, continuity: the unbroken rhythms of culinary practice as they unfold over time through bodies, voices, and rituals. The cookbook, read aloud or followed in silence, moves in rhythm with the breath of the cook. Instructions are taken in like inhalations, recipes performed like exhalations. In a symbolic sense, *El libro de Doña Petrona* is animated by breath – by voices that speak across generations, by annotations left in margins, by the sighs and laughter of shared meals. It is a book filled with the air of kitchens: fragrant, warm, noisy, alive.

In this elemental schema, air reminds us that cooking is not only a material act – it is also linguistic, performative, and aspirational. It rises. It invites. It connects. And in Doña Petrona's work, it circulates through word, image, and gesture, sustaining a vernacular community of knowledge that floats – sometimes invisibly, but always vitally – above the flame and the pot.

Conclusion

In this essay, I have proposed an elemental reading of *El libro de Doña Petrona* as a site where culinary practice, memory, gender, and cultural authorship intersect. By tracing the presence of fire, water, earth, and air in both literal and metaphorical registers, I have argued that Doña Petrona's cookbook is not merely a domestic manual but a work of vernacular epistemology – one in which embodied knowledge is articulated through technique and care. *El libro* appears instead as a text that continues to accumulate traces of use, marked by the handling and adjustments of those who have cooked with it over time, carrying forward ways of working and remembering in the kitchen.

Doña Petrona's work reflects a uniquely Latin American mode of culinary modernity – one shaped by class, race, and gender hierarchies but also by women's creative negotiation of these structures through the kitchen. Her performative authority as teacher, her use of illustrations, and her pedagogical tone together constitute a form of feminine modernity that is both aspirational and practical. In a context where culinary labour has often been dismissed as invisible or unskilled, Doña Petrona's cookbook makes that labour visible, intelligible, and transmissible.

The elemental approach has further shown how the material practices of cooking – stoking a flame, simmering a broth, kneading dough, whisking air – encode relational forms of knowing. These gestures, often relegated to the margins of history, become central to an understanding of how knowledge is produced, circulated, and preserved in domestic contexts. Ultimately, *El libro de Doña Petrona* offers a way of understanding cooking not only as work that holds together technique and relation but also as a philosophy of everyday life. It reminds us that culinary texts, particularly in their vernacular forms, are vital repositories of cultural meaning – where the work of the hands and the traces of use come together in acts of quiet, yet profound, authorship.

About the Author

Felipe Acevedo Riquelme holds a PhD in literature and is an Assistant Professor at Universidad de Concepción, Chile. His research explores the intersections of food, literature, gender, and knowledge.

Notes

1 Amy L. Tigner and Allison Carruth, *Literature and Food Studies* (Routledge, 2018), p. 3.
2 Juana Inés de la Cruz, *Carta Atenagórica y Respuesta a sor Filotea* (Red Ediciones, 2014), p. 64.
3 Megan J. Elias, *Food on the Page: Cookbooks and American Culture* (University of Pennsylvania Press, 2017), pp. 1–2.
4 Barbara Haber and Arlene Voski Avakian, 'Feminist Food Studies: A Brief History', in *From Betty Crocker to Feminist Food Studies: Critical Perspectives on Women and Food*, ed. by Barbara Haber and Arlene Voski Avakian (University of Massachusetts Press, 2005), p. 2.
5 Haber and Avakian, p. 4.
6 Haber and Avakian, p. 5.
7 Rebekah E. Pite, *Creating a Common Table in Twentieth-Century Argentina: Doña Petrona, Women, and Food* (University of North Carolina Press, 2013).
8 Michael Pollan, *Cooked: A Natural History of Transformation* (Penguin Books, 2014), pp. 56–61.
9 Begoña Alberdi, '*Mil fórmulas de cocina "La Negra"*: Labour, Gender, and Race in Argentina's Meat Industry, 1917–1940', *Journal of Latin American Cultural Studies*, 33.1 (2024), pp. 103–25.
10 Claude Lévi-Strauss, *Mitológicas III: El Origen de las Maneras de la Mesa* (Siglo XXI, 2003), p. 421.
11 Petrona C. de Gandulfo, *El libro de Doña Petrona*, 58th edn (Editorial Atlántida, 1963), p. 49. Subsequent references are cited parenthetically.
12 Pollan, *Cooked*, p. 146; Sonia Montecino, *La olla deleitosa: Cocinas mestizas de Chile* (Catalonia, 2015), p. 9.
13 Montecino, *La olla deleitosa*, p. 36; Luce Giard, 'Hacer de comer', in *La invención de lo cotidiano*, ed. by Michel de Certeau, Luce Giard, and Pierre Mayol, vol. 2 (Universidad Iberoamericana, 1999), p. 205.
14 Janet Theophano, *Eat My Words: Reading Women's Lives through the Cookbooks They Wrote* (Palgrave Macmillan, 2003), pp. 6 and 11.
15 David Le Breton, *Sabor del mundo: Una antropología de los sentidos* (Nueva Visión, 2007), p. 44; Kate Cairns and Josée Johnston, *Food and Femininity* (Bloomsbury, 2015), p. 68; Giard, 'Hacer de comer', p. 153.
16 Jane Bennett, *Vibrant Matter: A Political Ecology of Things* (Duke University Press, 2010), p. 41.
17 Montecino, *La olla deleitosa*, p. 19.
18 Elias, *Food on the Page*, p. 1.
19 Pite, *Creating a Common Table*, pp. 132–38.
20 Elias, *Food on the Page*, pp. 109–17.
21 Elias, *Food on the Page*, pp. 4–5; Theophano, *Eat My Words*, pp. 12–13, 46–47.

2
Air, Foams, Nitrogen, Helium, Smoke
The Use of Elements in Molecular Cuisine

Yanet Acosta

Air, the invisible element that makes cooking and living possible, became a protagonist in a new culinary typology called molecular cuisine.[1] The first dish created with this idea of making airy textures – giving air entity and prominence as an ingredient – was a white bean foam with sea urchins in 1994 at the restaurant elBulli in Spain. Since then, a line of cuisine has emerged in which various techniques are used to capture the element of air in a bite. Foams were produced in siphons – initially intended for whipping cream but reconfigured to create these new-generation sauces, which were followed by the so-called *aires*. Later, large bubbles were proposed, especially topping desserts like Juan Mari and Elena Arzak's Strawberry Bubbles,[2] which were first achieved through air and later through the inoculation of helium into a liquid, making the bubbles capable of floating and appearing as a small cloud or even an edible balloon like Helium Balloon with Apple Flavour by Grant Achatz in 2016.[3] The prominence of these elements was joined by smoke, either through the generation of aromatic smoke in a pipe that was released when a glass dome was removed from the dish for the diner (a creation by El Celler de Can Roca in 2004[4]) or through the smoke released by liquid nitrogen, which began to be used in the dining room for the on the spot preparation of sorbets or other types of granitas (the work of Luki Huber in 2018 for elBulli[5]).

At the time, these elements seemed more suited to a periodic table than a cookbook. Indeed, a dictionary made by the elBulli team to spread the language of the new cuisine – *Scientific Gastronomic Lexicon: The Keys to Understanding Cooking Today* – had a cover illustration inspired by the periodic table.[6]

This scientific approach was crucial in definitively distancing the new molecular cuisine from the traditional domestic realm. It elevated cooking to a different dimension, transforming it into a discipline closer to alchemy or even artistic genius. Before reaching the table or the kitchen, chefs began working in dedicated laboratories. elBulli Taller, located in Barcelona, was a spacious studio where ideas were tested and discussed. Arzak created a similar space, as did many other fine dining restaurants of the time. This new culinary style not only introduced cutting-edge equipment but also redefined how food was served – presented in spoons or on uniquely designed plates tailored to each dish. However, the most radical shift was the reinterpretation of ingredients: using familiar elements like air or water in new ways and introducing substances previously limited to industrial use, such as liquid nitrogen.

This image of the chef as a scientific and creative pioneer was further reinforced by international culinary congresses and the growing influence of mass media. Food photography also evolved, embracing a more artistic style that prioritized form and close-ups over full-plate presentations, often portraying chefs as solemn, visionary creators.

Air and Foams

In 1994, with a white bean foam and sea urchins, elBulli introduced a new culinary style that remains closely linked to the siphon today. Since 1990, Ferran Adrià had been obsessed with achieving a lighter texture and more intense flavour than possible in traditional mousses. He explained that foams were created to use only the juice or puree of an ingredient, without cream, eggs, or other additions.[7] Today, the siphon and foams have become symbols of what Arthur Lubow called the 'nueva nouvelle cuisine'.[8] *Le Monde 2* referred to it as 'investigative cuisine';[9] others called it 'molecular cuisine', using the term coined by Hervé This. Most Spanish newspapers at the time preferred 'avant-garde cuisine', as Elisabet Puiggròs Ruiz demonstrates in her study 'Analysis of the 21st-Century Gastronomic Movement in the Spanish Mainstream Press',[10] and Catalan journalist and Adrià's close friend Pau Arenós coined the term 'techno-emotional cuisine'.[11]

As Ferran Adrià explained:

> Foams are very well-known preparations today, which hundreds of chefs serve in their restaurants. It remains to be seen whether in twenty or thirty years they will continue to be so important. In relation to the controversy they have aroused, we still find it difficult to understand why there has been so much vehemence in the opinions that have been formulated. The current results allow us to affirm, without any hesitation, that there are good foams and others that are not so good, in the same way that there are more or less attractive mousses.[12]

He included several of these foams in his book: caviar with almond foam, black truffle juice cappuccino with almond foam, hot cottage cheese with caviar and radish chard foam, rock mussels with coriander foam, and vegetable stew in textures, in addition to the first one, dried bean foam with sea urchins. This foam was served inside a sea urchin shell and resembled the foam of the sea. After this experience and the resulting criticism of this type of 'non-food', he created a smoke foam in 1997, playing on the Spanish saying 'to sell smoke' (making empty promises).

Adrià's *Las espumas: Técnicas, tipos y usos* (*Foams: Techniques, Types, and Uses*) was the first book to widely share elBulli's pioneering technique of creating airs and foams.[13] Though no individual authors are credited, it was written by the elBulli team and features photographs by the restaurant's official photographer, artist Francesc Guillamet. The book was published by ICC (Innovación y Cualificación Culinaria), one of the most prominent suppliers of cutting-edge culinary tools at the time – selling everything from siphons to chemical additives and high-tech equipment for the new cuisine.

The book defines foam as follows:

> A recipe created by Ferran Adrià, cold or hot, based on the use of the iSi Gourmet Whip siphon. Initially, it consisted of making a very light-textured mousse with an intense flavour from a purée or gelatinous liquid. Over time, any recipe made with the siphon – even those including other ingredients such as egg whites, cream, or yolks – came to be known as a 'foam'.[14]

The book offers a step-by-step guide on how to use the siphon and introduces a structured method for creating foams, built around three core steps:

1. Choose the flavour.
2. Define the intended use.
3. Determine the desired density.[15]

The most revolutionary innovation, however, was the introduction of hot foams – made possible by a technique that initially seemed risky: placing the siphon in a double boiler (*bain-marie*) to heat its contents. This bold step opened the door to new culinary expressions, including one of the most iconic dishes of modern Spanish cuisine: the twenty-first-century potato omelette.

This dish, prepared in layers and served in a glass, was a 'deconstructed' version of the traditional *tortilla española*. At the top of this layered creation was a warm egg foam – a textural and conceptual reimagining of a beloved classic and a symbol of the avant-garde approach that defined elBulli.

In 2004, elBulli created its first 'air' – an evolution of foam in the quest for an even lighter sauce texture. The first air was made with carrots and an additive called Lecite, a form of lecithin used to create airy preparations. This innovation had already appeared in the public eye: in 2003, a photo of Ferran Adrià holding this carrot air featured on the cover of *The New York Times Magazine* with the title 'The Nueva Nouvelle Cuisine: How Spain Became the New France'. On the interior pages, the article was titled 'A Laboratory of Taste'.[16] The carrot air also featured in *Le Monde 2*, where Jean Michel Normand termed it: 'Certainly the lightest edible texture after the air we breathe. A variation of the foam technique'.[17]

Around this time, Adrià launched Texturas, a registered brand designed to make these seemingly impossible dishes accessible in other restaurants – and even at home. As the website states:

> elBulli Taller set out to expand the range of possible textures in cooking. As a result of this research, techniques such as foams, clouds, and others were developed, representing an evolution in its culinary style.
>
> The Texturas product line is essential for incorporating some of our most well-known techniques into your kitchen, such as hot gelatins, airs, gelatin caviar, or spherical ravioli.

> The products within the five families – Spherification, Gelification, Emulsification, Thickeners, and Surprises – are the result of a rigorous process of selection and testing. Texturas marks the beginning of a world of magical sensations that has continued to grow over the years.[18]

Nevertheless, the most essential tool for making foams is the siphon. In 1998, the Adrià brothers partnered with the Calabuig family to distribute it through the company ICC.[19]

The air disappears really quickly, so people combat this inconvenience with lots of new tools, like the Foam Kit by the company 100% Chef. The instructions for this product are:

> The perfect aerator for creating homogeneous, stable 'airs' in seconds – either continuously or on demand – in small or large quantities, all without using a hand blender.
>
> To operate, simply dilute a measured amount of soy lecithin or foaming agents such as BUBBLES with a liquid ingredient and pour it into a container that is taller than it is wide.
>
> Submerge the flat nozzle of the Foam KIT into the solution, press the button, and within seconds you'll be able to collect as much foam as you need. Turn it on or off with a single button press and control your foam production with precision.
>
> Work anywhere, with no need for plugs or cords, as the compressor runs on two 1.5V LR20 batteries, offering more than 96 hours of autonomy.[20]

This process represents a new kind of recipe, one closer to a scientific formula developed in a laboratory.

In fact, Arzak created a dedicated space called 'the laboratory' within the restaurant, where the team works collaboratively to develop new dishes. It's a bright, spacious room located in a separate apartment within the same building, equipped with an induction stovetop and a round table for discussing the day's culinary discoveries. Xabier Gutiérrez, who has worked with Juan Mari Arzak since 1980, is described as having the appearance of a 'mad scientist'.[21] Elena Arzak explains, 'The laboratory is the restaurant's core, where we work as a team. We like to describe our cuisine as author cuisine, Basque, creative, research-based, and avant-garde'.[22] In this laboratory, dessert bubbles were born – an innovation designed to add even more airiness to the plate and to make flavours subtler. The dessert called Strawberry Bubbles first appeared on Arzak's menu in 2005. Soon after, similar large bubbles began to appear in other restaurants, especially as toppings for desserts. For instance, in 2006, the restaurant Mugaritz introduced Cold Beet Bubbles.[23]

However, chef Andoni Luis Aduriz of Mugaritz went even further by solidifying air bubbles in a dish called Ortensia seca (Dried Cocoa Bubbles). In this technique, the presence of air becomes visible through the holes it leaves. Luis Aduriz, who was also obsessed with lightness, aimed to create 'a dough without dough, almost tasteless, so subtle that the body would hardly notice the difference between eating it or not'.[24] One such dish, titled Seasoned Lightness, was made only with kiwi and lemon juice, agar-agar, gelatine, and pepper. It had the appearance of a small, delicate, white pillow.

Lighter Than the Air: Nitrogen and Helium

And what is lighter than the air inside a bubble? Hydrogen and helium. These two elements became the focus of other chefs similarly obsessed with lightness. In 2016, Grant Achatz succeeded in creating floating bubbles by injecting helium into a liquid mixture, producing something that resembled a small cloud or even an edible balloon. The creation, called Helium Balloon with Apple Flavour, was made using inverted sugar. Achatz had spent seven years developing this idea, inspired by a child's birthday party.

For Achatz, the moment a diner interacts with the balloon – biting into it and then speaking in a helium-altered Mickey Mouse voice – evokes a return to childhood. He noted that such a surreal, playful experience in the setting of a three-Michelin-star restaurant creates a powerful emotional contrast.[25]

The first person to use liquid nitrogen for making instant sorbets was Hervé This, during his molecular gastronomy demonstrations in 1996. In 2003, British chef Heston Blumenthal began experimenting with the technique, and the elBulli team visited him at The Fat Duck to learn more about the process. This encounter sparked what they called the 'Blumenthal effect', leading elBulli to introduce several 'Nitro' creations on its 2004 menu.[26] The most famous of these creations was Dragon-Nitro, a spherical liquorice preparation. When guests placed it in their mouths, vapour would escape from their noses, mimicking a dragon breathing smoke.[27] At the same time, chef Dani García helped popularize the use of nitrogen in Spain with his chilled Andalusian soups, most notably his cherry gazpacho served with nitrogen granita of olive oil.

Smoke and Essential Water

Adding smoke flavour to dishes has been reproduced all over the world to the delight of diners. The first dish made with this technique was 'an octopus with La Vera paprika smoke' at El Celler de Can Roca in 2004. As the restaurant's website explains, 'The goal was to bring the smoke directly to the table and evoke emotional responses by appealing to olfactory memory'.

In 2004, elBulli introduced a simple yet symbolic 'cocktail' called Welcome Water. It consisted of a spherical ice glass filled with water, with a vanilla pod used as a straw. This is the favourite dish of Marc Cuspinera – chef at elBulli for several years – because of its simplicity: using just two ingredients that transformed at the very moment they were consumed.[28] He recounts it with enthusiasm, fascination, and passion thirty-four years after wanting to run away from 'that trap called elBulli', which went on to become the restaurant that revolutionized the forms, ingredients, and dishes of fine dining establishments around the world.

Eating Art

elBulli also transformed the way gastronomy was narrated through photography. Francesc Guillamet stumbled into food photography almost by accident. It happened in a bar in the 1980s, when someone he knew – a former DJ – talked him into it. That person was Juli Soler, who owned elBulli with Adrià. They had already crossed paths in Perpignan, where both had been browsing the latest photography and cooking books, and that's where their mutual understanding began.[29]

Up to that point, Guillamet had been photographing landscapes. What was originally supposed to be a few images for a book evolved into a catalogue that would document the dishes of a restaurant near the town of Roses in the Spanish province of Girona – a place so unexpectedly remote that reaching it feels like crossing into another world. This is where elBulli was born, thanks to a German couple who arrived in a camper van in the 1960s and decided to settle there.

The idea for such a catalogue came from Picasso's *catalogue raisonné* – a key inspiration for Ferran Adrià. But the photographic influence came from elsewhere: Reinhart Wolf's 1970s and 1980s books on New York architecture, Chinese and Japanese cuisine, and Spanish castles. Guillamet also drew from the visual elegance of *Vogue* photographer Irving Penn, photographer of the iconic *Frozen Foods with String Beans*. From that point on, a photographic journey began: Guillamet would go on to take thousands of photos and contribute to dozens of books. The first was *El sabor del Mediterráneo* (*The Taste of the Mediterranean*); the original layout drafts are displayed today in a showcase at elBulli1846.

Back then – times that Guillamet now says offered the best working conditions he ever had in terms of space and time – he photographed just three dishes per session, a far cry from the eleven he manages nowadays. The first photo of each dish was meant to clearly show the preparation – the restaurant's main goal – but the subsequent shots allowed the camera to 'interpret' the food. These were fresh, spontaneous images, styled not by professional stylists but by the chefs themselves.

Guillamet's work didn't just document the dishes from the elBulli kitchen; he turned food styling into a form of artistic expression, as his 2009 book *Comer Arte* (*Eating Art*) exemplifies. In fact, the elBulli Museum in Roses's Cala Montjoi holds 3612 photographs, most taken by him. Portraits of chefs working in this new style of cuisine also began to evolve. The shift is best explained by Ferran Adrià himself in an interview, where he recalls the moment a *New York Times* photographer came to elBulli to take his picture in 2003:

> The photographer arrived at Cala Montjoi around midnight, and the service was almost over. He took out his camera and said something I've always remembered: 'If you're talking about creation, you can't smile, because it wouldn't be credible'. That's why when I talk about creation and innovation, I'm serious, but if I talk about food or happiness, I'm not.[30]

Conclusion

'Disruptive' was the favourite word for Ferran Adrià in a decade that his colleague Santi Santamaría saw as a dangerous break with traditional cuisine and its traditional ingredients for a new 'chemical' cuisine, distant from Santamaría's values. He expressed this discomfort in his prizewinning 2009 book *La cocina al desnudo* (*Naked Cuisine*). He wrote, 'These are dangerous games. It's time to blow the whistle and end recess'.[31] That was a signal to know that in the future, more people would think in the same way. Between 2008 and 2012, especially with the economic crisis and even after the death of Santi Santamaría, his response to Adrià and others' disruptions started a movement against the use of artificial additives in haute cuisine.

However, Ferran Adrià's pursuit of lightness through foams and airs still holds a place in the food world. The culinary innovations to make air visible – created by chefs seen as geniuses, mad scientists, or alchemists and reinforced through a new style of photography – have added new dishes, vocabulary, and techniques to restaurant kitchens. Even the classical *Larousse Gastronomique* recognizes them: 'Nowadays, foam is used to describe all preparations made with a siphon'.[32]

This type of cuisine has become established under the name 'molecular' cuisine or gastronomy. The scientific image of this new way of cooking has taken hold not only in the media but also among scientific circles, where it formed a new discipline of research.[33] The result has been revolutionary: as 'restaurants around the world have started to adopt a more scientific approach in their kitchens, [. . .] several have become acclaimed as being among the best in the world'.[34] In this way, the new 'era of molecular cuisine' – as Davide Cassi called it – was inaugurated.[35]

About the Author

Yanet Acosta (PhD, information sciences) is a Professor at Rey Juan Carlos University, Madrid. She is the founder and director of the master's programme in gastronomic communication and journalism, The Foodie Studies, and is affiliated with the UNESCO Chair for Communication Research at Rey Juan Carlos. Recent scholarly publications trace how gastronomic communication is being studied in the Spanish academy (2023) and the roles of seafood in Mediterranean countries (2021). In addition, she is the editor of *Periodismo gastronómico, investigación y comunicación intercultural* (Universidad Rey Juan Carlos, 2021) and coeditor with Rebecca Ingram of *Food Studies en español: Investigación actual en gastronomía y comunicación* (Fragua, 2024). She wrote for ten years for Agencia Efe, the world's fourth-largest international news wire service, and is the author of three novels and a collection of short stories.

Notes

1 Hervé This, *Molecular Gastronomy: Exploring the Science of Flavor* (Columbia University Press, 2006).

2 Juan Mari Arzak and Elena Arzak, 'Pompas de Fresa', Madrid Fusión, 20 December 2005 <https://edicionesanteriores.madridfusion.net/es/receta/pompas-de-fresa/> [accessed 26 January 2026].

3 Emily Schell, 'Grant Achatz: "It Took Me Seven Years to Make the Edible Helium Balloon"', *El Mundo*, 14 February 2016 <https://www.elmundo.es/cultura/2016/02/14/56c0a5f746163f8b3c8b4571.html> [accessed 5 February 2026].

4 'Roca Timeline', El Celler de Can Roca <https://cellercanroca.com/los-hermanos-roca/roca-timeline/> [accessed 20 December 2024].

5 Luki Huber, *Diseños y esbozos para elBulli* (Planeta Gastro, 2018).

6 Alícia and elBulli Taller, *Léxico científico gastronómico* (Grupo Planeta, 2006).

7 Ferran Adrià, *elBulli (1994–1997)* (Ara Llibres, 2013).

8 Arthur Lubow, 'A Laboratory of Taste', *New York Times Magazine*, 10 August 2003 <https://www.nytimes.com/2003/08/10/magazine/a-laboratory-of-taste.html> [accessed 26 January 2026].

9 J. M. Normand, 'Ferran Adrià, l'alchimiste', *Le Monde 2*, 2 (25–31 January 2004), 24–33.

10 Elisabet Puiggròs Ruiz, 'Análisis del movimiento gastronómico del siglo XXI en la prensa generalista española' ['Analysis of the 21st-Century Gastronomic Movement in the Spanish Mainstream Press'], *Foodie Studies Magazine*, 4, 2019 <https://thefoodiestudies.com/analisis-del-movimiento-gastronomico-del-siglo-xxi-en-la-prensa-generalista-espanola/> [accessed 22 May 2026].

11 Pau Arenós, *La cocina de los valientes* (Penguin Random House, 2015).

12 Adrià, *elBulli*, p. 94.

13 Ferran Adrià, *Las espumas: Técnicas, tipos y usos* (International Cooking Concepts, 2004).

14 Adrià, p. 7.

15 Adrià, p. 11.

16 Lubow, 'Laboratory of Taste'.

17 Normand, 'Ferran Adrià', p. 31.

18 Albert Adrià, 'Texturas Albert y Ferran Adrià', Albert Adrià, 2026 <https://albertadria.com/id/texturas-albert-y-ferran-adria> [accessed 26 January 2026].

19 Cote Villar, 'El "divorcio" empresarial de Ferran Adrià y el creador de sus míticos sifones', *Vanitatis*, 17 December 2020 <https://www.vanitatis.elconfidencial.com/famosos/2020-12-17/ferran-adria-divorcio-empresarial-sifones-marc-calabuig_2874947/> [accessed 12 May 2026].

20 'Foam-KIT: Portable Foam Maker for Perfect Foam in Seconds', 100% Chef <https://100x100chef.com/shop/es/-aireadores-/188-foam-kit-8436020500527.html> [accessed 5 February 2026].

21 Juan Mari Arzak and Elena Arzak, *Arzak + Arzak* (Planeta Gastro, 2018), p. 53.

22 Arzak and Arzak, p. 69.

23 Arenós, *La cocina de los valientes*, p. 105.

24 Andoni Luis Aduriz, *Puntos de fuga* (Planeta Gastro, 2019), p. 213.

25 *Business Insider*, 'The Best Restaurant in America Serves Helium Balloons You Can Actually Eat', YouTube, 2017 <https://youtu.be/1PKvqdObaRw?si=H4GFjJ_IF6IlUgZR> [accessed 26 January 2026].

26 Ferran Adrià, Juli Soler, and Albert Adrià, *elBulli 2004* (RBA Libros, 2005).

27 Luke Huber's sorbets and granitas for elBulli also used liquid nitrogen (Huber, *Diseños y esbozos para elBulli*, p. 42).

28 Yanet Acosta, 'La épica de elBulli retratada por Francesc Guillamet', *Foodie Studies Magazine*, 8, 2024 <https://thefoodiestudies.com/la-epica-de-elBulli-retratada-por-francesc-guillamet/> [accessed 26 January 2026].

29 Acosta.

30 Isabel Gómez Melenchón, 'Ferran Adrià: salió la portada y fue la locura', *La Vanguardia*, 28 July 2019 <https://www.lavanguardia.com/comer/20190728/463731314560/ferran-adria-portada-he-new-york-times-magazine-el-bulli-verano.html> [accessed 12 May 2026].

31 Santi Santamaría, *La cocina al desnudo* (Temas de Hoy, 2009), p. 120.

32 *Larousse Gastronomique: The World's Greatest Culinary Encyclopedia* (Clarkson Potter, 2019), p. 461.

33 Hervé This, 'Molecular Gastronomy Is a Scientific Discipline, and Note by Note Cuisine Is the Next Culinary Trend', *Flavour*, 2.1 (2013) <https://doi.org/10.1186/2044-7248-2-1>.

34 Peter Barham and others, 'Molecular Gastronomy: A New Emerging Scientific Discipline', *Chemical Reviews*, 110.4 (2010), pp. 2313–65 <https://doi.org/10.1021/cr900105w>.

35 Davide Cassi, 'Science and Cooking: The Era of Molecular Cuisine', *EMBO Reports*, 12.3 (2011), pp. 191–96 <https://doi.org/10.1038/embor.2011.18>.

3
Elements, Flavours, and Food in Three Premodern Physiological Systems

Ken Albala

Modern biochemistry explains the sensation of flavour and smell as the impact on the human taste buds and olfactory receptors of molecules that are in turn made up of elements. Sourness on the tongue from a squirt of lemon juice comes from citric acid, or $C_6H_8O_7$ – comprised of carbon, hydrogen, and oxygen configured in a way that our senses recognize as a source of vitamin C, which our bodies, unusually among animals, cannot produce itself. Combined with lemon oil made up of $C_{10}H_{16}$ – limonene, which is a terpene, along with aldehydes (citral and citronellal) and esters like geranyl acetate – citric acid gives us the complex flavour we recognize as lemon. Or in simpler terms, what we taste, everything on the plate, every recipe, is simply a complex of elements that can be quantified. Moreover, the impact of these elements on our bodies can also be described in purely molecular terms.

Surprisingly, this is nothing new. The number of elements has proliferated, and our understanding of molecular structures is entirely new, but elemental theory has always been the cornerstone of gastronomy. In this paper, I will discuss three premodern elemental and physiological systems and their understandings of flavour, nutrition, and the interaction of food with the human body in ways that will reveal that cooking has always been thought of as an assemblage of elemental building blocks or a harnessing of natural processes. I begin with the ancient Greek system of humoural physiology, proceed to traditional Chinese medicine, and then continue to Indian Ayurveda.

Among the pre-Socratic philosophers, there was great debate about the nature of the universe and what it is made of. Thales of Miletus imagined that moisture was the fundamental element that gave rise to all matter and all living beings. Anaximenes, his successor, believed that air is the source of all other elements. Heraclitus of Ephesus pointed to fire, and Empedocles of Agrigentum insisted that earth be added to these as an element in its own right, giving us the four elements that stuck in ancient Greek science: earth, air, water, and fire.

The four elements can each be described in qualitative terms as well. That is, earth is cold and dry, water is cold and moist, air is hot and moist, and fire is hot and dry. Each has a different specific density, which is why earth (the heaviest element) is logically at the centre of the universe, water lies on its surface, air surrounds that, and fire circles around the heavens in the sun and stars. But all matter is made up of a combination of elements in varying degrees. That means that our bodies, as well as all the food we eat, are also comprised of elements and can be described with the same qualitative terms or primary qualities.

Within our bodies, four regulatory fluids correspond to the elements. These are called the humours. Blood, the most abundant, is described as hot and moist, phlegm is cold and moist, yellow bile is hot and dry, and black bile, or melancholia, is cold and dry. Health consists of a balance of these four humours, or *eukrasia*, although every individual is born with a predilection to abound in one particular humour – their complexion – thus making some people sanguine by nature, or ruddy and cheerful, and others phlegmatic, or lazy and apathetic. Cholerics are prone to outbursts of anger, and melancholics are depressed. The system has a built-in psychology, since the mind and body are directly connected as a psychosomatic whole.

As building blocks of nature, all foods we eat are made up of the same elements and can be described as heating/cooling or drying/moistening. We still employ vestiges of this system gastronomically when we say a martini is dry: it's obviously liquid, but its effect on the tongue is astringent and drying. Likewise, a hot chili pepper is not hot in terms of temperature, but it does burn in the mouth. The logic of dietary recommendation is that if an individual exceeds in any particular humour, they should eat foods opposite in quality. The logic of pharmaceuticals is the same, an allopathic system of curing with opposites. So a person with a hot and dry fever should eat a cold and moist cucumber to rebalance their humours. A person with a cold (another completely humoural term) should eat foods that are heating and moistening, like honey.

Perhaps not obvious is how foods were assigned humoural values in the first place. It was a simple matter of flavour. Sweet foods were considered the most nourishing and categorized as hot and moist, or sanguine. In this system, the foods that are closest to the human body are also those most easily assimilated into it; thus meat is the most nourishing food. Watery, insipid foods were thought to be cold and moist like the element water itself. So too were sour foods like lemons. Hot and dry choleric foods were those that burn and clearly heat the body, so anything spicy but also salty foods. Lastly, bitter foods like the earth itself are cold and dry.

These are all primary qualities, but all foods also have secondary qualities. They may be astringent and tighten up the passages of the body, like the styptic effect of tea on the palate. Or they may be aperient, widening the passages, or unctuous, smoothing them out. The Greeks were fairly obsessed with the process of digestion, so the consistency of food was just as important as its humoural makeup and should also be appropriate to the power of the stomach. People without much exercise have weaker systems and should eat foods lighter in texture, like chicken and fine white bread, while workers have powerful systems and can digest coarse and rough foods, like beef, cabbage, or beans.

And any food or drug may be tonic (tightening up the muscles), sudorific (making you sleep), expectorant (promoting salivation), cordial (good for the heart), diuretic (promoting the passage of urine), etc. But even without knowledge of these secondary properties, any individual could theoretically sense the body being too hot or cold, moist or dry, and could eat foods appropriately corrective to draw the humours back into balance. They could also choose foods that match their digestive strength. Dense foods make a delicate body ill because the whole system becomes clogged up, and light foods for a labourer will simply burn up, offering no nutritional value.

It should be mentioned here that food is only one of several ways that one can regulate the body. Exercise and rest, changing environments, emotional responses, and even sexual activity serve to heat/dry or cool/moisten the body. This system originated in the Hippocratic authors and was not only elaborated upon several centuries later by Galen of Pergamum but also added to by Muslim authors like Avicenna and Averroes. It was the Arabs who coined the term the six 'non-naturals', not because they are unnatural, but because they are external to the body and can be controlled and regulated.

Just as an individual corrects an imbalance with its opposite, likewise within an individual recipe, an ingredient that may exceed in one particular humour can be balanced by its opposite. Cold and moist milk will counteract the burning of a spicy dish. Hot mustard will cut through the cold and phlegmatic fattiness of pork. The gluey humours of fish were thought to be resolved by the acidity of lemon juice. There is no evidence that cooks devised recipes with these principles in mind, and it is just as likely that physicians looked at common culinary practices and defended or denounced them with humoural logic. But it is nonetheless possible that many common practices proliferated with the help of a medical rationale. It was believed, for example, that cheese after a meal would serve as a stopper on the top of a seething stomach, keeping foul vapours from ascending and causing dizziness and headaches. However, fruit after a meal would corrupt inside the body just as it does outside.

In the past, as today, there is also no evidence that people followed dietary advice, any more than they cooked directly out of cookbooks. Both are prescriptive literature rather than descriptive. But this system helps explain how people envisioned what they thought was going on in the universe, in their bodies, and perhaps even on the plate. The intersection of food and medicine is incontrovertible.

Humoural theory was passed down from the Greeks and Romans to the Medieval Muslim world. Much of the ancient scientific literature was translated into Arabic in the new Abbasid capital of Baghdad. From there, it was then translated from Arabic into Medieval Latin in Europe, in places like Salerno, and persisted into the seventeenth century. That means for roughly two millennia, foods were categorized according to an elemental understanding of how the universe works. It was only with the challenge of iatromechanical and iatrochemical theories in the Scientific Revolution that humouralism began to be challenged, not coincidentally at the same time modern haute cuisine was invented, though I hesitate to say that Western cuisine was liberated from dietetics (as Jean-Louis Flandrin put it).[1] They simply veered in different directions.

The ancient Chinese had a system that was in many ways similar but also based on elemental theory. How and why these similarities arose might be attributed to contact in ancient times or to common origins from some prehistoric way of thinking about the world. Or it might even be some form of Jungian collective subconscious symbolism. More likely, since we are all the same physical subjects dealing with similar external stimuli, it makes sense that we would come up with ways of explaining natural phenomena in similar ways. We all feel heat and cold, dryness and moisture, so logically we would all attribute these to the same empirically demonstrable sources.

Traditional Chinese medicine has five elements: fire, water, metal, earth, and wood. These are not so much building blocks as they are states of transformation in nature that explain the creation or destruction of matter. Sometimes they are described as five phases in the classic literature, such as the *Nei Ching*, or *Yellow Emperor's Classic of Internal Medicine*, by Huangdi. Here each element is linked to another, so in a cycle of generation, water makes wood grow, wood burned gives off fire, fire creates ash (a form of earth), earth is the source of metals, and metals heated flow like water. In a cycle of destruction, water puts out fire, fire melts metal, metal axes cut wood, wooden ploughs turn the earth over, and a dam made of earth stops water from flowing.

This all sounds quite poetic, but it is also used to describe physiological processes, since all nature obeys the same rules. And as in the Western system, Chinese medicine assigns specific nutritional and therapeutic properties to individual foods. In fact, one could argue that there is no culinary system that is more closely aligned with medicine than China. Cooking and feeding are always on some level correctives. Foods are categorized as increasing yin or yang. These are the two forces that govern the entire universe, yin being the female element (cold, soft, empty, dark) and yang being the male (hot, firm, full, light). If a person is diagnosed as having too much or too little yin or yang, they are prescribed foods that are opposite in value.

As in the Greek system, there are foods that cool, such as cucumbers, melons, and watery leafy greens, and others that heat, like ginger and lamb. Again, I think the similarities arise simply because these are the physical sensations these foods elicit. Likewise, the five basic flavours (*wuwei*) function in the same way. Sweet foods nourish the body and are a source of orthogenic *qi* – sort of like energy and immune response in the body. Sour foods tighten the body, just as they have a puckering effect on the tongue, which strengthens and tones the tendons. Bitter foods dry up superfluous humidity. Spicy foods help push food through the body and help circulation, and salty foods also resolve clogs and phlegm. The dietary logic is also a matter of correcting imbalances.

The five flavours in Chinese medicine are also derived from the elements: sweet comes from the earth, sour from wood, bitter from fire, spicy from metal, and salty from water. (Incidentally, that is completely different from the Western system.) Nonetheless, foods can be categorized according to their flavour and their effect on the body, though there are also neutral foods, like rice, beans, cabbage, and nuts. Cooking is a matter of balancing the ingredients and cooking methods with the state of the individual's body.

Ayurvedic medicine is also based on an elemental theory and is also a matter of prescribing food in order to balance the body. The foundational text in this system is the *Caraka Samhita*, which describes five elements that, in combination, give rise to three basic forces which govern all bodily processes. In Sanskrit, these are called *doshas*. The elements are fire, air, earth, water, and space. Space and air combine to form the *vata dosha*, which is the principle of movement in the body. Fire and water combine to form the *pitta dosha*, which is the principle of digestion, and the *kapha dosha*, the structural principle that holds everything together. Like the complexion in the West, every individual is born with a *prakriti*, a temperamental imbalance of one of the *doshas*. So *vata* individuals are thin and bony, talk and eat quickly; we might call them highly strung. People with too much *pitta* have hot digestive systems, which makes them prone to anger – and foods that heat the body, like spices, are harmful to them. *Kapha* people are bulky, have slow digestive systems, and are very stable but slower. And naturally, a person with too little *kapha* would be fragile and weak.

Every food in the Ayurvedic system is classified as reducing or increasing any particular *dosha*. Sweet foods increase *kapha*, as well as those that are heavy and oily. They are used to balance and nourish *pitta* and *vata* in the body. By contrast, sour foods are heating and aggravate *pitta* and *kapha doshas*. Salty foods are also heating and can harm those imbalanced with *pitta* and *kapha*. Pungent foods are heating and drying, which aggravates *pitta* and *vata* but helps balance *kapha* because it promotes circulation. Bitter foods are cold and dry and reduce excess *kapha* and *pitta*, and astringent foods act in the same way. Thus in Ayurveda, flavour is key to understanding how a food will behave in the body and is linked to the dominant element.

The specific recommendations in these three systems do not align neatly and sometimes are completely opposite. Spicy or pungent foods in all three systems are considered heating, which makes sense, but sour foods are cooling in the Greek system, toning in the Chinese system, and heating in the Indian system. This would suggest that there are no real connections or mutual influences in these three systems; their similarities spring from the fact that humans all experience sensations in similar ways, even though we don't agree on how those sensations affect the body.

In any case, the specific details aside, it is evident that all three systems are based on an elemental theory that translates directly into flavours, bodily functions, and ultimately dietary recommendations which influence what people cook and eat.

As an epilogue, I would like to add one more premodern system into consideration. It is a little more difficult to make sense of, given the lack of classic written sources and the fact that most of it was first seen through the eyes of Spanish conquistadors, who were intent on destroying the Indigenous religion and all its systems of belief. I am speaking of Aztec Mexico. It has been argued, most notably by George Foster, that the Hippocratic/Galenic system was transported lock, stock, and barrel to the New World and survived among *curanderas* whom he studied in Mexico.[2] Bernard Ortiz de Montellano made a plausible argument that there was an entirely separate Indigenous system which simply coincidentally categorized foods in similar ways, describing them as heating and cooling.[3] I would contend that both are somewhat true. There was definitely an Aztec medical system, and Greek medicine was simply laid on top superficially and added new features. For example, the Aztec system did not categorize foods as drying or moistening: the Greek system via the Spanish added that. This is much like how Christianity was simply laid on top of Aztec religion so as to accommodate it, changing the names of local deities to Christian saints and altering Indigenous holidays to be vaguely Christian (like *Día de los Muertos*). The same can be said of cooking as well: some Aztec dishes are completely Native, like guacamole, while others look European with spices and nuts but are fundamentally Aztec – a *mole poblano*, for example.

The same can be said of the dietary recommendations that can still be gleaned from contemporary practice. They are an amalgam of Aztec theory with a veneer of Hippocratic medicine. Again, their similarity is the result of common human experience, just as it is with the Chinese and Indian systems.

About the Author

Ken Albala is Tully Knoles Endowed Professor of History at the University of the Pacific in Stockton, California. He has published thirty books, the latest of which is *The World Atlas of Fermented Foods and Drinks*.

Notes

1 Jean-Louis Flandrin, 'Introduction: The Early Modern Period', in *Food: A Culinary History*, ed. by Jean-Louis Flandrin and Massimo Montanari (Columbia University Press, 1999), pp. 349–73 (pp. 363–64).
2 George M. Foster, *Hippocrates' Latin American Legacy: Humoral Medicine in the New World* (Gordon and Breach, 1994).
3 Bernard Ortiz de Montellano, *Aztec Medicine, Health and Nutrition* (Rutgers University Press, 1990).

4
Elemental Uptake in Food
From Terroir to Terror

Sabine E. Apitz

In a process as old as life itself, microbes and plants evolved to extract nutrients they required to live and grow from the earth – from the air, water, rocks, and, as they developed, soils. These processes fed the microorganisms and plants – their extraction of chemicals, respiration, and excretion then also changed the geochemistry of the earth. Animals evolved to extract what they need not only from air and water but also from the consumption of plants and each other in intricate food webs. Thus, all animals, including us, depend for sustenance upon the ability of plants to extract and convert chemicals from the earth – the food we gather from the land and water around us nourishes us. Plants not only generate organic nutrients such as starches, sugars, and proteins that we need to grow but also accumulate the elements essential to our functioning. These elements taken up by plants feed them and are transferred up the food chain to all life.

Macronutrients comprise the bulk of what we need to provide energy, structure, and function; these include carbohydrates, proteins, and fats. These are organic compounds – built primarily of carbon (C), oxygen (O), and hydrogen (H), making up about 95 percent of plant dry matter. Major nutrients, nitrogen (N), phosphorus (P), and potassium (K), are necessary in large amounts. Secondary elements or nutrients, silicon (Si), calcium (Ca), magnesium (Mg), sulphur (S), chlorine (Cl), and sodium (Na), comprise 4 percent of plant dry matter. Trace elements (minor elements, microelements, or micronutrients) exist at much lower levels, often parts per million. Boron (B), chlorine (Cl), copper (Cu), iron (Fe), manganese (Mn), molybdenum (Mo), sodium (Na), zinc (Zn), and (less certainly) cobalt (Co) are essential to both plants and animals. Animals also require chromium (Cr), iodine (I), selenium (Se), and possibly fluorine (F), barium (Ba), and strontium (Sr).[1] However, to paraphrase Paracelsus, the dose makes the poison – these are toxic at higher levels.[2] Other elements such as arsenic (As) and mercury (Hg) are nonessential trace elements which also may be taken up by plants.[3]

Soils are the source of the essential minerals that support plant growth and thus our nutrition. Elements can be deposited on plant surfaces or taken up by roots; these can be transferred to various tissues. Fish bioconcentrate metals from water and food; some accumulate up the food chain. Plant mineral deficiencies can impact plant growth and yields. Not all of an element that an organism is exposed to can be taken up; bioavailability is controlled by many factors. Crops produced on soil with poor phytoavailability of essential elements, if they grow at all, can cause human nutrient malnutrition, illness, and even death, as can a lack of variety in the diet.[4] Two-thirds of the world's population have insufficient intake of one or more minerals, mostly Fe, Zn, I, S, and to a lesser extent, Ca, Cu, and Mg. Our diet can also provide us with too much of an essential nutrient or of elements not necessary to life and can poison us. Finding the Goldilocks level of the countless elements, essential and toxic, in our diet is a moving target in a changing world.

Historically, if we ate a varied and whole food diet, we were able to provide the trace elements essential to our health. Modern agricultural practices allow us to grow food in greater abundance; increased yields are feeding billions. However, although larger yields have brought more calories, there have been significant reductions in many essential nutrients in food crops. In Finland, for example, although the consumption of vegetables has risen, average daily mineral and trace element intakes have dropped.[5] Changes in crop varieties seeking higher yields, growth rates, and pest resistance have reduced the uptake of essential nutrients; faster growth and bulkier plants generate a greater volume of less nutrient-dense food.[6]

Evaluating the role of farming practices on the nutritional quality of foods is complex; studies focusing on countless differing aspects of farming practices make drawing sweeping conclusions and cross-study comparisons challenging. However, meta-analyses suggest that a critical factor is soil health, defined as the capacity of soil to function as a living ecosystem that sustains plants, animals, and people.[7] (Standard criteria to evaluate such 'health' are still lacking.[8]) Natural soils can have rich biodiversity, especially in root zones where plants and the fungal, microbial, and invertebrate components of ecosystems have complex and symbiotic relationships, exchanging chemicals for often mutual benefit. Ecosystem engineers, such as earthworms in cooler climates and termites in warmer climates, enhance soil biodiversity. Fungi in association with the root zone exchange sugars from plants for nutrients they extract from soil, and microbes affect the biogeochemical cycles that affect micronutrient bioavailability.[9] Although soils can be supplemented and plants can be bred to take up more nutrients, the biodiversity in soils controls plants' ability to take them up.

Modern agricultural practices, such as ploughing and applying pesticides and fertilizers, significantly reduce soil biodiversity, disrupting these symbiotic processes.[10] Ploughing removes invertebrates, breaks down soil and microbial structure, and enhances the loss of soil organic matter essential to nutrient uptake; agrochemicals also reduce biodiversity.

Farms that reduce these practices, whether organic or not, grow more nutrient-dense vegetables. Much of the yield loss attributed to organic farming can be increased with organic-rich soils fertilized with compost and/or manure. These plants have higher micronutrients as well as higher levels of plant chemicals which have important antioxidant and anti-inflammatory properties.[11] The practices which produce less nourishing foods (and less biodiverse soils) can be reduced without sacrificing the yields required to feed ever-growing populations.

By focusing on bulk and calories, we may have created quantity over quality. How we tend our gardens matters – we must attend not only to the plants but also to the vastly complex soil ecosystems they depend upon to be healthy and nutritious.

What Can We Digest?

Would it be enough for us to eat more nutrient-rich foods? Although a lack of dietary diversity is a major cause of inadequate micronutrient consumption, poor nutrient absorption in our digestive systems is also a major driver of malnutrition. The amount of, for example, Fe that can be absorbed from plant food depends on what else is eaten in the same meal, suggesting that not only what is eaten but also with what can be critical to nutrition.[12]

Gut microbiota are essential to mineral absorption in the gastrointestinal tract, producing enzymes that increase mineral bioavailability. These personal ecosystems can be enhanced by what we eat – they benefit from vitamins, micronutrients, and fibre.[13] These interactions support essential bodily functions. The consumption of probiotics, or beneficial bacteria, enhances mineral uptake.[14] Prebiotics, dietary fibres indigestible to humans but essential for our intestinal microbiome, enhance the absorption of essential minerals. Some of this microbial action may be beneficial outside our bodies; fermented foods also increase the bioavailability of minerals, as does the soaking and pre-sprouting of seeds.[15]

How we tend to our own personal gardens, in our guts and, perhaps, in our fermentation crocks, is essential to nutrition – feeding and supplementing our symbiotic ecosystems in turn feeds us.

A Taste for the Elements?

The concept of terroir suggests that the soil, water, and weather of a place impart special properties and flavours to our crops. Terroir and elements of soil are invoked as drivers of flavour for wine, as well as other foods.[16] The concept of 'minerality' in wine tasting suggests that certain minerals from the soil impart taste. However, any minerals in wine are only distantly related to soil components, in trace quantities, and tasteless when compared to the many flavoured organic compounds.[17] On one hand, although soil type and composition may affect nutrient and water availability, soil geochemistry does not in itself seem to impart a taste as much as does climate.[18]

On the other hand, when a variety of strawberries were grown in the same geography and climate but in different soils, higher-quality soil produced more nutritious berries.[19] Root-associated microbes may be significant in developing the flavour of fruits and seeds, so perhaps we taste microbes rather than the soil itself.[20] Soil microbial systems are enhanced by soil organic matter, which enhances nutrient uptake. As soil mineral content can also affect microbial populations, these elements may be, albeit indirectly, affecting flavour. Grapes, and thus wine, show trace metal signatures that are unique to the soil they were grown in.[21] Although clear correlations between mineral content and flavour have not been measured, relationships may still exist.

'Nutritional intelligence' is often attributed to nonhuman animals.[22] As omnivores, we eat a varied diet, getting different nutrients from different foods – do our tastes guide us? Taste intensity (e.g. fattiness, saltiness, sweetness, umami) can be loosely correlated with a higher probability of nutrient density.[23] However, this correlation may be confounded in a world of ultra-processed food and, given a preference for calorie-dense foods, can lead to poorer health outcomes in an ultra-processed foodscape.[24] Most studies have focused on macronutrients, but when food pairing and portioning were examined, a preference for meals with 'micronutrient complementarity' and an avoidance of nutrient redundancy was observed.[25] Food choice is complex, and it is difficult to determine whether a subject chooses a food due to its nutritional value and, if so, if this 'knowledge' is based upon an intrinsic understanding or cultural learning, which might be quite vulnerable in a rapidly changing, or ultra-processed, food world.[26]

The seeking out of minerals lacking in the diet is attributed to many wild and domestic animals. Why not humans? If we are deficient, do we crave them and seek out specific nutrients? Pica – a human 'craving for and consumption of non-food items, including the ingestion of earth' and other items – has been shown to be associated with a risk of low Fe and Zn.[27] Those who suffer from deficiencies seem to know and seek sources of minerals, despite the potential risks of dirt eating.[28]

Elemental Excess

Elements as Pollutants

Plants, with the help of their symbionts, gather essential elements; these nourish us. However, the same or other elements can also be toxic, sometimes at very low levels. The mechanisms by which plants take up essential elements can also take up others. Many elements exist naturally in soils at very low levels or in nonavailable forms, but humans concentrate and release many elements. Agriculture, industry, roads, waste, mining, energy production, and many other processes introduce elements to soil, water, and plants. Some elements can inhibit plant growth or interfere with the uptake of essential elements, and essential and toxic elements may enhance or reduce each other's toxicity, uptake, or other effects.[29] These interactions are complex and not fully understood, but elements both naturally in the soil and concentrated by human activities can at times pose risks to human health.

How Does Your Garden Grow?

Although many people grow their own food, where it is grown can affect trace element levels in produce. For example, garden-grown lettuce from urban Montreal had higher levels of heavy metals than those bought in local grocery stores.[30] In other urban areas, with stronger and closer contaminant sources, these results might differ. Gardens may be affected by contamination from industry, agriculture, transport, and urban deposition, as well as from contaminated mulch or compost. Decades ago, I struggled to explain to members of my 'organic' community garden collective why the 'free' donations of telephone poles or railroad sleepers, both treated with toxic materials to reduce degradation, were not appropriate as garden edging, even if the impacts were not immediately visible.

Some gardeners have less choice. Crops grown near organic-rich waste sites in Uganda had metal levels exceeding consumer safety limits.[31] Similarly, crop plants in an area impacted by mine tailings in Brazil demonstrated elevated levels of some metals but reduced levels of the essential element Zn.[32]

A common source of fertilization is sewage sludge. Sewage systems collect contaminants from many sources. Although degradable organics may be broken down and pathogens killed, sludge retains and concentrates nondegradable elements and forever chemicals from pharmaceuticals and personal care, households, agriculture, and industry. Only a few are monitored or regulated, based upon generic criteria. With repeated applications, these can accumulate in soils and the food chain, though uptake is dependent upon the nature of the metal, sludge, soil, and crop properties.[33] Although waste reduction and circularity, which seek to make resources out of waste, are important to sustainability, potential contamination and its impact on crop growth and food quality must be considered.

The lines between a family garden, market garden, and commercial field may be tenuous – in a global market, we cannot assume that exposure to contamination is limited to where something is grown or that we are safely removed from sources of contamination. The waste from our consumption, even if shipped far from home, may, in some form, return to us in our food. We all have a stake (beyond the obvious moral one) in ensuring that pollution and food safety are addressed everywhere and for all.

Gathering Food

A great source of pleasure for me comes from foraging in the countryside on my walks. However, location matters. Plants near industrial or waste areas should be avoided, as well as areas impacted by road traffic, chemical spraying, or animal fouling. Although the banning of leaded fuel has reduced lead risks on roadsides, automobile brake linings, tyres, and combustion release a range of toxins, from soot and microplastics to toxic organics and trace metals, posing respiratory and dietary risks.[34] For example, Cu, Pb, and Cd were measured in washed and unwashed wild roadside blackberries in Kent in the UK. Single servings of these berries were not found to exceed guideline values, but larger quantities of a kilogram per day might exceed allowable doses. While Pb could be reduced by washing berries, it did not significantly reduce Cu or Cd; metals were both loosely sorbed on surfaces and more tightly bound or taken up into cells, from surfaces or soil.[35] These low risks may be offset by the positive social and nutritional effects, but foraging should be done with the environment in mind, and produce should be washed.

Wild-grown mushrooms are a culinary delicacy and forager's delight, but they also have a propensity to accumulate radioactive K (^{40}K) and caesium (^{137}Cs and ^{134}Cs). These exist at low levels naturally but are enriched due to radioactive fallout from bombs and nuclear disasters such as the Chernobyl power station meltdown in 1986.[36] Fallout decreases with distance from the source, but long radionuclide half-lives mean that wild mushrooms and other foods harvested nearby will remain of concern for centuries.[37] In nearby Scandinavia, where mushrooms are abundant, mushroom eaters, such as deer, have elevated levels of radionuclides.[38]

That impacts from nuclear fallout are first depositional and then taken up in the wider food chain is supported by data from radionuclides in local Japanese food after the Fukushima meltdown in 2011. Initially, radioactivity was observed in vegetables, mushrooms, and fruits and berries. In the following months, these levels dropped, with peaks seen in animal products due to their consumption of these food items. This was particularly strong in the meat of boars, who feed on mushrooms, illustrating that elements move through the food chain.[39] Although it is reported that radioactivity in Fukushima's food is back to pre-accident levels, there is still stigma.[40] Even when government agencies suggest that risk – for example, for Pacific seafood – is low, distrust remains, along with an expectation that precautionary monitoring continues.[41]

What Does the Future Hold?

Food security is a situation that exists when all people, at all times, have physical, social, and economic access to sufficient, safe, and nutritious food that meets their dietary needs and food preferences for an active and healthy life.[42] Aspects of food security include food availability, access, utilization, and stability over time, as well as agency and sustainability.[43] Even with the most optimistic projections, climate change is expected to have extreme and interacting impacts on food security via the breakdown of food systems, including crops, livestock, and fisheries, as well as disruptions in food distribution linked to global warming, drought, flooding, and precipitation variability and extremes. These changes will particularly impact already vulnerable populations.[44] Further climate change-linked drivers of crop yield loss include increased ground-level ozone, pollinator loss, soil degradation, coastal inundation, and salt intrusion, while loss of coral reefs and mangrove forests, as well as ocean temperature rise, acidification, overfishing, and oxygen loss, will affect fishery and other seafood yields.[45]

There are many interdependent changes – including fertility, water, yield, nutrient concentration and bioavailability, and increased food antinutrient content – that affect our ability to access sufficient nutrition for global populations.[46] Unsustainable practices such as food waste, large-scale monoculture, excess fertilizer and agrochemical use, and deforestation and habitat destruction all increase greenhouse gas production and reduce biodiversity, further affecting food system sustainability.

These changes will exacerbate inequality, child malnutrition, and nutrition-associated diseases, adversely impacting children and adults, particularly in Africa and Asia.[47] Our ability to equitably feed and nourish a still-growing population will be one of the major future challenges. Failure to do so will cause untold suffering but also significant socio-political and economic upheaval.

In terms of micronutrients, a significant driver of mineral deficiencies will be reductions in the availability and access to fruits and vegetables. Higher atmospheric CO_2 will also reduce food quality; the protein and mineral content of cereals will decrease due to the dilution effects of faster growth without greater nutrient uptake. The concentration of many micronutrients in crops is predicted to decrease significantly. More than a billion more people will suffer increases in existing deficiencies under these conditions; the decline in Zn content alone is projected to lead to millions of additional people with Zn deficiency.[48]

Higher CO_2 will decrease protein and micronutrient content in rice, subjecting six hundred million people with rice as a staple to the risk of micronutrient deficiency by 2050.[49] Further, higher temperatures and CO_2 will supercharge the microbes that cause As deposition on rice, so As toxicity in this staple crop is predicted to rise significantly unless measures are taken to avoid this effect, which will lead to increases in bladder and lung cancer, heart disease, and diabetes.[50] Heat-induced changes in microbial communities are likely also to shift the dynamics of other essential and toxic elements in the root zones of crops.

Elemental Eating

There is an urgent need for strategies to prevent or reduce these impacts. The first and most urgent is to limit greenhouse gas emissions – immediately. However, the political will to make these essential changes globally remains in doubt, and many changes are now irreversible. Many mitigation strategies will be required to reduce the risk of micronutrient malnutrition: these include postharvest fortification, micronutrient supplementation, and biofortification of foods and staple crops; plant breeding and genetic approaches to increase nutrient content and stress tolerance of plants; as well as dietary changes, including a reduction in animal protein consumption, increased consumption of insect protein, and the promulgation of guidelines for changes in food systems and nutrition.[51] Clearly, the trade-offs between crop yield and nutrition as well as nutrient bioavailability must be considered. The nutritional benefits of animal-sourced foods must be balanced against their environmental footprint and impacts on biodiversity.[52] Caring for the planet and all our futures requires that we care for the waters, soils, air, plants, and the tiny microbial, fungal, and invertebrate helpers that ensure that our food contains the nutrients we need. Caring as well for our personal ecosystems will help ensure that we make the most of all the nutrients available.

Humans are far more tolerant of risks they have chosen – people drive and smoke, although those activities carry clear and relatively high risks, but risks from contaminated food are generally neither chosen nor in our control. We react to such risks, even if lower than those we choose to accept, with terror. We want to trust that our food nourishes and not poisons.

Although there are many other pollutants, the focus of this paper has been on essential and nonessential elements taken up by plants and in our food. Different elements behave differently in their uptake, accumulation, transport, and risk, and it is important to understand these processes to make informed decisions. Understanding the flow of different elements to and from soil (and air, sediment, and water) to the food chain and our bodies through pathways from sorption and uptake to bioaccumulation and biomagnification helps us ensure we maximize our access to nutrients while avoiding, if we can, elemental toxins.

Knowing where (and when) one's food comes from is important, when one has the luxury of choice. Once in the food chain, toxic elements do not easily go away. Understanding these issues, the cook (and eater) may better navigate the challenges and choices necessary to go about minimizing risk while maximizing nutrition, but many issues still require research and systemic or policy changes.[53] As engaged and sceptical cooks, eaters, and voters, we must dig a little deeper than the headlines, evaluating what is important and what we can and cannot change as we seek nourishing, safe, just, and joyful food in a changing and uncertain world.

About the Author

With degrees in chemistry and oceanography and an international career in various aspects of environmental science, **Sabine E. Apitz** has focused on the evolving question of how we transparently and meaningfully use cross-disciplinary science to more sustainably evaluate and manage human impacts and demands on ecosystems, building links in the chains between what we can measure as scientists and what we want to achieve as a society.

Notes

1 Food and Agriculture Organization of the United Nations (FAO), 'Trace Elements in Soils and Agriculture', *FAO Soils Bulletin*, 17 (1979), p. 69.
2 Spyros N. Michaleas and others, 'Theophrastus Bombastus Von Hohenheim (Paracelsus) (1493–1541): The Eminent Physician and Pioneer of Toxicology', *Toxicology Reports*, 8 (2021), pp. 411–14.
3 M. J. McLaughlin, D. R. Parker, and J. M. Clarke, 'Metals and Micronutrients – Food Safety Issues', *Field Crops Research*, 60 (1999), pp. 143–63.
4 M. A. Olivera and P. J. Gregory, 'Soil, Food Security and Human Health: A Review', *European Journal of Soil Science*, 66 (2015), pp. 257–76.
5 Päivi Ekholm and others, 'Changes in the Mineral and Trace Element Contents of Cereals, Fruits and Vegetables in Finland', *Journal of Food Composition and Analysis*, 20 (2007), pp. 487–95.
6 Donald R. Davis, Melvin D. Epp, and Hugh D. Riordan, 'Changes in USDA Food Composition Data for 43 Garden Crops, 1950 to 1999', *Journal of the American College of Nutrition*, 23 (2004), pp. 669–82.
7 John Doran, 'Soil Health and Global Sustainability: Translating Science into Practice', *Agriculture, Ecosystems & Environment*, 88 (2002), pp. 119–27.
8 K. M. Nienstedt and others, 'Development of a Framework Based on an Ecosystem Services Approach for Deriving Specific Protection Goals for Environmental Risk Assessment of Pesticides', *Science of the Total Environment*, 415 (2012), pp. 31–38.
9 Diana H. Wall, Uffe N. Nielsen, and Johan Six, 'Soil Biodiversity and Human Health', *Nature*, 528 (2015), pp. 69–76.
10 David R. Montgomery and Anne Biklé, *What Your Food Ate: How to Heal Our Land and Reclaim Our Health* (Norton, 2022), p. 400.
11 David R. Montgomery and Anne Biklé, 'Soil Health and Nutrient Density: Beyond Organic vs. Conventional Farming', *Frontiers in Sustainable Food Systems*, 5 (2021), article 699147.
12 Olivera and Gregory, 'Soil, Food Security'.
13 Noushin Hadadi and others, 'Intestinal Microbiota as a Route for Micronutrient Bioavailability', *Current Opinion in Endocrine and Metabolic Research*, 20 (2021), article 100285.
14 Viktor Bielik and Martin Kolisek, 'Bioaccessibility and Bioavailability of Minerals in Relation to a Healthy Gut Microbiome', *International Journal of Molecular Sciences*, 22 (2021), article 6803.
15 Davis, Epp, and Riordan, 'Changes in USDA'.
16 'The Impact of Soil and Climate on Wine Flavor', Wiens Cellars, 2025 <https://www.wienscellars.com/the-impact-of-soil-and-climate-on-wine-flavor-2> [accessed 29 March 2025].
17 Wendy V. Parr and others, 'Minerality in Wine: Towards the Reality behind the Myths', *Beverages*, 2018, p. 19.
18 Cornélis van Leeuwen, Jean-Philippe Roby, and Laure de Rességuier, 'Soil-Related Terroir Factors: A Review', *OENO One*, 52 (2018), pp. 173–88.
19 Kim Woodford, 'A Look at the Effects of Terroir on Taste. A Whole Pot of Soup Can Taste No Better Than the Sum of Its Parts, Right?', *Daily World*, 10 March 2018.
20 Ryohei Thomas Nakano, 'Taste of Microbes: The Terroir Explained by Rhizospheric Microbes', *New Phytologist*, 243 (2024), pp. 1639–41.
21 Salvatore Pepi and others, 'Relationship among Geochemical Elements in Soil and Grapes as Terroir Fingerprinting in Vitis vinifera L. cv. "Glera"', *Chemie der Erde*, 77 (2017), pp. 121–30.

22 Beth Burritt, ‘Mineral Nutrition: Are Animals Nutritionally Wise?’, Utah State University Cooperative Extension, 2012 <https://extension.usu.edu/behave/files/Mineral_Nutr.pdf > [accessed 31 March 2025].
23 Christophe Martin and Sylvie Issanchou, ‘Nutrient Sensing: What Can We Learn from Different Tastes about the Nutrient Contents in Today’s Foods?’, *Food Quality and Preference*, 741 (2019), pp. 185–96.
24 Djin Gie Liem and Catherine Georgina Russell, ‘The Influence of Taste Liking on the Consumption of Nutrient Rich and Nutrient Poor Foods’, *Frontiers in Nutrition*, 6 (2019), article 174.
25 Jeffrey M. Brunstrom and others, ‘Human Nutritional Intelligence Underestimated? Exposing Sensitivities to Food Composition in Everyday Dietary Decisions’, *Physiology & Behavior*, 263 (2023), article 114127.
26 Jeffrey M. Brunstrom and Mark Schatzker, ‘Micronutrients and Food Choice: A Case of “Nutritional Wisdom” in Humans?’, *Appetite*, 174 (2022), article 106055.
27 Diana Miao, Sera L. Young, and Christopher D. Golden, ‘A Meta-Analysis of Pica and Micronutrient Status’, *Am. J. Hum. Biol.*, 27 (2015), pp. 84–93.
28 ‘What Is Dirt Eating?’, WebMD, 2024 <https://www.webmd.com/mental-health/what-is-dirt-eating> [accessed 4 April 2025].
29 Minori Uchimiya and others, ‘Chemical Speciation, Plant Uptake, and Toxicity of Heavy Metals in Agricultural Soils’, *J. Agric. Food Chem.*, 68 (2020), pp. 12856–69.
30 William Hendershot and Patricia Turmel, ‘Learned Discourses – Is Food Grown in Urban Gardens Safe?’, *Integr. Environ. Assess. and Manag.*, 3 (2007), pp. 463–64.
31 Florence Barbara Awino and others, ‘Occurrence of Trace Metals in Food Crops Grown on the Mbale Dumpsite, Uganda, and Human Health Risks’, *Integr. Environ. Assess. and Manag.*, 16 (2019), pp. 362–77.
32 Daniel Gomes Coelho and others, ‘Evaluation of Metals in Soil and Tissues of Economic-Interest Plants Grown in Sites Affected by the Fundão Dam Failure in Mariana, Brazil’, *Integr. Environ. Assess. and Manag.*, 16 (2019), pp. 596–607.
33 Nuno Nunes and others, ‘Review of Sewage Sludge as a Soil Amendment in Relation to Current International Guidelines: A Heavy Metal Perspective’, *Sustainability*, 13 (2021), article 2317.
34 Raihan K. Khan and Mark A. Strand, ‘Road Dust and Its Effect on Human Health: A Literature Review’, *Epidemiology & Health*, 40 (2018), e2018013.
35 Lauren K. Chamberlain and others, ‘Heavy Metal Contamination (Cu, Pb, and Cd) of Washed and Unwashed Roadside Blackberries (*Rubus fruticose L.*)’, *Integr. Environ. Assess. and Manag.*, 20 (2024), pp. 2107–15.
36 Pavel Kalač, ‘A Review of Edible Mushroom Radioactivity’, *Food Chemistry*, 75 (2001), pp. 29–35.
37 Kalač, ‘Review of Edible Mushroom Radioactivity’; ‘Foods Subject to Controls of Radioactive Contamination Related to Chernobyl’, AGRINFO, 14 February 2024 <https://agrinfo.eu/book-of-reports/foods-subject-to-controls-of-radioactive-contamination-related-to-chernobyl/> [accessed 4 April 2025].
38 Kalač, ‘Review of Edible Mushroom Radioactivity’.
39 Alexey V. Nesterenko, Vassily B. Nesterenko, and Alexey V. Yablokov, ‘12. Chernobyl’s Radioactive Contamination of Food and People’, *Chernobyl: Ann. N.Y. Acad. Sci.*, 1181 (2009), pp. 289–302.
40 Sachiko Hirakawa and others, ‘Surveys of Food Intake Just after the Nuclear Accident at the Fukushima Daiichi Nuclear Power Station’, *Food Hyg. Saf. Sci.*, 58 (2017), pp. 36–42.
41 Pepi and others, ‘Relationship among Geochemical Elements’.
42 FAO and others, *The State of Food Security and Nutrition in the World 2020: Transforming Food Systems for Affordable Healthy Diets* (FAO, 2020), p. 320.
43 HLPE, *Food Security and Nutrition: Building a Global Narrative Towards 2030* (HLPE, 2020), p. 112.
44 Alisher Mirzabaev and others, ‘Severe Climate Change Risks to Food Security and Nutrition’, *Climate Risk Management*, 39 (2023), article 100473.
45 Richard D. Semba and others, ‘The Potential Impact of Climate Change on the Micronutrient-Rich Food Supply’, *Adv. Nutr.*, 13 (2021), pp. 80–100.
46 Victor Owino and others, ‘The Impact of Climate Change on Food Systems, Diet Quality, Nutrition, and Health Outcomes: A Narrative Review’, *Frontiers in Climate*, 4 (2022), article 941842.
47 Carlo Agostoni and others, ‘Interlinkages between Climate Change and Food Systems: The Impact on Child Malnutrition – Narrative Review’, *Nutrients*, 15 (2023), p. 41; Amanda Ruggeri, ‘How Climate

Change Could Affect Arsenic in Rice', *BBC*, 25 April 2025; Jessica C. Fanzo and Shauna M. Downs, 'Climate Change and Nutrition-Associated Diseases', *Nature Reviews Disease Primers*, 7 (2021), p. 90; Owino and others, 'Impact of Climate Change'.

48 Owino and others, 'Impact of Climate Change'.

49 Carla S. Santos, Ephrem Habyarimana, and Marta W. Vasconcelos, 'Editorial: The Impact of Climate Change on Nutrient Composition of Staple Foods and the Role of Diversification in Increasing Food System Resilience', *Frontiers in Plant Science*, 14 (2023), article 1087712.

50 Ruggeri, 'How Climate Change'.

51 Santos, Habyarimana, and Vasconcelos 'Editorial'.

52 Zaigham Shahzad and Hatem Rouached, 'Protecting Plant Nutrition from the Effects of Climate Change', *Current Biology*, 32 (2022), R721–R35.

53 Shahzad and Rouached, 'Protecting Plant Nutrition'; Thalia Morrow Sparling and others, 'Intersections of Climate Change with Food Systems, Nutrition, and Health: An Overview and Evidence Map', *Adv. Nutr.*, 15 (2024), article 100274.

5
Considering Fire and Water
Cooking and Drinking in Steerage during the North Atlantic Crossing, 1845–1855

Andrea Broomfield

William Smith, in *An Emigrant's Narrative; or, A Voice from the Steerage* (1850), recounted horrific conditions aboard the *India* on its eight-week journey from Liverpool to New York in winter 1846–47. Ireland was in the worst throes of *an Gorta Mór*, or the Great Hunger, and Smith was crowded 'tween decks with hundreds of starving steerage passengers. Due to an eleven-day delay in the ship's departure, Smith witnessed passengers consume up to 'half their provisions, without the [financial] means of obtaining more'.[1]

Well before boarding, Smith set the scene for the misery that followed. During a vicious storm three weeks out, 'it rained so heavily the whole day we could not make fires on deck to cook our victuals with', Smith recounted. Even if it were possible to cook in the downpour, ship crews battened the hatches leading down to the 'tween decks to avoid taking on excess water. Given the oppressive heat and the vile stench below, the captain attempted to open the aft hatchway a few hours at a time, but for five days, these suffocating, sick passengers 'had nothing but cold water' with their biscuit, 'for, having no means of cooking [. . .] such provisions as could not be eaten without cooking, could not be eaten at all', Smith wrote.[2]

The situation grew increasingly desperate. Wind lulls replaced gales, and the ship drifted aimlessly. Smith and others had depleted their food stores and were subsisting on the daily pound of sea biscuit ('full of maggots') that ships were required by law to distribute. Biscuits, however, along with the water, were running out, so the captain reduced the ration from two quarts to one quart a day per adult and one pint per child. At five weeks, the water ration was reduced 'from a quart to a pint per day', and the biscuit ration again halved.[3] Many were already sick from diseases exacerbated by cold and malnourishment, but now fever raged through the ship. In dire straits, the captain summoned two passing ships, neither of which could bring the *India* relief.[4]

Literally dying of thirst by the time the ship began its final leg of the voyage, Smith captured prosaically a gruesome irony: 'Our situation was truly deplorable; the bitter thought that unless we soon got relief we must perish for want of water, while we were surrounded by a sea of it'. What water did remain 'was in such a state of decomposition, that no human being could drink it, until forced by being placed in similar circumstances with ourselves'.[5]

Smith's recollections of the North Atlantic crossing reflect the experiences of some two million Irish emigrants who sailed to North America between 1846 and 1855, with 1.5 million of them landing in the United States.[6] Even if they were blessed with fast voyages and fair weather, steerage passengers learned quickly that surviving the crossing came down to two elements: water and fire. Most provisions required rehydration and cooking, but fires were too few and often prohibited. Meanwhile, potable water was never enough, almost always foul and a disease carrier. In 1847 alone, some 25,000 to 30,000 of the 215,000 Irish who attempted to settle overseas perished in their attempt.[7] The majority died on board ship or in quarantine, succumbing to waterborne illnesses and/or to complications accompanying dehydration and malnutrition.

The 1840s and 1850s were complicated, significant decades in North Atlantic maritime history, especially regarding technological advances that transformed the 'ocean railway' from sail to steam. Furthermore, tension between *laissez-faire* economics and demand for government regulation intensified, particularly when, in 1845, *Phytophthora infestans* infected Irish potato crops and placed unprecedented stress on a shipping industry unprepared to transport millions of emigrants. Roughly seventy-five thousand people left Ireland in 1845 alone.[8] They boarded barely seaworthy transient cargo vessels and several older first- and second-generation American packet ships.

While historians have paid significant attention to the predictably horrific conditions aboard unseaworthy cargo vessels, so-called coffin ships that primarily sailed from Irish ports, less attention has been paid to American packets, the preferred way to cross the Atlantic if an Irish emigrant could afford the financial cost of crossing the Irish Sea to depart from Liverpool. American ships (primarily packets) accounted for up to 90 percent of the foreign ships docked there. By 1848, they captured ever-larger proportions of the passenger trade.[9]

Given that American packets were reputedly the best-made, safest, and fastest ships sailing the North Atlantic, why would an Irish emigrant not prefer them over other vessels? Were they not the best choice? The answer is ambivalent at best, succinctly captured by Foster in *Work and Wages; or, The Penny Emigrant's Guide*: 'As regards mere ships, Liverpool ranks highest; as regards treatment of passengers, lowest'. Furthermore, Foster considered, 'The advantages of going to New York are the more abundant choice of the best ships, – with, however, the worst treatment'.[10]

This paper considers Foster's observations as they pertained to fire and water, for the logistics of staying fed and hydrated on purportedly the best vessels afloat underscore the truth of Foster's observations and enhance our understanding of the role these touted vessels played between 1845 and 1855. Passengers' acts mandated by the British Parliament between 1842 and 1855 and by the US Congress between 1817 and 1855 were designed to mitigate passenger suffering when it came to diet and hydration, but they proved ineffective. Foster was a catalyst for Britain and the United States passing the strictest regulations in 1855, but that push came too late. Steamships, starting with the Inman Line in spring 1852, ultimately ended the worst of passenger suffering by taking the unprecedented step of admitting steerage passengers aboard its ships for just a few pounds more than it cost to take a sailing ship.

While many Famine and Irish histories devote a chapter to the North Atlantic crossing, they primarily concern the most egregious injustices aboard vessels not designed to carry passengers in the first place. When it comes to packets, however, historians tend to either bunch the history of their involvement in the emigrant trade together with that of other emigrant vessels (such as freighters) or neglect careful consideration of emigrants and instead concentrate on the luxuries that packets offered first-cabin passengers. Information is particularly vague on how packets stored and distributed water and provisions and how steerage passengers cooked.[11] However, by foregrounding fire and water, I seek to shed more light on a traumatic history. While the Irish were vulnerable to gross governmental ineptitude and cruelty, so too were they vulnerable, in the worst and most extreme ways, to the elements.

American Packets: Background and Context

Packets were typically three-masted, square-rigged ships that carried passengers, cargo, and the mails (bound in packets, hence the name). The first private packet companies began in New York City, starting in 1818 with the Wright, Thompson, Marshall, and Thompson Line, popularly known as Black Ball. Unlike other existing packet services, Black Ball committed to sailing monthly from New York and Liverpool on advertised departure dates, regardless of weather or how much cargo was in its ships' holds.[12] Given this unprecedented commitment to schedules and speed, Black Ball was immediately popular with the merchant class, and the company's success encouraged competitors along the US East Coast. The ships carried an abundance of sail to help with speed, and westbound voyages averaged between twenty-seven and fifty-one days, with eastbound voyages typically faster, between sixteen and thirty-one days.[13] The first generation of packets, built between 1816 and 1825, averaged 350 tons.[14] Around 1825, these vessels were replaced by 'a larger and more uniform type', around 500 to 800 tons.[15] Until the mid-1840s, American packets were at the height of their importance and withstood competition from the first generation of steamships. Not only did packets depart on time, but they offered first-cabin passengers sumptuous accommodations and 'excellent cuisine'. They 'made little effort to secure steerage passengers' because the 'tween decks space in the upper hold 'could be used just as profitably for fine freight' as it could for steerage passengers.[16]

However, as steamships gained stature, sailing packets lost status. To stay solvent, companies lowered first-cabin ticket prices and began accepting ever-increasing numbers of steerage passengers. Packets sailing when the worst effects of the blight began to be felt in autumn 1845 might be compared to restaurants that were once famous for luxury but that began fading into sad shadows of their former selves. Edward Knight Collins's Dramatic Line, for example, was famous in its heyday for gorgeously appointed large ships, but they began falling into disrepair. Its 895-ton *Garrick*, for example, had become by 1854, in the words of Captain R. W. Foster, 'most shamefully neglected and allowed to go to destruction in hull and rigging' even though it was still sailing.[17]

Macabre, perhaps, but *an Gorta Mór*'s onset might have been seen by packet companies as a 'godsend'. To accommodate swells of desperate emigrants, ship carpenters reoutfitted older vessels for westbound voyages by taking the 'tween decks – or more specifically, the upper hold between the ship's main deck and lower deck – and throwing up rough pine berths. Other companies followed suit. The Copes Line reduced fares and began attracting steerage passengers whose numbers per ship increased from 64 in 1840 to 133 in 1845 and 214 in 1850, while its westbound first-cabin fares fell from £35 to £30 in the mid-1830s and to £20 in the mid-1840s. Steerage fares fluctuated between £3 and £4 and resembled other American packet lines in this price range.[18]

Meanwhile, packet companies also responded to demand by building three-deck ships ranging from 1000 to 1600 tons and creating within them the equivalent of a third and fourth class, with steerage designed 'between the upper deck and the main deck, and the lower tween decks between the main deck and the lower deck'.[19] Size, however, seldom came with more amenities or better service, and sometimes it meant the worst service and passenger abuse, particularly regarding water, food, and cooking.

Water Logistics

When it came to bringing aboard, storing, and dispensing water on packets, the chaos and failures were arguably as pronounced as what emigrants experienced on crowded cargo vessels leaving Ireland. Certainly, American packets flaunted or ignored the passengers' acts governing ships departing from UK ports.[20] From 1842 through 1855, the acts addressing the water ration remained unchanged: 'a supply of not less than three quarts of water' to 'be issued to each passenger per day', a point to which I return, as so much misery emanated from it.[21]

Water quality was never guaranteed safe from contaminants and inevitably was 'brackish or foul'.[22] An 1848 investigation 'disclosed that at nearly half the Irish ports ordinary river water was taken aboard for the passengers to drink'.[23] Liverpool ships were supplied with water from the privately controlled Harrington Water Company or the Liverpool Waterworks Company. Bootle claimed (without verification) that its water was 'abundant and excellent'. However, it was impossible in the mid-1840s to be certain of water's purity, irrespective of what companies or city officials claimed.[24] In Liverpool, water was pumped into ships' casks by means of dock hydrants, ideally on departure day. But prior to filling, the casks lay there in a filthy jumble of crates, baggage, and livestock.[25]

Repeatedly, the passengers' acts regarding water were overlooked or manipulated, resulting in emigrants' constant thirst and misery but also exposure to infectious disease. Although the chief emigration officer at Liverpool was supposed to test the water, he typically took the ship captain's word that the ship was compliant. Isaac Brown, in 'Ten Trips across the Atlantic' (1848), recalled that prior to embarkation, if the emigration officer or an agent 'does at all come on board the ships sailing out of Liverpool, it must be into the cabin to see the captain, for he is never seen anywhere else below'. The Office of the Emigration Agency, Brown surmised, 'is very likely a place of government patronage'.[26] Repeatedly, contaminated water and failed inspections contributed to infectious disease outbreaks at ports and on ships.

During the worst years of *an Gorta Mór*, 1846–49, water casks were invariably old and so leaky that water often ran short by mid-voyage, even when a ship made good time.[27]

William Smith's *India* voyage was so desperate because the captain and/or emigration officials apparently ignored laws that dictated both water storage and quantity. Shortly after *India* sailed, it sprang a leak and was returned to dock. Smith watched as water casks were removed so the leak could be repaired. By 1847, a new passengers' act stipulated that when 'distressed vessels' returned to dock, provisions and water were to be replenished.[28] But such did not happen. A month into the voyage, it was discovered that four of the casks, 'by oversight or neglect, had not been refilled'.[29]

Even if the water was pumped into watertight casks in fair condition, it would usually end up contaminated by the casks themselves because most had previously held other substances, including turpentine, oil, animal hides packed in salt to prepare them for the tanner, rum, molasses, or wine.[30] While the 1842 Passenger Act dictated that casks be 'sweetened by being soaked in quicklime' before filling, it is dubious to what degree 'sweetening' helped – if it took place at all.[31] To make the water more palatable and safer, it needed boiling, but steerage passengers were limited in access to fire, and the linkage between illness and dirty water was not fully understood. Epidemiology was only in its early stages in the 1840s. Regardless, had water been boiled, diseases such as diarrhoea, dysentery, and cholera that routinely raced through steerage might have been slowed. Water storage on passenger ships would not be remedied in any meaningful fashion until the 1850s, when newer vessels installed iron tanks in the hold. These fed water, by means of pipes and a pump, to the poop deck, although even then, these innovations 'did not always work to the benefit of the hapless souls in steerage'.[32]

Cooking Logistics

While a ship cook was, at least theoretically, on hand to oversee the stoves and meal preparation by 1849, until ships were required to serve passengers cooked meals in appointed dining areas, passengers were left to their own devices or at the mercy of the cooks. It is thus to fire and cooking that I turn.

Fire caused the gravest threat to wooden vessels, and hence, regulations surrounding its use on board were strict, including where, when, and how long cooking fires burned. Typically, they were permitted between 7 a.m. and 7 p.m. when the weather was fair. Aside from their quarters on the 'tween decks, steerage passengers were only permitted to use space on the main deck to cook and wash. Packets in the 600- to 1000-ton range typically included on their main deck a deckhouse to berth the crew and to operate the galleys. On some ships, a steerage galley sat adjacent to the galley that fed the crew and cabin passengers. Basil Lubbock characterized them as cramped spaces, 'hemmed in by the cow-house, sheep, pig and poultry pens', whose meat and milk supplied first-cabin passengers and officers. The galley 'was generally open to the weather, a mere range of fires, above which a rail with sliding hooks served to suspend the cook pots over the flames'.[33]

While some newer packets had functioning stoves, most in the mid-1840s offered steerage passengers nothing more than communal grates housed on either side of the foredeck: 'rudimentary boxes lined with bricks, a crude form of barbecue' and poorly protected from waves and rain. Robert Whyte characterized his vessel's 'fireplaces' as 'shaped something like an old-fashioned settee – the coals being confined by two or three iron bars in front'.[34]

Typically, the ship carpenter or second mate allowed groups of twenty to thirty passengers on deck at a time.[35] While this regimen allowed for order, it inevitably created anxiety for those trapped below waiting for their turn to cook. Certainly, brawls around the fires were often reported, including Sarah Mytton Maury's account regarding her 1845 voyage aboard New Line's *Hottinguer*.[36] This 993-ton ship adhered to the recommended one grate per one hundred passengers. For its 397 steerage passengers, it offered four grates.[37] The daily news at the cabin passengers' breakfast table, recounted Maury, concerned the steward's reports of the previous night's fight in steerage, one often started over food: 'What did they fight about?' 'About the fire; same thing they always fight about, – all will cook at a time'.[38] Despite four grates, competition for space and time at the fire was intense because hunger exacerbated people's anxiety and bad tempers, as did the ever-present threat of water, be it rain or, as one steerage passenger told Maury, the 'spray, and the wind' that put out fires.[39]

Complicating matters further was the distance from the booby hatch or main entrance from the 'tween decks to the grates, situated well forward on the main deck. 'How often I have witnessed dozens of young girls saluted by a green wave as they tried to reach the galley', wrote the narrator of 'An Old-Time Emigrant Ship'. They sometimes ended up in a heap, 'on top of each other in the lee scuppers in a most helpless condition and washed the food which they were taking to have cooked into the sea to the delight of the seagulls'.[40]

Given the stress caused by the uncertainty of when one might be able to prepare a meal, many strategies were hatched among passengers and complicit captains and crews. William Cobbett urged steerage passengers to carry 'a bottle of brandy' as a bribe 'to be judiciously administered'. The cook 'would bid you toss your money into the sea; but he will suck down your brandy; and you will get many a nice thing prepared by him, which you never would get, if it were not for that brandy'.[41]

Water and Fire: Slack Thirst or Cook?

Although it is unclear how such manoeuvrings happened if passengers were trapped below during a storm, some likely found the means to follow another piece of Cobbett's advice. Bring along apples as well as brandy, he wrote. 'Your wife will sit up in her berth, in very rough weather, and make the puddings in a large tin pan. [. . .] The cook will boil them for you; he will bake or broil cakes for you, and boiling water is all that you want for the slops of various description' as well as warming drinks.[42]

How vital warm drinks and hot food would be to passengers who spent much of their time shivering, damp, sometimes soaked, and often ill. The most common provisions that the Irish carried aboard ship were oatmeal and potatoes, both of which greatly benefitted from being cooked in boiling water, as was true of most ship-issued rations.[43] Starting in 1842 and increasing in amount and variety in 1849, passengers were to receive weekly, along with water, breadstuffs (navy bread or biscuit) and/or bread substitutes, especially potatoes, flour (maize or wheat), and oats.[44] When the horrors of 1847 resulted in thousands of passengers dying of starvation-related diseases because they were surviving solely on that ration, legislation increased the amount and variety of food doled out, but the water remained at three quarts despite the fact that the additional food still required water as a cooking medium and/or ingredient. Clause 24 of the 1849 Passengers' Act mandated two and a half pounds of bread or navy biscuit, along with one pound of flour, five pounds of oatmeal, two pounds of rice, two ounces of tea, and a half pound each of sugar and molasses. Potatoes could be substituted for oatmeal or rice.[45]

However, not only was it sometimes impossible to reach a fire to boil water or cook, but a lot of times, passengers had to choose between slaking their thirst or cooking their food, and thus, I reach the most damning of injustices that literally came down to the elements. In short, there was seldom enough potable water; a mere twelve cups per adult and six per child (or less) was a soberingly small amount to work with. Two powerful men, Sir Stephen Edward de Vere (1812–1904) and his cousin, Vere Henry Louis Foster (1819–1900), made that forced choice between cooking or drinking crystal clear to government officials and a scandalized public.

Sailing to Quebec alongside his Irish tenants in 1847 on what he described as a 'better-regulated' vessel, de Vere wrote:

> The supply of water shipped on board was abundant, but the quantity served out to the passengers was so scanty that they were frequently obliged to throw overboard their salt provisions and rice (a most important article of their food), because they had not water enough for both the necessary cooking, and the satisfying of their raging thirst afterwards.

Passengers remained 'for days together in their dark close berths, because they thus suffered less from hunger, though compelled, at the same time', de Vere repeated, 'by want of water to heave overboard their salt provisions and rice'.[46]

Roughly three years after de Vere's voyage, Foster submitted his own scandalous reports regarding an October–December 1850 voyage aboard Black Star's new packet, *Washington*. He exposed the cruelty that Commander A. Page and his crew wielded over passengers whose lives literally depended on them. When *Washington* left Liverpool, Foster recorded that 'the whole 900 and odd passengers were called forward at once to receive their water pumped out into their cans from barrels on deck'. However, water distribution 'was twice capriciously stopped by the mates of the ship who, during the whole time without any provocation, cursed and abused and cuffed and kicked the passengers and their tin cans'. Only sixty received their water ration on the first day, while on the next, the rest received theirs, albeit with more physical abuse. While water was more frequently distributed than provisions, the uncertainty as to when and how much would be allotted further increased the passengers' terror and misery.[47]

Foster laid bare the realities of what transpired on many packets, irrespective of tougher passengers' acts, and he did so in ways that powerfully reinforced his observations about ships in his *Emigrant's Guide*. The 1849 act – on paper – was treated by many as a breakthrough, with Clause 26 for the first time stipulating that 'no ship carrying

100 passengers shall clear out without a person rated in the ship's articles as passengers' cook, and a proper cooking apparatus on board'.[48] But as Foster reported, the situation regarding stoves and cooks on the *Washington* amounted to abuse. He also heard passengers complaining of 'favouritism shown by the passengers' cooks to those who give them money or whiskey, and who consequently get five or six meals cooked daily'. On 25 November, Foster reported that a twelfth child died. As with the others, this child 'died of dysentery for want of proper nourishing food and was thrown into the sea sewn up, along with a great stone, in a cloth'.[49]

Some packet ships did perform admirably when it came to the laws mandating rations and cooking areas, including those in Enoch Train's well-regarded Diamond Line. However, the Black Star Line operated some of the worst vessels, including the *Atlas* under Captain Osborne, whose 415 passengers fared 'considerably worse' than those on the *Washington*, wrote Foster. The *William Rathbone* likewise sported gross irregularities with the distribution of rations.[50]

Conclusion

Foregrounding water and fire in the history of Irish migration during *an Gorta Mór* likewise foregrounds the futility of the passengers' acts, despite modest improvements made over twelve or so years. Had various commissions and Parliament members more forcefully dealt just with the matter of water – demanding that ships stock enough, that it be stored in new waterproof casks, that passengers be granted double the amount doled out, that the doling occur daily – the journey would have been easier for millions to endure, and many lives would have been saved.

Tragically, scientists, engineers, and government officials did address the seriousness of water, but their efforts were ignored or fell short. In one of the final and most damning investigations to go before Parliament, initiated by John O'Connell in March 1854 regarding 'the recent causes of loss of life on board emigrant ships', experts repeatedly addressed the problem of water. Dr O'Doherty, a ship physician who made ten crossings between 1850 and 1854, testified that 'the water allowance was almost invariably short, with the initial deficiency often aggravated by bursting of bad casks'. He deplored the 'quality and the cooking of the food – it "is not what you would give your servants" – and he described American vessels on which the bread left over from one crossing was furnished again for the next'.[51] Meanwhile, engineers and scientists at the behest of emigration commissioners experimented with condensers that could provide water if a ship were to run short, and they worked with distillation and purification machines.[52] However, the commissioners confined the use of such distilling and purification machines to steam vessels, presumably because of the absence of engineers on sailing ships.[53]

Steamships ultimately rendered the passengers' acts redundant, particularly 'the invention of a new kind of steamship: not quite as fast as the mail steamers, but as safe and well ordered, and – the crucial point – much cheaper to run than any steam vessel yet on the ocean'.[54] The Inman Line broke precedent and began taking aboard steerage passengers in 1852–53.[55] Although life for this class aboard steamships was uncomfortable and crowded, the water was abundant and filtered. The significantly shorter voyage, if nothing else, ended complications that arose from sailing for weeks when one was already dehydrated and malnourished, if not starving.

In 1912, Peter Roberts, author of *The New Immigration*, made a salient observation regarding steerage passengers and their creature comforts. As with the packet ship companies before them, steamship companies 'study the comfort and convenience' of cabin passengers who 'cross the ocean for pleasure', while the needs of steerage passengers 'have had the least possible attention'.[56] Although Irish Americans are several generations removed from *an Gorta Mór*, more so than when Roberts recorded his observations in 1912, they still resonate with descendants of Famine emigrants, especially Roberts's observation that 'the dread of hunger and of thirst, the terrible ravages of ship fever and epidemics, the absence of conveniences which common decency demands [. . .] still hang as a pall in the [descendants'] memories, and more than half a century of life in the United States has not been able to obliterate the impression. The throes of that voyage – made two generations ago – are still rehearsed with a feeling which can only be explained by the abject misery endured'.[57] A minute focus on fire and water brings a particular force to Roberts's observations, as it did to Foster's.

About the Author

Andrea Broomfield is a Professor of English at Johnson County Community College in Overland Park, Kansas, US. She is the author of *Food and Cooking in Victorian England: A History* and *Kansas City: A Food Biography*, among other books and articles.

Notes

1 William Smith, *An Emigrant's Narrative* (William Smith, 1850), pp. 5–7.

2 Smith, p. 8.

3 Smith, pp. 12, 18.

4 Smith, p. 13.

5 Smith, pp. 14, 17.

6 Stephen Fox, *Ocean Railway* (Harper Perennial, 2003), p. 169. Exact numbers of those arriving in North America are nearly impossible to calculate. Prior to 1847, the first year the New York Commissioners of Emigration began to regularly differentiate Irish emigrants from others arriving from the UK, such differentiations were not made or were sloppily done. See Terry Coleman, *Passage to America: A History of Emigrants from Great Britain and Ireland to America in the Mid-Nineteenth Century* (Hutchinson, 1972), pp. 297–98.

7 Oliver MacDonagh, *Pattern of Government Growth 1800–1860: The Passenger Acts and Their Enforcement* (Gregg Revivals, 1993), p. 187.

8 Patrick Fitzgerald and Brian Lambkin, *Migration in Irish History, 1607–2007* (Palgrave Macmillan, 2008), p. 176.

9 MacDonagh, *Pattern of Government Growth*, p. 223.

10 Cathy Joynt Labath, 'Vere Foster's Penny Emigrant Guide, "Emigration to North America"', *Irish in Iowa*, 2001–06 <https://www.celticcousins.net/irishiniowa/pennyemigrantguide.htm> [accessed 30 April 2025].

11 Cian T. McMahon devotes significant attention to 'coffin' ships: see *Coffin Ship: Life and Death at Sea during the Great Irish Famine* (New York University Press, 2022). Edward Laxton's *Famine Ships: The Irish Exodus to America, 1846–51* (Henry Holt, 1997) likewise covers the worst voyages. Robert Albion's 1965 *Square-Riggers on Schedule: The New York Sailing Packets to England, France, and the Cotton Ports* (Princeton University Press, 1965), a definitive packet-ship history, is indicative of the limited attention

scholars have paid to steerage-class passengers aboard sailing packets. As Albion notes, 'We shall not examine the steerage as minutely as we have the cabin because it was not as characteristic of the packets in their heyday' (p. 247). Furthermore, information on packet ships' interior arrangements – their galleys, water tanks, and filtration systems – is difficult to locate. Packets were 'built with the same integrity, and the same governing rules, as clippers of the day', William Crothers argued, but they 'did not generate the interest or captivate the imagination of the public as the clippers had done', especially regarding the history of emigrants relying on them for voyages to the United States during *an Gorta Mór* (*American-Built Packets and Freighters of the 1850s* (McFarland & Company, 2013), p. 6).

12 Albion, *Square-Riggers on Schedule*, p. 20.

13 Seija-Riitta Laakso, *Across the Oceans: Development of the Overseas Business Information Transmission, 1815–1875* (Studia Fennica Historica, 2016), p. 111.

14 Tonnage refers to 'old tonnage measurement', defined as a vessel's cubic capacity based on the American rule of measurement in effect until 1865.

15 Albion, *Square-Riggers on Schedule*, p. 79.

16 Albion, p. 247.

17 Qtd in Basil Lubbock, *Western Ocean Packets* (James Brown & Son, 1925), p. 119.

18 John Killick, 'Transatlantic Steerage Fares and Migration, 1815–1860', *Economic History Review*, 67.1 (2014), p. 179.

19 Crothers, *American-Built Packets*, p. 333.

20 UK and the US passengers' acts are complicated because many legal battles concerned who governed a vessel on the high seas. However, regardless of which nation owned a ship, when it docked at a UK port, it was subject (at least on paper) to the British passengers' acts. MacDonagh's scholarship (*Pattern of Government Growth*) on the complications remains the best.

21 'The Passenger Acts', *Australian* (Sydney, NSW), 25 July 1842, Trove, National Library of Australia <https://trove.nla.gov.au/newspaper/article/37116643> [accessed 30 April 2025].

22 MacDonagh, *Pattern of Government Growth*, p. 151.

23 MacDonagh, p. 50.

24 Coleman, *Passage to America*, p. 114.

25 Smith, *Emigrant's Narrative*, pp. 17–18.

26 MacDonagh, *Pattern of Government Growth*, p. 224; Isaac Brown, 'Ten Trips across the Atlantic', *The People: Their Rights and Liberties Their Duties and Their Interests*, 1.12 (1848), p. 91.

27 MacDonagh, *Pattern of Government Growth*, p. 151.

28 MacDonagh, p. 201.

29 Smith, *Emigrant's Narrative*, pp. 17–18.

30 Kevin Brown, *Poxed and Scurvied: The Story of Sickness and Health at Sea* (Seaforth Publishing, 2011), p. 129; MacDonagh, *Pattern of Government Growth*, p. 50; Melvin Maddocks, *Atlantic Crossing* (Time-Life Books, 1981), p. 122.

31 MacDonagh, *Pattern of Government Growth*, pp. 50, 151.

32 Maddocks, *Atlantic Crossing*, p. 122.

33 Lubbock, *Western Ocean Packets*, p. 102.

34 Laxton, *Famine Ships*, p. 29; Robert Whyte, 'Ocean Plague', in *Irish Famine Migration Narratives: Eye-Witness Testimonies*, ed. by Jason King, vol. II of *History of the Irish Famine*, ed. by Christine Kinealy, Gerard Moran, and Jason King (Routledge, 2019), p. 93.

35 Laxton, *Famine Ships*, pp. 12, 29.

36 MacDonagh, *Pattern of Government Growth*, pp. 49–50.

37 Albion, *Square-Riggers on Schedule*, pp. 247–49; Sarah Mytton Maury, *Englishwoman in America* (Thomas Richardson and Son, 1848), p. 5.

38 Maury, *Englishwoman in America*, p. 22.

39 Maury, p. 27.

40 Denis Kearney, 'An Old-Time Emigrant Ship: Across the Atlantic on a Blackball Liner', *San Francisco Call and Post*, 7 April 1895, p. 15.

41 Maury, *Englishwoman in America*, p. 50; William Cobbett, *Emigrant's Guide* (Mills, Jowett, and Mills, 1829), pp. 106–07.
42 Cobbett, *Emigrant's Guide*, p. 115.
43 Lubbock, *Western Ocean Packets*, p. 102.
44 'Passenger Acts'. For ticket prices and rations, 'passenger' meant 'adult passenger'. Two children under fourteen and three children under seven equaled one passenger. Those under one year were not counted. See MacDonagh, *Pattern of Government Growth*, p. 150.
45 'Passengers' Act, 1849', Rootsweb, Ancestery.com, 1997–2018 <https://freepages.rootsweb.com/~pbtyc/genealogy/Gazette/Shipping_Legislation_etc/Passengers_Act_1849.htm> [accessed 30 April 2025].
46 'Steven de Vere to T.F. Elliot, 30 November 1847', in *Irish Famine Migration Narratives*, ed. by Kinealy, Moran, and King, II, p. 48.
47 I relied on excerpts of Foster's diary reprinted in Laxton's *Famine Ships*: see pp. 194–99.
48 'Passengers' Act, 1849'.
49 Qtd in Laxton, *Famine Ships*, pp. 198–99.
50 Laxton, p. 200.
51 Qtd in MacDonagh, *Pattern of Government Growth*, pp. 270, 278–79.
52 Fox, *Ocean Railway*, p. 182; MacDonagh, *Pattern of Government Growth*, goes into detail about water requirements in the passengers' acts in 1855: see p. 247.
53 MacDonagh, *Pattern of Government Growth*, p. 247.
54 Fox, *Ocean Railway*, p. 172.
55 Fox gives the most lucid history of the Inman Line: see pp. 168–95.
56 Peter Roberts, *The New Immigration* (Macmillan, 1912), p. 1.
57 Roberts, p. 1.

6
The Flavour of Bitterness in Rituals of Trauma
From the Jewish Passover Seder and Beyond

Jonathan Brumberg-Kraus

Can bitterness be therapeutic in meal rituals that address historical trauma in communal stories of identity and origins? In my Jewish tradition of the Passover seder, bitter foods punctuate the bitter (or bittersweet) story of oppression and redemption: the paradigmatic Exodus from enslavement in Egypt. We taste *maror* (the bitter herb) because the Egyptians 'embittered our lives'; we eat the bittersweet 'Hillel sandwich' (which combines a bitter herb and sweet *charoset* between two pieces of matzah, the unleavened 'bread of affliction') – 'to sweeten the bitter'. With trauma an important component of Jewish identity and a fundamental component of the human condition and the stories we tell about ourselves, it makes sense to taste bitter foods at meal rituals to evoke those stories, as in Jewish Passover seders, Christian Eucharistic meals, and American Thanksgiving dinner rituals. Yet bitter foods are not central components of the latter two rituals. The bread and wine of Communion are not notably bitter but rather convey their symbolism for the body and blood of Christ 'given for us' in their visual and tactile sensory qualities: the solid earth-toned and savoury substance of the bread (like flesh), the flowing liquid red of the wine (like blood). The tastes are hearty and substantial (except in relatively tasteless Communion wafers), the wine sweet, intoxicating, and celebratory. Consumption combines the visual, tactile cues to the trauma of Jesus's death with the joyful, sustaining, uplifting flavours of the promised Resurrection and salvation – not *bitter*sweet per se, but still a complex oxymoronic sensation of sorrow and pleasure. Thanksgiving meal foods are hearty, savoury, stuffed, and often sweet: stuffed turkey, sweet potatoes, pumpkin and pecan pies, and tart-sweet cranberry sauce, with desserts and sides doubled or tripled compared to ordinary meals to symbolize the sweet abundance of the fall harvest with Native American foods. And there's the rub, the conspicuous absence of bitter flavours and the bitter history that brought Americans to this day: the genocide of Native Americans and colonization and appropriation of their land and the enslavement of Black Africans upon whose suffering our country was built. While the tart-sweetness of cranberry sauce could substitute sour for bitter to evoke the not wholly sweet experience of American history, it typically doesn't. We don't usually call attention to the discordant sweet sourness of Native American cranberry sauces garnishing our turkeys, revealing instead a collective American amnesia for the painful parts of our national history.

Memory meals are selective about what and how to remember. We can recall the bittersweetness of historical trauma almost completely with the contrasting flavours on our palates in the Jewish Passover seder; perhaps sense them by contrasting tastes with the tactile and bloodlike visual sensations of Christian Communion; or omit the bitter from our conscious memories altogether in American Thanksgiving meals as usually practiced. Thus, while bitterness or bittersweetness may have an important role in the stories of identities to tell at meal rituals, they are but one of several available tools people choose when they use meals to recall their stories.

To argue for using bitter or bittersweet foods to tell stories of identity to integrate both trauma and triumph, I will discuss (1) how bitterness works on our emotions and memory, moving us from flavour to memory, cognition, and mood; (2) whether meals evoke memory for its own sake (i.e. of deceased or absent family, friends, and places) or cultivate some specific pedagogical agenda; (3) how some modern Jewish meal trauma rituals surprisingly depart from the traditional Passover seder bitter herb memory practice; (4) how to put the bitter and bittersweet back in new rituals of trauma and reconciliation beyond Jewish stories and rituals (e.g. the idea of 'bitter medicine' especially in Native American and Palestinian herbalist traditions); and (5) how to make rituals of inclusive tables at which colonizers and the colonized can live together in peace.

How Bitterness Works on Our Emotions and Memory

Why stress the elemental flavour of bitterness, to which we are 'naturally' averse? Its primary function seems to be to trigger a disgust reflex, a first defence in our mouths protecting us from ingesting poisons that can hurt or kill us. Jennifer McLagan, who wrote the book *Bitter*, calls it 'the world's most dangerous flavor'.[1] Yet when we and our palates have matured, we tend to acquire a taste for bitter things in small doses and/or in the right combination with other flavours, 'like coffee, bitter alcohols, and chocolate'.[2] So what is the element of bitter, and how exactly does it affect our emotions and memory? For McLagan, 'Bitter is both taste and a flavor'.[3] That is, bitter is not just taste on our tongues:

> Smell, temperature, color, texture, and how the food feels in our mouth all relay a sense of bitterness to our brain. The pungency of arugula and horseradish can evoke a taste of bitter, as can the astringency that you find in celery, or the tannins in tea and cooked apricots. These sensations are delivered not through our taste buds, but via our somatosensory system, which includes touch, temperature, and texture.

Moreover, what people call bitter varies because of 'cultural, environmental, experiential, and genetic factors'. Some taste bitterness more strongly than others; some will taste rutabagas as bitter, others as sweet. Thus, 'when it comes to bitter, our understanding of it – even our experience of it – differs more widely than that of any of the other basic tastes'.[4]

McLagan roughly distinguishes between five-ish kinds of bitter: (1) the taste of bitter greens (greens in the chicory or endive family); (2) the 'liquid bitterness' of coffee, bitter alcohols (like Aperol), and tea; (3) 'pungently bitter', the typical characteristic of plants in the brassica family; (4) 'subtly bitter', especially in some fruits; and (5) what she calls 'surprisingly bitter', her catchall term for (a) plant foods not in the chicory or brassica families (white asparagus, cardoons, and celery), (b) 'dark' and burnt flavours (chocolate, toast, and caramel), (c) 'forbidden bitter' foods (e.g. tobacco consumed as food), and (d) 'very bitter' foods (fenugreek and bitter melon).[5] The taste of bitter greens like chicory, endive, and escarole is what most call bitter and seems most like the 'bitter herbs' to which the Biblical and rabbinic Passover observances originally referred. Now, especially among Ashkenazi Jews, it's the 'pungently bitter' horseradish root consumed as the *maror*/bitter herb. The pungently bitter brassicas – such as brussels sprouts, cabbage, kale, rutabaga, turnip, broccoli, mustard, horseradish, and arugula – typically have a sulphurous smell when cooked but also bear the medicinal benefits of the glucosinolates they contain.[6] We sense pungently bitter brassicas not only by tasting them but also by smelling them in the mouth and nasal passage and feeling them as a stinging sensation (which some might call spicy or hot). The quality of liquid bitters speaks for itself, as in some of their names: 'Angostura bitters' and 'Amaro'. Among 'subtly bitter' foods, McLagan includes fruits like grapefruit, the peels and piths of citrus fruits, the tannins in red currants and apricots, walnuts (from their skins), and extra virgin olive oil. The antioxidants in some have health benefits. Among her 'surprisingly bitter' foods, the 'forbidden bitter' of tobacco is patently unhealthy and not to be served to children, points McLagan makes when offering two tobacco-based recipes, tobacco chocolate truffles and tobacco panna cotta.[7] The 'very bitter' flavoured fenugreek and bitter melon are well-known in Indian culture for their medicinal value and their extreme bitterness. And the 'dark' burnt bitterness of chocolate, toast, and caramel, despite their bitterness, are generally considered appealing flavours, in contrast to the mixed or, for some, off-putting connotations of the pungently bitter brassica.

I doubt McLagan intends her categorization of 'bitter' to be rigorous or scientific; rather, she demonstrates the wide range of flavours people recognize as bitter. It also underlines certain important things they have in common: these bitter foods often have medicinal benefits; several are definitely 'adult flavours', acquired tastes connoting a sort of maturity; and all, especially the pungently bitter, have a powerful smell component, reinforced by other sensory components that make bitter more than just a taste but rather a complex flavour.

Smelling as well as tasting bitterness makes bitter flavours have such a significant impact on memory and emotion. That is, the 'smelling taste' of bitter flavours is both 'orthonasal olfaction' (smelled directly through the nose before and when first ingesting the food) and 'retronasal olfaction' (aromas circulating in the nasal cavity smelled while the food is in the eater's mouth). Smell has a crucial special link to memory. As McLagan puts it:

> Information from our taste buds is transmitted to our brain by an indirect route, whereas smell goes directly to our brain, registering in the hippocampus, where we store our emotions and memories, especially memories of places. *Odors create*

> *primal and long-lasting memories, which is why smell evokes a powerful emotional response. A smell can make us recall a place, a person, an event, or a feeling.* Is it possible to discuss the links between food and memory without invoking Marcel Proust's experience with a madeleine that led him to write a very long book? Probably, but it is worth noting that Proust points out that simply seeing a madeleine didn't trigger his memory:
>
>> The sight of the little madeleine had recalled nothing to my mind before I tasted it; perhaps because I had so often seen such things in the meantime, without tasting them, on the trays in pastry-cooks' windows, that their image had dissociated itself from those Combray days to take its place among others more recent . . .
>
> It was only when he smelt the madeleine mixed with the tea did Proust recall his long-forgotten childhood at Combray. [. . .] Like Proust, we all have memories, many of which remain buried until we encounter that distinctive odor again. These odor memories don't change the taste of what we eat, but they have a strong emotional power. So, while our taste buds can be retrained with repeated exposure, if the smell of the food is linked to a bad or traumatic memory, we may never like the food. Can a food smell bitter? Burnt toast, caramelizing sugar, roasting coffee, the medicinal odor of bitter alcohols, and the sulfurous, earthy smell of brussels sprouts, all trigger an expectation of bitter in our brain. Taste scientists argue that bitter is simply a taste, but for the cook, bitter also has a smell.[8]

Suppose we can reverse the process and, through repeated exposure to tasting bitter foods as something pleasurable and healing, retrain the traumatic memories associated with bitter experiences into something healing and assimilable, if not exactly pleasurable. In a psychological study of the effects of tasting sweetness and bitterness on moods, Rachel Herz, Caitlin Cunningham, and Theresa White suggest that, while tasting bitterness tends to promote hostile moods, 'the effect of bitter taste on hostility was eliminated if participants were given the opportunity to first evaluate how agreeable they were. [. . .] Priming a positive social self-schema counteracts the transient hostility towards others that bitter tastes may induce'.[9] Thus, if you're primed to think about bitter flavours as pleasurable experiences, as I'm attempting here, or to eat bitter herbs in the tasty Hillel sandwich at a joyful social gathering of a Passover seder, it can affect your mood positively.[10] If bitter flavours experienced in appealing combinations or balanced with other flavours (as medicinal, in pleasant company, and as an achievement of maturity over time) can be enjoyed – without suppressing the undertones of the bitter – why can't we do the same with bitter memories? Especially if those memories are not wholly or completely bitter, as if eating a bitter melon by itself rather than as part of a complex and balanced dish.

Memory for Its Own Sake or Memories with a Pedagogical Agenda

We use food at ritual meals especially to evoke 'memories of things past', like Proust's madeleine, through the taste, smell, or sight of certain foods. We can refine these memories into different categories. Some memories connect with ancestors, family, and friends separated from us by death or distance. We 'miss' them simply because they are physically not there, or at least not immediately visible. In a past paper on the Jewish harvest festival of Sukkot, I discussed the custom of inviting *ushpizin*, 'guests', the Biblical ancestors like Abraham and Sarah and the invisible mystical attributes of God, the *sefirot*, often depicted in pictures on the wall of the sukkah or suggested by the actual human guests invited to eat in the sukkah.[11] Some practice an improvised ritual asking participants to name an absent person they wish were there eating with them. But often the evocation of such memories causes some pain or nostalgia, literally the 'pain of going back', *nostos* + *algia*, which may be either the pain of remembering people, places, and times one has lost – from the passage of time and the displacements by choice or compulsion, or memories of pain *per se* – shared pain from our pasts, directly experienced in our bodies, or constructed in the narratives of our group memories. Such are the memories evoked by the bitter herb at the Passover seder, 'because the Egyptians embittered our lives' when we were enslaved, or the painful suffering of Jesus's betrayal and death heard in the words of institution recited when consuming Communion. Finally, in the recollection of the taste of bitterness *per se*, not primarily a metaphor for the pain of loss or absence, but more in the sense of a verbal 'remembrance', an explicit mention (like *le-hazkir* in Hebrew, literally to 'cause to remember') calls attention to the bitter taste in one's mouth. But that too has certain very specific connotations, as in the bitterness of poison that can kill us (hence our natural gag or disgust reflex), the bitterness of medicine (often a smaller dose of the poison) that momentarily hurts us to heal us, or bitterness as an essential component of balance, offset by other tastes like sweet, salty, sour, and pungent, as in Ayurvedic eating.[12] If bitterness is not often pleasant by itself, in combination with other sweet and/or savoury flavours, as in coffee, dark chocolate, Guinness stout, Aperol spritzes, or Chinese dishes with bitter melon, the whole is more delicious than its parts taken separately.

In such combinations, we enjoy the taste of complexity, of opposites pleasingly balanced, of richness and nuance. As I have argued in another paper, when we can enjoy the convergence of different, even opposing flavours in our mouths, we are more predisposed to hold conflicting and complex ideas and emotions in our minds.[13] We've tasted complexity and know that it is good, to paraphrase Psalm 34. 8. It's especially hard for us to hold together cognitively the coexistence of evil and good, suffering and salvation, in ourselves and in the world we live. That, as I've argued in many places, is the point the late nineteenth-century Hasidic Rabbi Yehudah Aryeh Lieb of Ger, also known as Sefat Emet, makes in his interpretation of eating the Hillel Sandwich at the Passover seder, which combines the bitter herb (typically horseradish among Ashkenazi Jews or a bitter green like romaine lettuce among Sephardi and Mizrachi Jews) with the sweet charoset fruit paste between pieces of matzah. He says:

> Hillel used to combine the Passover lamb, unleavened bread, and bitter herbs together. Because the bitterness of the exile is not a separate thing, [. . .] but rather the whole order by which the Holy One of Blessing directs us is all one way. As it is said, He causes death, brings life, and implants salvation. The word *maror*, 'bitter herb', is [numerically equivalent to] '*mot*', death. For in exile, we are considered as if we're 'dead'. But this is necessary for there to be new life and salvation for generations after. And therefore, we need to eat the bitter herb because we are now in exile, and by eating the bitter herb, we can sweeten the bitterness on this night. And thus, when swallowing the bitter herb, we will not feel the taste of the bitter herb more than the matzah, but the bitter herb is sweetened now.[14]

The modern Jewish Biblical commentator Avivah Gottlieb Zornberg notes that the cognitive difficulty of holding good and evil together is a deeply holy moment we experience upon realizing, in our process of learning, that we are wrong:

> Since good and evil are so thoroughly intermingled, it is precisely in the experience of slippage – of getting it wrong, of being embroiled in the inevitable mix of good and evil, falsehood and truth – that the theological moment is experienced. The moment of the most vulnerable humanity, when one stumbles over the complexity of things, brings one into greatest intimacy with God.

Remarkably, she too characterizes this experience as a 'pungent *taste*'.[15]

Departures from Tasting the Bitter

This is not necessarily the way that some of the most notable recent Jewish efforts deal with Jewish trauma in meal rituals. For example, Jewish Holocaust cookbooks, such as Cara De Silva's *In Memory's Kitchen* and June Hersh's *Food, Hope & Resilience*; the cookbook *Shavuot of Longing*, published by families and friends of the victims of the 7 October 2023 Hamas attack; and empty-table-setting installations organized in both Israel and abroad are all contemporary efforts to respond to Jewish trauma.[16] These cookbooks focus on the absence and loss of the people missed by remembering the foods they liked to cook or eat, just as table installations emphasize their physical absence from shared tables. In other words, they evoke memories that connect with ways of life with ancestors, family, and friends who have been lost because they are separated from us by incarceration, death, or distance.

Thus, the Jewish female inmates imprisoned by the Nazis in Theresienstadt gathered to recall recipes for the central European dishes they used to be able to cook and eat, recipes Mina Pachter recorded in the cookbook that was found and which Cara De Silva published and popularized. There are two memory meal rituals here, both in what the women themselves did to preserve their human dignity in the face of their dehumanization in camp conditions and in De Silva's memorialization of them and their food memories as acts of spiritual resistance. For the women in Theresienstadt, 'the recipes were not recalled fondly; rather, they were a painful fantasy for the authors of the recipes'.[17] As

De Silva puts it, 'Recalling recipes was an act of discipline that required them to suppress their current hunger and to think of the ordinary world before the camps – and perhaps to dare to dream of a world after the camps'.[18] De Silva preserves the cookbook as is, though many of the recipes are missing ingredients or correct proportions of ingredients, not intending them to be cooked by readers of her book.

In contrast, in *Food, Hope & Resilience: Authentic Recipes and Remarkable Stories from Holocaust Survivors*, June Hersh explicitly states how she made sure the survivors' recipes were cookable:

> Most of these cooks prepare from instinct, taste and smell or by *shitteryne*, a little of this and a little of that[.] How much is an eggshell of matzo meal or a glass of oil [. . .]? Is a *bisel* a teaspoon or a pinch? And what about those recipes written in shorthand or scribbled as illegible notes stained with cooking oil and sauce? [. . .] That's when I knew I needed to meticulously test and re-test every recipe and write each precisely, providing you with clear directions, accessible ingredients, consistent format and exact measurements. [. . .] All recipes were reviewed and approved by the contributor.[19]

The cookbook is organized around interviews with Holocaust survivors, followed by their own recipes or, in her new edition, similar recipes contributed by celebrity Jewish chefs like Michael Solomonov.[20]

Also intended to be cooked are recipes collected as a response to the losses experienced by Israeli Jewish families and friends of those killed or taken hostage in the 7 October 2023 massacre. The Hostages and Missing Families Forum published *Shavuot of Longing: Their Recipes on Our Table*, a 'special recipe book' dedicated 'to our brothers and sisters'. In it, they 'gathered their favorite dishes, unearthed their family recipes, and bound all our longing into this book for our loved ones'. The focus of the book is on the pain of their loved ones' absence, evoking in the title the holiday of Shavuot:

> It's hard for us to imagine going through another holiday without them, as these occasions magnify the void we feel. [. . .] They are not here with us yet, but in mind and spirit, we are there with them. Our hearts ache as we feel the anguish of captivity experienced by our loved ones still held in Gaza. Shavuot in Hebrew means 'weeks', but we've been endlessly yearning for their return for many months now. How we yearn for the day when we will be able to prepare the recipes found in the book with them.

Until then, the editor

> invite[s] you to prepare a recipe from the book, take a photo, and upload it to social media with the hashtag: #Shavuot_of_Longing. This way we can continue to tell their stories, raise awareness of their situation, and demand the immediate return of all of our brothers and sisters.[21]

Thus the book's main focus is not on the food elements of the recipes but rather on the victims and hostages themselves, whose photographs, along with brief descriptions of their lives and likes, precede each recipe remembered for them (most actually composed by professional chefs). Conspicuously absent are remarks calling attention to bitter foods and flavours *per se* in the recipes. Rather, it's sweetness or other flavour memories explicitly evoked in the preface for the '75 recipes, near and dear to our hearts. From the pastries they start their day with (along with a cup of coffee) to a hot bowl of soup in winter, to desserts that sweeten their hearts'.[22] Similarly, the table-setting installations for the missing hostages focus on the painful absence of the specific people themselves rather than flavour-cued pain in food rituals, which connotes more the internalization and integration of pain as an element of the wholeness of our personal and collective experiences. This point is not to disparage the memory practices in Holocaust and post-7 October cookbooks and table rituals but rather to note their difference and departure from the Jewish meal memory practice of tasting the bitter herb in the Passover seder ritual.

Putting the Bitter and Bittersweet Back in New Rituals of Trauma and Reconciliation

That said, I want to put the bitter and bittersweet back in new rituals of trauma and reconciliation, inspired by the Passover consumption of the bitter herb and the Hillel sandwich, in applications beyond Jewish stories and rituals.

I have glimpsed ways that bitterness and trauma are fused with loving memories of place and people in Palestinian cookbooks, in Native and African American food stories, and in contemporary efforts to reimagine Thanksgiving meals.[23] Consider in particular the idea of the healing power of bitter medicine, an important implication of the multivalent flavour of bitterness. Two recent books by the Native American *curandera* (traditional healer) Felicia Cocotzkin Ruiz and Lebanese American Layla Feghali turn to their respective cultures' herbalist traditions to reconnect to their human and other-than-human ancestors (Feghali refers to some of the latter as 'plantcestors'), from whom violent displacement by settler colonists and continued impacts to this day have separated them.[24] Notably, both Ruiz and Feghali dig deep into the edible and medicinal features of plants they discuss and, while not focusing on their bitter flavour *per se*, do in fact pay as much attention to how they taste, smell, feel, and look as they do to the memories and stories of family members they evoke. Such a sensory approach to memories of both loss and healing suggests to me the intentional deployment of bitter foods, particularly Native American foods like chocolate, chokeberries and chokecherries, and cranberries in new rituals to evoke bittersweet experiences of both rupture and some sort of reconciliation – what Native American Patty Krawec calls 'becoming kin'.[25]

Table Rituals at Which Colonizers and the Colonized Can Live Together in Peace?

It is one thing to remember the painful loss of our own family and kin but quite another to remember and live with the painful legacy of our history with others who wronged us or whom we wronged. I'm thinking particularly of American history, but I also have in mind the history of Israel/Palestine. How can colonizers and the colonized, settlers (many themselves displaced from their homelands) and Indigenous people live together in peace? Are there ways that table rituals like Thanksgiving dinner or diasporic cookbooks can foster the coexistence of people violently separated from one another, their lands, and their ancestral homes?

Regarding creating new American rituals that do this, Masha Gessen, a queer nonbinary Russian Jewish journalist living in the United States, and Patty Krawec, an Anishinaabe woman from Lac Seul First Nation, have some instructive recommendations. Gessen recommends 'narrative therapy':

> People may tell what they did and what was done to them, they may tell their stories repeatedly in specific settings and formats, and they may produce a record of the telling. [. . .] By creating new rituals, it gives voice to people who have not been heard in public. By focusing on the acts of telling and listening, it may challenge assumptions about justice as the act of meting out punishment. [. . .] In the framework of collective trauma, moral injury has been inflicted on society as a whole, not only on the victims but also on the perpetrators and the bystanders—not in the same way, to be sure, and not in equal measure, but it is all of society that has been injured. The goal of reckoning is moral restoration.[26]

Krawec describes this sort of moral restoration as 'becoming kin', a healing practice for settlers to reframe themselves as kin to those they've harmed:

> So many people are still dancing the shut-eye dance, refusing to see or hear the harms of settler colonialism, something I hope is now obvious to you [. . .]. Doing the work of picking up your own bundle, of returning to yourself, will help you move from being a settler to becoming kin. [. . .] We walk this path together, as the Two Row Wampum, that original treaty that the Haudenosaunee offered the settlers, lays out: each in our own ways, but together. Nii'kinaaganaa. We are related.[27]

So, for example, if one were to reconstruct American Thanksgiving meal rituals to these ends, as my son Max Brumberg-Kraus and I are doing together, to compose a new American Thanksgiving Haggadah, the stories we tell and the symbolic foods we eat will work to these goals. In broad strokes, this would mean including a diversity of foods and a diversity of stories from the voices of other Americans with experiences different from our own. These foods and stories should reflect the diversity of the people and cultures of America, respect the biodiversity of the other-than-human persons with whom we share this land (e.g. the Three Sisters – corn, squash, and beans), and acknowledge those not there in person in our choice of representative foods and the words we speak formally and informally at the table. And to accentuate the complexity and affective impact of our ritual, of the very real and painful parts of our histories in becoming kin, bitter and bittersweet foods are essential.

Thus, I'll end with a description of a ritual I've imagined for it, inspired by a story Krawec puts in a postscript to her book. Her story:

> Nanaboozhoo was known for many things, including wisdom, and people often sought him out, wanting to learn from him. One day, a particular Anishinaabe asked Nanaboozhoo how he got so smart, and Nanaboozhoo told him about smart berries. 'Oh, where are these smartberries?' the Anishinaabe wanted to know. He begged and pleaded, and eventually Nanaboozhoo agreed to take him for a walk to where the smart berries were. After a while, Nanaboozhoo stopped. Before them on the path was a scattering of berries on the ground.
>
> 'There they are', he said, and in a flash, the Anishinaabe scooped up the smart berries and ate them all. Just as quickly he spat them out. 'These are rabbit pellets!' he cried angrily. 'See?' Nanaboozhoo said. 'You're smarter already'.[28]

Now eat some raw cranberries dusted in cocoa powder and then sweetened dried cranberries. It's hard to retell a story at odds with the ones you grew up with and that shaped who you are. What is telling an inclusive story of Thanksgiving like? It is like eating smartberries. Shouldn't we have eaten the tart berries before revealing the punchline? 'Yuk, these are rabbit pellets'.

See, you're smarter already!

About the Author

Jonathan Brumberg-Kraus is a Professor of Religion at Wheaton College in Massachusetts. His most recent book is *Gastronomic Judaism as Culinary Midrash.*

Notes

1 Jennifer McLagan, *Bitter: A Taste of the World's Most Dangerous Flavor, with Recipes* (Ten Speed Press, 2014); Jennifer McLagan, 'The Bitter Truth', interview in Nicola Twilley and Cynthia Graber, *Gastropod*, 22 September 2015 <https://gastropod.com/the-bitter-truth/> [accessed 20 November 2024].

2 McLagan, *Bitter*, p. 2.

3 McLagan, p. 3.

4 McLagan, p. 4.

5 McLagan, pp. 9, 53, 97, 135, 175, 211.

6 McLagan, p. 98.

7 McLagan, pp. 233–34.

8 McLagan, p. 241 (my emphasis).

9 Rachel S. Herz, Caitlin M. Cunningham, and Theresa L. White, 'Advancing Investigations into the Effects of Sweet and Bitter Tastes on Agreeableness and Hostility', *Food Quality and Preference*, 121 (2024), p. 105254, doi:10.1016/j.foodqual.2024.105254.

10 As our guest at a recent Passover seder, Rachel Herz did a version of her experiment (at our request). I personally attest to the impact of this prosocial priming on counteracting any hostile or unpleasant mood that drinking the bitter solution she provided might have provoked. On the contrary, it made me very happy and agreeable!

11 Jonathan Brumberg-Kraus, 'Sukkot: The Paradigmatic Harvest Festival', in *Celebration: Proceedings of the Oxford Symposium on Food and Cookery 2011*, ed. by Mark McWilliams (Prospect Books, 2012), pp. 56–65.

12 Usha Lad and Vedant Lad, *Ayurvedic Cooking for Self-Healing* (Ayurvedic Press, 1997).

13 Jonathan Brumberg-Kraus, '"Truly the Ear Test Words as the Palate Tastes Food" (Job 12:11): Synaesthetic Food Metaphors for the Experience of the Divine in Jewish Tradition', in *Food and Language: Proceedings of the Oxford Symposium on Food and Cooking 2009*, ed. by Richard Hosking (Prospect Books, 2010), pp. 42–51.

14 Rabbi Judah Aryeh Leib Alter (S'fat Emet) in Rabbis Judah Aryeh Leib Alter and Solomon Rabinowich, *Hagadah Shel Pesah:`im Sefat Emet* (Mahadurah hadashah mehuderet u-metukenet, 1965), p. 133, originally written in the late nineteenth century; my translation.

15 Avivah Gottlieb Zornberg, *The Hidden Order of Intimacy: Reflections on the Book of Leviticus* (Schocken, 2022), pp. 118–19 (my emphasis).

16 Cara De Silva, *In Memory's Kitchen: A Legacy from the Women of Terezín* (J. Aronson, 1996); June Hersh, *Food, Hope & Resilience: Authentic Recipes and Remarkable Stories from Holocaust Survivors* (History Press, 2023), a revision of *Recipes Remembered: A Celebration of Survival* (Museum of Jewish Heritage, 2013); *Shavuot of Longing: Their Recipes on Our Table*, ed. by Orli Peli-Bronshtein (Foody/Be'eri Printers, 2024); Philissa Cramer, '200-Seat Empty Shabbat Tables Are Set for Israeli Hostages in Tel Aviv', *Jerusalem Post*, 21 October 2023 <https://www.jpost.com/israel-news/article-769451> [accessed 20 November 2024]; Zoe MacDiarmid, 'Jewish and Student Organizations Assemble an Empty Shabbat Table Memorial for 229 Hamas Hostages', *Huntington News*, 1 November 2023 <https://huntnewsnu.com/73481/campus/jewish-and-student-organizations-assemble-an-empty-shabbat-table-memorial-for-229-hamas-hostages/> [accessed 20 November 2024].

17 Talya Slaw, 'On Holocaust Cookbooks: Fourth Generation Jews and the Re-Creation of Jewish Food Culture' (unpublished thesis, University of Kansas, 2018), p. 10.

18 De Silva, *In Memory's Kitchen*, xvi.

19 Hersh, *Food, Hope & Resilience*, p. 14.

20 Hersh, pp. 247, 262.

21 Peli-Bronshtein, *Shavuot of Longing*, p. 3.

22 Peli-Bronshtein, *Shavuot of Longing*, p. 3. Technically, coffee is bitter, but many Israelis tend to prefer their coffee very sweet.

23 Such efforts include Reem Assil, *Arabiyya: Recipes from the Life of an Arab in Diaspora* (Ten Speed Press, 2022); Layla K. Feghali, *The Land in Our Bones: Plantcestral Herbalism and Healing Cultures from Syria to the Sinai – Earth-Based Pathways to Ancestral Stewardship and Belonging in Diaspora* (North Atlantic Books, 2024); Joudie Kalla and Jamie Orlando Smith, *Baladi: A Celebration of Food from Land and Sea* (Interlink Books, 2018); Reem Kassis, *The Palestinian Table* (Phaidon Press, 2017); Robin Wall Kimmerer, *Braiding Sweetgrass: Indigenous Wisdom, Scientific Knowledge and the Teachings of Plants* (Milkweed Editions, 2015); Felicia Cocotzin Ruiz, *Earth Medicines: Ancestral Wisdom, Healing Recipes, and Wellness Rituals from a Curandera* (Roost Books, 2021); Sean Sherman, *The Sioux Chef's Indigenous Kitchen* (University of Minnesota Press, 2017); Michael W. Twitty, *The Cooking Gene: A Journey through African American Culinary History in the Old South* (Amistad, 2017); Debnita Chakravarti, 'Feel Good Reel Food: A Taste of the Cultural Kedgeree in Gurinder Chadha's What's Cooking?', in *Reel Food: Essays on Food and Film*, ed. by Anne L. Bower (Taylor & Francis Group, 2004), pp. 17–26.

24 Feghali, *Land in Our Bones*, pp. 41–43.

25 Patty Krawec, *Becoming Kin: An Indigenous Call to Unforgetting the Past and Reimagining Our Future* (Broadleaf Books, 2022).

26 Masha Gessen, 'Why America Needs a Reckoning with the Trump Era', *New Yorker*, 10 November 2020 <https://www.newyorker.com/news/our-columnists/why-america-needs-a-reckoning-with-the-trump-era> [accessed 19 November 2020].

27 Krawec, *Becoming Kin*, pp. 186–87.

28 Krawec, p. 189.

7
Elements of Time
Calendrical Complexity and Culinary Continuity in Southern Italy

Anthony F. Buccini

(Post-)Modern vs. Traditional Cuisines and Discourse Networks

Between the culinary cultures of the most wealthy and globalized nations, there still exist notable differences at all levels of their culinary grammars (Buccini 2016: 112ff.). And yet, hand in hand with the process of globalization goes a strong drift toward a more homogenized, international approach to food. Here I'm not thinking of the ever-increasing array of ingredients and dishes that are habitually consumed in cities throughout the world – pasta and pizza, sushi and ramen, hummus, tacos, etc. – but rather some basic attitudes and ideas that pervade postmodern approaches to cookery and consumption. Manifested to the most extreme degree in the United States but present to varying degrees in all so-called advanced societies, a clear shift has occurred away from traditional ways of thinking about alimentation and cuisine. One sees this shift, for example, in a lessening of the role of religion in culinary behaviour, an ever-stronger commodification of all things culinary and particularly of culinary knowledge, an increasing preoccupation with forms of culinary competition, and with that an insatiable appetite for culinary novelty and a high appreciation of innovation, to name just some of the most conspicuous recent trends. At the heart of all these developments is, I believe, a shift in the nature of the discourse networks in relation to food in those affluent, globalized societies, specifically a shift regarding the transfer of culinary knowledge away from familial and proximate community discourse networks to increasingly far-ranging and impersonal discourse networks, rapidly expanding on the role of the printed word through the electronic media of television and internet. In effect, what we see over the past century or more is a massive broadening and acceleration of socio-culinary developments that before modern times were limited almost wholly to the elite echelons of some societies, which accords both with the affluence of the 'first-world' nations and the intensification of both the global commerce in foodstuffs and the explosive expansion of communications.

In previous work, first in Buccini 2016, I have termed these affluent culinary cultures of our days, as well as those of some elites of the past, *exocuisines*, given their openness to and reliance on the exchange of culinary knowledge through impersonal discourse networks. These stand against *endocuisines*, in which culinary information is largely or wholly exchanged through intimate and local discourse networks, with direct generational transfer playing a pervasive role. We might therefore also refer to endocuisines as traditional cuisines, which, as a result of their locally focused discourse networks, exhibit a high degree of interconnectedness between cuisine and other tradition-bound cultural domains and a very strong inclination toward conservatism. Whereas members of an exoculinary culture prize innovation, novelty, and cosmopolitan sophistication, those in an endoculinary culture prize adherence to traditional foodways and thus deep local knowledge. Thus, a fundamental difference between current exocuisines and traditional endocuisines is the conception of cuisine in the former as largely an expression of individuals' taste and knowledge and in the latter as a shared communal institution that continues knowledge accrued by the ancestors of the community as a whole, and thus cuisine serves as a key element of their in-group or ethnic identity.

Cuisine, as a cultural domain interconnected with other cultural domains in complex ways, reflects broad sociocultural developments, and my proposed typology of cuisines can easily be related to the division more recently proposed by Henrich in his book of 2020 between what he calls 'WEIRD' – an acronym for 'Western, Educated, Industrialized, Rich and Democratic' – societies and other societies both past and present around the world. Henrich's ambitious goal is to explain how and why the WEIRD societies have come to dominate the globe, a topic far outside this paper, but two elements of his analysis have direct relevance here. First, as he found a bias in socioculturally related psychological studies which has long obscured the fundamental division which he proposes (xii), we similarly find a skewing in the analysis of culinary behaviour arising from the fact that the field of food studies is itself a product of exoculinary culture, a field largely practiced by food scholars and authors for the edification of audiences, all of whom are immersed in exoculinary cultures. Consequently, the modes of thinking about food on which endocuisines are based are, to a considerable degree, alien and even incomprehensible to those engaged in first-world food discourse. Second, at the heart of the strangeness of endocuisines for WEIRD, exoculinary food scholars, authors, and consumers are deeply rooted patterns of thought regarding tradition and the relationship between the individual and the communal: the manifestations in ingredients and dishes of traditional cuisines may often be appreciated, embraced, and exploited for harvesting to enrich individuals' globalized cookery, but the sociocultural framework that has produced and maintained traditional cuisines is incompatible with exoculinary culture and is, even when praised, generally examined only superficially and often viewed as 'primitive'.

Indeed, despite the widespread appreciation of traditional foods among people of an exoculinary persuasion, one can observe some of the (post-)modernist, WEIRD discomfort with and even hostility toward tradition in some academic food writing. One less obvious manifestation of this discomfort is the commonly asserted idea that all cuisines are 'hybrids', a claim which renders the idea of culinary hybridity essentially meaningless by throwing all degrees of contact-induced change into one and the same pot, from, for example, Mexican cuisines, which reflect a deep intertwining of autochthonous and colonialist foodways at all levels – a genuine hybrid in my view – to Italian cuisines, which have certainly taken up over the centuries a number of essentially peripheral innovations through various cultural contacts but at the core show a rectilinear developmental path throughout (Buccini and Dahlstrom 2020: 4). While defining 'hybrid' and 'hybridity' may be a dry, academic endeavour, the issue takes a dark turn when the idea that all cuisines are hybrids is used to refute representatives of traditional cuisines when those representatives speak of authenticity and cultural appropriation from their own endoculinary perspective.

Hostility to tradition comes to expression in, for example, the iconoclastic view that 'Italy's regional cuisines are not only modern invention, they may even have been created in the Americas' (Pilcher 2013: 427; for discussion, see Buccini 2024a: 1ff.). Like culinary hybridity, the notion of invented tradition is often used as a sort of intellectual weapon by WEIRD scholars to dismiss in a glib way culinary traditions that have complex histories. I do not deny wholly the existence of invented traditions, if by that term one intends some cultural or specifically culinary element which is of recent origin and thus has not gone through the process of being passed down through multiple stages of generational transfer but is falsely claimed to have done so, presumably for commercial or political reasons. Given, however, that all traditions have in a real sense been invented, the question arises of how we are to separate the somehow genuinely bogus pseudo-tradition from a tradition that has undergone change or updating so that it better reflects new concerns or attitudes of the community that embraces it. The clearly pejorative application of the term 'invented tradition' or 'modern invention', implying some measure of deceit, to a set of traditional cuisines by an outsider with limited knowledge of those cuisines and an exoculinary perspective should be troubling.

Traditions are not immutable; indeed, if they are part of a living culture, they cannot be, for, just as linguists are keenly aware that all living languages are necessarily always undergoing change, so too must we recognize that all living cuisines are undergoing change. That living cultures change is therefore a trivial observation. The interesting questions are, What is the rate of that change? What is the structural significance of that change? What are the social parameters of that change? and What is the original cause of that change? (Buccini and Dahlstrom 2020: 5).

Tomatoes, Pasta, and False Iconoclasm

Central to the claim that Italian regional cuisines are somehow a modern invention are the roles of pasta and the tomato in contemporary Italian cookery, and with these foods, there comes naturally a focus on Southern Italy. With regard to the tomato, I will only say here that this Columbian Exchange item, introduced to Spanish-ruled Southern Italy ultimately from Mexico via Spain in the sixteenth century, was rapidly taken up as a food by the poor in the Mezzogiorno and beyond, as demonstrated by multiple learned Italian texts of that century which decry the poor nutritional qualities of this fruit while also noting that people do in fact eat it (Buccini 2006: 135). The rise of the use of the tomato in cookery was a facet of the long-standing Italian love of umami flavouring, and for the poor of Southern Italy, tomato sauces became an affordable substitute for the expensive meat broth that was an entrenched feature of festive cookery and the quotidian cookery of the wealthy, among other things, as a way to dress pasta. This development represents no culinary revolution but rather the exploitation of a new, easy-to-grow ingredient to supplement a long-established culinary practice (Buccini 2011, 2025).

With regard to pasta, we note that both scholarly and popular food writers have generally accepted the theory that it was the Arabs, in the wake of their conquest of Sicily in the Middle Ages (ninth–tenth century), who introduced this food to Italy; a broad form of the theory posits Arab influence in the spread of all forms of pasta, while a narrower version limits Arab influence to the spread primarily of dried pasta. Enthusiasm for this theory is, however, in inverse proportion to the strength of the historical and linguistic evidence proffered in its support: only the most superficial and prejudicial examination of the established facts is compatible with the Arab theory, as I have argued in previous publications (Buccini 2013, 2015); here I will briefly review some of the material most relevant to this paper from Buccini 2024a: 7ff. and 2024b.

Whereas no single name of a basic form of pasta in Italy can, on the basis of linguistic and philological grounds, be traced back to Arabic, almost all are native words from regional languages/dialects or more recently coined in standard Italian. Of even greater significance for the history of this food in Italy is the fact that among the names of the most basic forms of pasta in Southern Italy are two which are unambiguously Greek in origin – i.e. they are first attested already in Classical Greek sources and, moreover, have solid Greek etymologies (Buccini 2024b). In this regard, it must be recalled that significant portions of continental Southern Italy and Sicily were fully part of the Greek world in antiquity as a result of Greek colonization, which began already in the eighth century BCE, with the Greek language surviving in some areas well into the Middle Ages and even in small pockets to the present day. The pasta names in question are:

1. Gk. *lágana*, which survives today in a large area of western, continental Southern Italy in various dialect forms and denotes a short broad noodle cut from a rolled-out sheet of dough. In much of the eastern part of the continental south, this pasta form used in the same kinds of prepared dishes is found, but the term *lágana* has been replaced there by the term *lasagna*.
2. Gk. *itria*, reflexes of which survive in parts of Sicily and on the continent in southern Puglia (where the Greek language survived most robustly) and in a band of territory around the border between Campania and Basilicata and continuing northward through mostly mountainous terrain with poor roads and no major cities, a classic relic zone for linguistic and other cultural features. The dialect reflexes of *itria* – *tria* in Sicily and southern Puglia, in Bari *triddë*, and in Campania and Basilicata *triilli*/*trilli* or *triiddi*/*triddi* (reinforced with a diminutive suffix) – almost all denote a more or less long stringlike pasta.

For those who are committed to denying the antiquity of pasta in Greece and Southern Italy, both *lágana* and *itria*, on the basis of some classical textual references, are interpreted as 'cakes' or 'pastries', for they are indeed in some sources described as being fried and dressed with honey and seeds. Other texts strongly imply, however, that a habitual way of preparing them was boiling; given that the words themselves originated in reference to shapes, one must conclude that in classical times, the primary classification of bits of dough was their form, with the method of their cooking being secondary (Buccini 2024a: 7–8).

The evidence – historical and geographical, textual and linguistic, and ultimately also culinary – makes it overwhelmingly clear that Southern Italian *lágana* were a part of local cuisines long before the rise of Islam and the Arab expansion into Sicily in the ninth century AD. Does it seem reasonable to ignore the clear relationship between these two pasta forms and to insist, as all proponents of the Arab theory of pasta do, that the Arabs brought stringlike pasta under the name *itriyya* to places in Italy where the Greeks, those who coined the term some time well before classical antiquity, had made their homes starting in the mid-first millennium BC?

In various presentations over the past fifteen years, I have suggested that the native Southern Italian term for stringlike pasta – *vermicelli*, nowadays widely replaced by the recent Italian term *spaghetti* – was itself a Southern Italo-Romance replacement for the even older Greek term *itria*. Last year, I learned that my claim has found confirmation in a fragment of an early medieval medical text brought to light in 2015 by philologist Vincenzo Ortoleva, where we find the phrase 'et hedria id est uermiculissi'. Here *hedria* is a known variant of *itria*, and *uermiculissi* can most reasonably be interpreted as a mistaken reading or copying of *vermiculi (e)lissi*: this fragment, dated to the seventh century AD, reads then 'and *itria*, that is boiled vermicelli' (Ortoleva 2015: 202ff.). In other words, *vermicelli* did replace the term *itria* in most of Southern Italy, and this form of pasta was undoubtedly known there already two centuries before the Arab expansion! One must conclude that *itria*, like *lágana*, have been consumed in the Mezzogiorno since classical times.

That *lágana* as pasta were common fare in the Mezzogiorno in classical times need not merely be inferred from the broader evidence. Rather, it is directly attested in Horace's (65–8 BCE) satire I.vi, where he says that in the evening he likes to stroll to his home in Rome and eat a pot of 'leeks, chick peas and *lágana*' ('inde domum me ad porri et ciceris refero laganique catinum') (Horace 1926: 84–85; cf. Buccini 2013: 97). Though generally interpreted as being a stew of leeks and chickpeas served with fried flat bread, Horace's meal is to anyone steeped in Southern Italian culture a clear reference to a dish featuring the aforementioned short broad noodles which is to this day consumed throughout Southern Italy, cooked and served together as described in the satire.

The broader context of this food reference is important, for a poet of Horace's talent chooses what he includes with care. The satire is an extended commentary on the place of the lowborn in Roman society and on the possibilities for advancement available to the ambitious. Horace, a native of Venosa – a Roman colony in the midst of Greek-influenced Sabellic/Lucanian territory – and son of a freedman, expresses the limitations of his own ambition and with that his glad acceptance of who his humble parents were and who he is and what he is genuinely content with. It is here that he makes a series of references to non-elite culture, including two to food: first, his pleasure in going to market and asking the prices of 'greens and grain', and second, his pleasure in dining on *lágana e ceci*. For his audience, Horace is surely counting on these as illustrating humble culinary tastes, but in addition, it must also be the case that his invocation of the dish is an expression of his identity with and pride in being a Southern Italian, in being his freedman-father's son from Lucania, and he is counting on his audience recognizing the reference. In other words, *lágana e ceci* was a dish laden with symbolic meaning no less than flavour, as it still is for many Southern Italians.

The Culinary Calendar and Cultural Identity in Southern Italy

It is well known that in earlier times, pasta was eaten in Italy far less often than it is today, though it is also clear from the onset of the transmission of vernacular texts there (*c.* thirteenth century) that it was widely consumed in a multitude of forms and preparations already in the Middle Ages, especially but not exclusively by the bourgeois and elite sectors of society. The commercial and ultimately industrial manufacture of dried pasta in the modern period was a major step in increasing the frequency of consumption, and yet, for the poor, even in the Mezzogiorno, pasta was hardly an everyday food. For those who claim that Italian regional cuisines are a 'modern invention', this is a key point. Pilcher, for example, claims with some justification (particularly for the desperate period leading to mass emigration in the late nineteenth century) that 'well into the twentieth century, peasants subsisted on monotonous porridges of maize, chestnuts, broad beans, or rice', from which he concludes the modern invention was under (Italian-)American influence: 'It seems that contemporary Italian cuisine emerged largely from festival foods such as *maccheroni*, which became items of everyday consumption in the United States' (2013: 427). Of course, it is absurd for him to divorce completely the foodways of the very poor from those of the less poor and bourgeoisie, but even more troubling is his suggestion that 'festival foods' are somehow not an integral part of a cuisine in general. Here a myopically exoculinary and perhaps particularly WEIRD-American perspective is expressed. In the United States, virtually all holidays among the mainstream's well-off are celebratory occasions when one might eat foods drawn from any cuisine, depending on the recipes and ingredients to which one has access. And though I have argued elsewhere that celebratory occasions are, indeed, openings in all culinary systems through which external influences can often easily enter, in an endocuisine, there is typically a far greater degree of filtering of such influences trickling down the social scale, not just through economic constraints but also through adherence to rules of the culinary grammar.

In several papers, I have discussed a core aspect of Southern Italian cuisine—namely, the culinary calendar and its fundamental organization around a three-way division between (1) ordinary days, (2) celebratory or joyous holidays, and (3) solemn holidays (esp. Buccini 2012, 2024c). This division is connected to the Roman Catholic Church's dietary rules, which traditionally demanded abstinence from terrestrial animal products, either partial (meat only) or total (meat, dairy, eggs, etc.), on the fairly numerous solemn days, which included extended periods of abstinence during Lent and Advent; also required was some measure of feasting on celebratory holidays. The yearly pattern is recapitulated each week: Sunday is a celebratory occasion, Friday (and formerly also Wednesday and even Saturday) is a solemn day with abstinence, and the other days are ordinary. These rules are today much relaxed and even so now ignored by many, though many Southern Italians continue to observe them, some for religious reasons but some also just as the rhythm of their culinary culture.

Cross-culturally, celebratory holidays are typically occasions when people consume foods that are ordinarily beyond the celebrants' financial means and/or require time-intensive preparations: the poor might emulate the foodways of the bourgeoisie, the bourgeoisie those of the rich, and the rich may look outside their own surroundings and emulate the ways of foreign elites. Thus, the trickle-down pattern of external influences is found even in strongly traditional and locally oriented cuisines. In Southern Italy, this has meant that the celebratory holiday (and Sunday) meals for the lower socioeconomic circles have often featured special forms of pasta and muscle meats, things enjoyed habitually by their wealthier neighbours but too expensive for regular consumption for the poor. That these foods were well known, always longed for, but only infrequently enjoyed by the poor hardly means that they somehow stood outside the cuisine, even specifically for the poor; from a psychological standpoint, the very opposite is true, as borne out by the deep, old obsession with *maccarune e carne* in non-elite Southern Italian culture.

The aforementioned solemn holidays have received little attention in food studies precisely because they are so peripheral to the mainstreams of increasingly nonreligious WEIRD societies. In most traditional societies, however, they are of great importance. In Southern Italy, such occasions are all tied to the Catholic faith, with some observed broadly and others being local (village, parish) in nature, but it has long been recognized that there is a connection between some elements of Catholic observance and pre-Christian practice, as, for example, in the case of Christmas (24–25 December), which is widely thought to be in part a continuation of the pagan Roman winter solstice (21/22 December) celebration (Saturnalia, 17–23 December) (Ogilvie 1969: 98). If we consider the other principal Catholic solemn holidays with this in mind, we can see that there is in fact a partial but suggestive correspondence with the four major points in the solar calendar. Indeed, of the generally observed solemn holidays, there is one for each season: in early summer, there is the Feast of Saint John the Baptist (24 June), which also nearly coincides with the summer solstice (20/21 June). In spring, the Feast of Saint Joseph (19 March) occurs very close to the spring equinox (20 March). Turning to the autumn season, however, the major solemn occasions occur quite a bit later than the autumnal equinox (21/22 September), namely All Saints' Day (1 November) and All Souls' Day (2 November). Of course, Christmas Day is one of the two universally observed joyous Catholic holidays (alongside the moveable Easter), but its vigil, Christmas Eve, though also in practice a joyous occasion, is in formal terms one of the most solemn religious days in the calendar – i.e. one of the few days when strict abstinence from all animal products is required.

The autumnal deviation between principal solemn holidays and the equinox suggests that there is something else involved here, and indeed, in Southern Italy, All Saints' and All Souls' Days fall just before the sowing of the most important crop, wheat. From this perspective, we must reconsider the summer holiday, Saint John's Feast, which should be considered together with another nonlocal holiday, namely the Feast of Saint Anthony of Padua (13 June): together, these occasions coincide with the naturally variable beginning of the harvest of wheat. Less obvious to us is the connection between Saint Joseph's Day and agriculture, but in rural Southern Italy, this time is when – if things go as they should – wheat plants emerge from below ground and take recognizable form, *c.* 130 days after autumnal sowing. The hibernal holiday period does not coincide with any work in the cultivation of wheat, and I would suggest that this in itself is important: this relative break from major farmwork allowed for the extended nature of the Saturnalia holiday, which continues in the Christian conception of Christmas as a festive period (24 December–6 January), while also marking the crucial stage of vernalization – when the wheat is dead and buried – and anticipating the new growing season powered by the reborn sun.

All holidays are commemorative in nature and thus look back to the past in some measure, but the two most important Catholic celebratory holidays, Christmas Day and Easter, are focused on (re-)birth and are inherently forward-looking. As part of a yearly cycle, the principal solemn holidays can also be seen as anticipatory, especially those associated with harvesting, but their focus is more on the past, especially the autumnal days remembering the dead. I suggest that, considering also the clear connection of these occasions to the central crop, wheat, and the land that yields it, these solemn holidays are first and foremost intended to remind Southern Italians of where they came from – i.e. their ancestors and the lands that nourished them.

Whereas one celebrates the joyous holidays with meals featuring meat and often elaborate, rich pasta preparations, the solemn holidays are days of abstinence – perhaps a Christian distancing from pagan sacrificial practices – and by necessity more focused on the grain products themselves. Of course, fried and baked wheaten-dough products, both sweet and savoury and often shaped in symbolic ways, are prominent for all these holidays, and in some cases, there is unambiguous evidence that these holiday foods have direct analogues that were enjoyed already in classical antiquity (Buccini 2024a: 6). In this regard, I call attention also to the decidedly archaic use of wheatberries associated in particular with the autumn and spring holidays, paired with legumes or fruits on solemn occasions (and with an animal product, *ricotta*, for Easter), a practice which clearly reflects Greek traditions of classical times in both the Mezzogiorno and Sicily (Buccini 2019: 81ff.).

Also associated with the principal solemn holidays in Southern Italy are two forms of pasta, and not coincidentally, they are the aforementioned *lágana* and *tria* or *vermicelli*. More widespread than the consumption of wheatberries is the consumption of the former during the autumn and spring events, and in many places, it is specifically the previously discussed *lágana e ceci* which by tradition must be consumed. Given Horace's highly suggestive invocation of this dish, it seems inconceivable that we are not dealing here with a culinary expression of local or regional identity that has been maintained for more than two thousand years. That this dish has been served with little variation over all this time is supported by the fact that, considered more broadly, there is a strikingly conservative character to all the dishes traditionally associated with Southern Italy's solemn feasts: they are all humble in character and generally still today eschew the inclusion of ingredients introduced to Italy in the Middle Ages by the Arabs or as a result of the Columbian Exchange. Here we see a striking parallel with the culinary aspect of funerary practices in other cultures (Buccini 2024a: 9–10).

With this in mind, I call attention to the fact that Saint John's Day and especially Christmas Eve are occasions that call for dishes of *vermicelli*. Though nowadays, spaghetti with clam sauce has become a very popular Christmas Eve *primo*, it is evident that far more widespread in even the recent past was a dish of *vermicelli* dressed with anchovies. In my (Campanian) family, this dish would be made on other (ordinary) occasions with the New World *peperoncino*, but on Christmas Eve, the dish has always been spiced only with black pepper, known in Italy since ancient times. If we consider, moreover, the deep love of umami seasoning in Italy from garum-happy Roman times on, it is hard not to see a dish of *vermicelli* and anchovies as a quintessential culinary expression of Southern Italian cultural identity.

However frequently or infrequently Southern Italians have eaten pasta, it can be no coincidence that the two forms discussed here are so deeply entrenched in the complex cultural institutions that lie behind the solemn holidays, with multiple ties to culinary practice dating back to pre-Christian times. Without doubt, much in Southern Italian cuisine has changed over the past two thousand or more years, but if we are looking to date its 'invention', the core of that process occurred in the first millennium BCE in the course of the interactions of Sabellians, Greeks, and Romans.

About the Author

Anthony F. Buccini is currently an independent scholar who formerly taught at the University of Chicago in the Departments of Germanic Languages and Literatures and Linguistics. He received his BA from Columbia University and his PhD in Germanic linguistics from Cornell University (PhD 1992); he also studied and later conducted research as a Fulbright Scholar at the Katholieke Universiteit Leuven. In linguistics his primary areas of interest are historical linguistics, dialectology, and language contact, with particular emphasis on the Germanic, Romance, and Celtic languages. In food studies, his research focuses on the culinary history of the Mediterranean and the Atlantic World, as well as broader theoretical and methodological issues. He is a two-time winner of the Sophie Coe Prize in Food History (2005, 2018). https://anthonybuccini.com.

References

Buccini, Anthony F. 2006. 'Western Mediterranean Vegetable Stews and the Integration of Culinary Exotica', in *Authenticity in the Kitchen: Proceedings of the Oxford Symposium on Food and Cookery 2005*, ed. by Richard Hosking (Prospect Books), pp. 132–45.

———. 2011. 'Continuity in Culinary Aesthetics in the Western Mediterranean: Roman *Garum* and *Liquamen* in the Light of Local Survival of Fermented Fish Seasonings in Japan and the Western Mediterranean', in *Cured, Fermented and Smoked Foods: Proceedings of the Oxford Symposium on Food and Cookery 2010*, ed. by Helen Saberi (Prospect Books), pp. 66–75.

———. 2012. '*Chi vuol godere la festa, digiuni la vigilia*: On the Relationship between Fasting and Feasting', in *Celebration: Proceedings of the Oxford Symposium on Food and Cookery 2011*, ed. by Mark McWilliams (Prospect Books), pp. 66–75.

———. 2013. 'Lasagna: A Layered History', in *Stuffed and Wrapped Foods: Proceedings of the Oxford Symposium on Food and Cookery 2012*, ed. by Mark McWilliams (Prospect Books), pp. 94–104.

———. 2015. 'The Merchants of Genoa and the Diffusion of Southern Italian Pasta Culture in Europe', in *Food and Markets: Proceedings of the Oxford Symposium on Food and Cookery 2014*, ed. by Mark McWilliams (Prospect Books), pp. 54–64.

———. 2016. 'Defining "Cuisine": Communication, Culinary Grammar and the Typology of Cuisine', in *Food and Communication: Proceedings of the Oxford Symposium on Food and Cookery 2015*, ed. by Mark McWilliams (Prospect Books), pp. 105–21.

———. 2019. 'Neapolitan *Pastiera* and the Religious Significance of Wheatberries', in *Seeds: Proceedings of the Oxford Symposium on Food and Cookery 2018*, ed. by Mark McWilliams (Prospect Books), pp. 79–88.

———. 2024a. 'Collective Memory, Culinary Continuity, and Solemn Repasts: Lagana, Itria and the History of Pasta in Southern Italy', Dublin Gastronomy Symposium 2024 <https://arrow.tudublin.ie/cgi/viewcontent.cgi?article=1350&context=dgs> [accessed 27 January 2026].

———. 2024b. 'The Fata Morgana of Arab *Itriyya*: Historical Linguistic and Dialectological Perspectives on the Diffusion of Pasta.' Paper delivered (in preparation for publication) at the Institute of Historical Research, School of Advanced Study, University of London, Food History Seminar, February 2024.

———. 2024c. 'Spaghetti – *Never on Sunday . . . Well, Almost Never*: Structure and Rules in an Endocuisine', in *Food Rules & Rituals, Proceedings of the Oxford Symposium on Food and Cookery 2023*, ed. by Mark McWilliams (Prospect Books), pp. 50–60.

———. 2025. 'Some Mediterranean Culinary Isoglosses: Perry's "Olive Line" in the Broader Context of the Umami-Enhancer Isogloss Bundle', in *Gardens, Flowers, and Fruit: Proceedings of the Oxford Symposium on Food and Cookery 2024*, ed. by Mark McWilliams (Equinox), pp. 61–69.

———, and Amy Dahlstrom. 2020. 'Culinary Change, Disruption, and Death: Do Traditional Cuisines Have a Future?', Dublin Gastronomy Symposium 2020 <https://arrow.tudublin.ie/cgi/viewcontent.cgi?article=1179&context=dgs> [accessed 27 January 2026].

Henrich, Joseph. 2020. *The WEIRDest People in the World* (Picador).

Horace. 1926. *Satires. Epistles. The Art of Poetry*, trans. by H. Rushton Fairclough. Loeb Classical Library 194 (Harvard University Press).

Ogilvie, R. M. 1969. *The Romans and Their Gods* (Norton).

Ortoleva, Vincenzo. 2015. 'Un frammento inedito di un non identificato trattato di medicina tardolatino', *Revue d'Histoire des Textes*, 10, pp. 197–214.

Pilcher, Jeffrey M. 2013. 'Taco Bell, Maseca, and Slow Food: A Postmodern Apocalypse for Mexico's Peasant Cuisine?', in *Food and Culture: A Reader*, ed. by Carole Counihan and Penny Van Esterik (Routledge), pp. 427–36.

8
Feast of the Four Elements
La Pêche à Pied/Coastal Foraging in Mont-Saint-Michel Bay

Mary Margaret Chappell

Even if you don't know Le Mont-Saint-Michel or its exact location in France, you are probably familiar with its shape and architecture. The towering, fortified abbey perched atop a rocky island is a UNESCO World Heritage Site and one of the most instantly recognizable monuments in the world.[1] Le Mont owes much of its renown to the bay that surrounds it, which is home to the greatest tides in continental Europe. Hundreds of years of silting in the bay now mean that Le Mont is only encircled by the bay's tidal waters a handful of times each year, during the equinoctial tides, or *grandes marées*. As their name implies, these super tides occur around the spring and autumn equinoxes, when the earth and moon align with the sun, compounding their gravitational pull on the seas. During these *grandes marées*, the water rises higher than usual over the tidal flats, and Le Mont becomes a true island once more, if only for a few hours. It is a dramatic sight to behold – and a major draw for tourists.

For locals who live on or near Mont-Saint-Michel Bay, the real allure of the *grandes marées* isn't when the tidal waters are high, but when they are low. The same natural and astronomical elements that bring the tide right up to the fortified walls of Le Mont also draw it farther out, exposing vast expanses of mud, sand, and rocks that are teeming with wild oysters, mussels, clams, winkles, shrimp, scallops, lobsters, squid, fish – even the elusive abalone. That delectable, gatherable treasure is all for the taking for those who head out into the bay for *la pêche à pied* (coastal foraging).

The Bay and Its Tides

Earth, water, wind, and sun are the natural elements that define the Mont-Saint-Michel Bay, amplify its tides, nurture its ecosystem, and make it such a unique and bountiful place for coastal foraging. The bay owes its massive tidal range first and foremost to the shape of the earth – i.e. the geography in and around it.[2] Tucked beneath the Cotentin Peninsula on the English Channel, the shallow, fan-shaped bay culminates in a funnel-like estuary which amplifies the Channel's already significant tidal flow. Refracted currents from the Cotentin Peninsula in Normandy push even more water into it, and the end result is a 10- to 15-metre (33- to 50-foot) tidal range and an intertidal zone of over 250 square kilometres (96 square miles).[3] With each ebb and flow of the twice-daily tides, millions of gallons of seawater bathe that intertidal zone in plankton and nutrients that feed a variety of fish and shellfish, including the farmed oysters from Cancale and the AOP (Protected Designation of Origin) Mont-Saint-Michel bouchot mussels.

Sun and wind play determining, if ephemeral, roles in the bay's tides as well. High-pressure weather systems – a.k.a. sunshine and fair weather – depress sea levels and cause extra-low tides. Low-pressure systems and storms mean higher tide levels than predicted. Onshore winds from the north/northwest drive rising tides farther into the bay, pushing them to reach higher levels. Offshore winds from the north and east blow the water out of the bay, drawing water levels to their lowest ebbs. At any time during a tide cycle, a front can blow in or the wind can shift, and the tidal range will vary accordingly. When these meteorological events occur during an equinoctial tide, a super tide can suddenly become a mega tide (not an actual scientific term) that defies all astronomical and oceanographic predictions. These mega tide events are what send waves crashing over seawalls at high tide and, alternately, lay bare stretches of seabed that haven't been uncovered for years. They are what Mont-Saint-Michel tourists live for – local coastal foragers, too: for the lower the tide goes, the more marine life is exposed, and the more time foragers have to catch, gather, and dig for it.

Generally speaking, however, a typical *grande marée* in Mont-Saint-Michel Bay has enough of a tidal range to allow gathering or catching all but the deepest-dwelling intertidal marine life. Moreover, the immensity of the bay's intertidal zone, its gently sloping seabed, and its profusion of edible sea creatures are what make coastal foraging in the bay so accessible and attractive.

The Bay and Its Bounty: Three 'Elemental' Case Studies

The four natural elements that determine and affect the tides in Mont-Saint-Michel Bay can also function as an entry point for examining the particularities of specific *pêche à pied* areas. The water's tidal flow is the elemental constant these locations have in common. Otherwise, each can be said to have a dominant element that can make – or break – a *pêche à pied* outing. An examination of three locations in the bay – Cherrueix, Cancale, and Port-Picain/Port-Mer on the Brittany side – offers a good overview of how the elements directly affect coastal foraging conditions.

Earthly Expanses: Cherrueix

Cherrueix, a town that lies at the southern base of Mont-Saint-Michel Bay, is one of the most popular spots for *la pêche à pied* in the area. At each *grande marée*, hundreds of people convene at La Laronnière or La Chapelle Sainte Anne, the two main entry points to the bay's foreshore. The sheer earthly vastness of the intertidal zone before them is a sight to behold. Practically speaking, the miles and miles of uncovered foreshore stretch the time needed to forage and turn the excursion into an all-afternoon outing.

High noon or a little after is the usual meeting time for the groups that go out on the Cherrueix tide flats together. A quick picnic by the car or on the grassy embankment is the *de rigueur* fortification for the excursion. And it is quite an excursion: six kilometres (three and a half miles) of sand, mud, and tide pools to cross to get to the lowest edge of the intertidal zone. Most people make the trip on foot, but some get driven out in tractor beds outfitted especially for the occasion.

'It's not so much the trip out that's hard but the trip back, when the baskets and buckets are full and you have to carry them', explains Pierre Lemonnier, who lives right by the entrance in La Laronnière and has been chauffeuring *pêcheurs* to low water with his tractor for twenty-five years. 'They sometimes complain about the 10 euros I charge, but they always say it's worth it on the way back'.

Once foragers reach the lower tidal zone, everyone disperses to search for edible treasure. Some rake for clams. Others stalk the farthest reaches of the tide for wild oysters and scallops that live on the seabed. A few adventurous souls head further out to mussel farms, where the pools that form around the stakes are a favourite hideout for crabs and shrimp.

It is a heady sensation to be out so far at the bottom of the sea. But time and tide wait for no one, as the saying goes, and the moment the tide turns, it's time to turn back. With a mean speed of six kilometres (three and a half miles) per hour, the rising water can quickly overtake anyone on foot who stays out too long.

Wind-Sensitive Ebbs: Cancale

In Cancale, on the eastern side of the bay, oysters have been big business since the sixteenth century, when the town furnished the royal tables of François I.[4] Coastal foragers come to Cancale primarily, if not solely, for oysters, both the native flat European oysters (*Ostrea edulis*) and the farmed, crinkle-shelled Pacific oysters (*Magallana gigas*) that now propagate wild. There are plenty to be had, but most of the foreshore is covered in private oyster beds that are off limits to coastal foragers. Recreational oystering is therefore only possible on the lowest intertidal zones beyond the beds. Here, the wind and its effect on the tide are paramount. Every incremental tidal ebb counts. A single centimetre can uncover a swathe of foreshore that is ten metres (thirty-three feet) wide and reveals hundreds of wild oysters. It can be pouring rain and blowing a gale, but as long as an offshore wind is blowing from the north or northeast, the tide will ebb beyond the beds, and gathering oysters will be as easy as hunting for Easter eggs. Onshore winds from the west and northwest have the opposite effect. They can prevent the tide from ebbing to the edge of the oyster beds at all, even during the greatest equinoctial tide forecasts.[5] Those who know the oyster beds well won't even bother going out if an onshore wind is disagreeably brisk or a northwesterly storm is forecast. It's too much effort, and there's too much of a risk of returning *bredouille*, or empty-handed.[6]

Sun-Ruled Shellfish: Port-Picain and Port-Mer

Nestled on the western side of the Pointe du Grouin headland just north of Cancale, Port-Picain and Port-Mer have got it all: rocks for crabs, lobsters, winkles, and abalone; sand for razor clams and sand eels; silty mud for cockles, clams, and oysters; and clear water shallows for *le bouquet* (common prawns). The south-facing beaches are coastal foraging idylls where guides teach beginner *pêche à pied* courses and schools bring groups of students to learn about marine life.[7] There are even outdoor cafés and public restrooms.

When the air is warm, the sun is bright, and the breeze is soft, Port-Picain and Port-Mer offer an embarrassment of riches to be dug, caught, or gathered. Burrowing bivalves (cockles, clams, razor clams) rise closer to the surface and leave telltale holes that are easy to spot. Crabs and lobsters are lured out of the deep by the seasonal promise of tide pools filled with food, and winkles cluster on the rocks to graze on the algae bloom.

Dark skies, chilly temperatures, and stormy conditions are another story, however. When the weather and water are cold and rough, crustaceans make themselves scarce, seeking safety and surer food sources in deeper waters. Cockles and clams remain hidden beneath the mud and sand. The two coves are also subject to the same tidal variables mentioned previously, and onshore winds can prevent the tide from ebbing low enough to expose the most propitious foraging areas.

'Everything depends on the weather there', grumbles Roland Audic, a retired oyster farmer who knows all the coastal foraging areas of the region well. 'Then, when it's sunny, there are too many people on the beach, and they scare everything away. The best tides for Port-Picain and Port-Mer are in the late fall, when all the tourists have gone, but the water is still warm, and the oysters and clams have fattened up after their reproductive period'.

The Siren Call of *Les Grandes Marées*

'We live by the rhythm of the tides', says Olivier Mocé, an avid *pêcheur à pied* whose home is right beside the foreshore in Cherrueix. 'Invariably, our conversations always return to the last *grande marée* – or the next'.

Olivier goes so far as to schedule his vacation days during *les grandes marées* each year, something that's easy to do by consulting an *annuaire de marées* (tide table). Local tide tables are compiled by port (the nearest are Cancale, Granville, and Saint-Malo) and widely available as small booklets and online apps. The tables display tide forecasts as monthly charts with columns for the times and heights of each day's tides along with the tide's coefficient. Tidal coefficients are calculations only used in France, and they are extremely handy for anyone boating, fishing, or foraging in the bay area.[8] Numbers ranging from 20 to 120 act as a kind of tidal shorthand, with 20 representing the smallest tidal range and 120 the absolute highest. A tidal coefficient of 120 happens only once every eighteen years when an equinoctial tide occurs on the exact day of the equinox. A 100 coefficient is considered the mean tidal range for an equinoctial tide, so any coefficient over 100 is the golden ticket for Olivier and the *pêcheurs à pied* in the bay area.

Jeanne Grossin is another *pêche à pied* devotee who plans her life around the equinoctial tides. When the 27-year-old real estate agent from Cancale learned the tidal coefficient for a warm, sunny day in April would be 108, she immediately cancelled all her appointments for the afternoon. She borrowed a bucket and a rake, grabbed her boots, and raced to Cherrueix so that she could slosh around on the tidal flats and forage for clams and oysters. 'There's absolutely no meeting whatsoever that would make me miss it, not even a house closing', she declared.[9]

Jeanne caught the *pêche à pied* bug from her father, Manu Grossin, a retired fisherman who learned to forage the coast when he was very young. While his grandmother and great-grandmother worked their oyster beds, they would let him gather winkles and net shrimp in the tide pools nearby. By the age of 10, Manu was selling his winkles to restaurants on the harbour for pocket money. The long days and nights of a commercial fishing career never diminished his love for *la pêche à pied*, either. 'Sometimes I'd come in after 24 hours on the boat, eat a sandwich, then put my boots back on and head out to *la marée*', he admits.[10]

Marie-Josephe Quéma, an oyster farmer who passed away in April 2025 at the age of 103, felt much the same way. 'Maman would work a full day tending the oysters, then walk to the water's edge at low tide to see what she could find or catch', explained her daughter, Marie-Christine Masson. 'She would poke around the rocks and oyster basins and usually find a lobster with a dogfish wrapped around it. The dogfish was waiting for the lobster to moult so it could eat it, but instead we would get both for dinner'.[11]

Then there is Andrea Invidia, an Italian with a vacation home in Cancale who plans his stays in Brittany around the *grandes marées*. During the three- or four-day periods when the tidal coefficient is over 100, he's out on the mudflats every afternoon gathering oysters and clams. Some of these he eats, but most of them he stocks in the basins of an obliging oyster farmer so he can take them back with him to Italy and share them with friends.

La Pêche à Pied, a Cultural Phenomenon

Even with these examples, the cultural scope and significance of *la pêche à pied* is as hard to explain as the term itself. The literal translation of *pêche à pied* is 'fishing on foot', which sounds both oxymoronic (how can you fish on foot?) and clunky in English. Yet that 'on foot' is crucial to the official activity. The French Office for Biodiversity defines *la pêche à pied* as 'a fishing activity carried out without the use of a boat or other motorized equipment, without losing contact with the ground, and without any breathing equipment allowing the person to remain submerged'.[12]

'Coastal foraging' is the generally accepted expression (and one I use here) in English, but it doesn't have the same breadth of meaning or common usage. In English, you wouldn't say, 'Hey, do you want to go coastal foraging?' but you would say, 'Hey, do you want to go clam digging?' – or any other species-specific verb. *La pêche à pied* offers a wonderfully concise way of expressing all the clam digging, oyster gathering, abalone prying, winkle picking, shrimp scooping, scallop grabbing, and lobster trapping that can happen at low tide. It even includes snaring the stray sole, squid, or conger eel to be found in a tide pool or beneath a rock (because no true *pêcheur à pied* would pass up one of those).

According to a survey carried out by the Observatoire de l'Environnement en Bretagne, over 450,000 people forage the intertidal zones along the coast of Brittany each year. Or as the Saint-Malo/Baie du Mont-Saint-Michel tourism website proclaims (somewhat hyperbolically), 'Everyone who lives in Brittany has a friend or family member who is a *pêche à pied* enthusiast'.[13]

What was once a mostly professional or subsistence activity practiced by men and women of nearby communities is now a major recreational pastime that spans generations and class divisions. In less than a hundred years – historians date the advent of recreational coastal foraging to the establishment of paid leave in 1936 – *la pêche à pied* has become as much a part of the seaside experience as beachgoing or boating.[14] Today, newspapers herald the *grandes marées* with features on *la pêche à pied* and informational articles that remind readers of the rules and regulations that are strictly enforced. Most beaches are outfitted with signs that attractively and concisely detail regulatory information, and all tourist outlets distribute *pêche à pied* brochures and fishing rulers.

Throughout the region, supermarkets sell tide chart booklets next to the registers. Newsagents stock clam rakes. Hardware stores display an array of baskets and tools in the front aisles. There are postcards that feature *la pêche à pied*, T-shirts emblazoned with *la pêche à pied*, and *pêche à pied* baskets tricked out with stylish flourishes (that would never last long out on the muddy, wet tidal flats).

But is this commercialized, aspirational aspect of *la pêche à pied* truly representative of its popularity and longevity? According to a study by the Réseau Littoria in 2021, '55% to 65% of coastal foragers in France are over 50 [. . .] teenagers and young adults are mostly absent, though young children are often present in family groups with their parents and grandparents'.[15]

Luckily, there are exceptions to these findings. There's Jeanne Grossin, who takes great pleasure in inviting her friends along to go dig for clams and gather oysters in the bay. There's also Vincent Cobat, a teacher from Saint Pierre de Plesguen who learned how to hunt for abalone from a retired fisherman. He has since taught a group of 30- and 40-something friends how to do it and now organizes outings for all of them. Their desire to pass on their passion and experience is what will keep the traditional, ancestral practices alive.

C'est Gratuit. C'est Frais. Et C'est BON!

'It's free. It's fresh. And it's GOOD!' Those were the first things Jeanne Grossin could think of when I asked her about why she loved *la pêche à pied* so much. She went on to talk about the primal hunter/gathering euphoria of foraging for your own food, the exhilaration and pleasant exhaustion of being out on the bay all day, and the interesting people she has met, but savouring her catch after the fact was first and foremost in her mind. This is perhaps what makes *la pêche à pied* so wildly popular and so uniquely French. Ultimately, the entire experience is centred around the sharing and enjoyment of food.

'We picnic together, we open a few oysters and clams out in the bay together, we have a drink with oysters before we leave together, then we get together later in the day or the week to eat our catch', is how Olivier Mocé summarizes the traditional *pêche à pied* experience. Vincent Cobat's group of friends always pool their catch of abalones and whatever else they find, then choose a time to enjoy them together in the days that follow. Preparations tend to be simple, like the area's cuisine: lots of butter, a little parsley, and wine, plus the garlic, shallots, and onions that grow so well on the farms that line the bay. In truth, the seafood is so fresh it needs little to no adornment or accompaniment aside from fresh bread, salted butter, and plenty of chilled white wine.

Sharing is also an integral part of *la pêche à pied* culture on the bay. The experience, the knowledge, the best foraging areas, even the catch: it's all generously and enthusiastically passed along to family, friends, and even friendly strangers. There are more than enough fish and shellfish in the bay to go around. Rare are the days when a *pêcheur à pied*, even a beginner, returns from an outing without a surplus. The legal catch limits are generous. Each forager is allowed up to 60 oysters, 150 clams, or a whopping 300 cockles or mussels per day.

'There's a lot of sharing that goes on, especially with older neighbours who can no longer go out', explains Olivier Mocé. 'They often get a small basket of clams and oysters from a neighbour or friend'. When I take my leave of Pierre Lemonnier after talking to him about his *pêche à pied* experiences, he asks if I'd like a few oysters – even though I barely know him. I thank him but decline, not because I'm being polite, but because I'll have plenty waiting for me at home. On the village-like backstreets of the Cancale harbour where I live, my neighbours and I swap our catch when we return from our excursions as well:

– You want some oysters? Here, how about a dozen?
– Do you like razor clams? Have some for you and your guests.
– I have too many winkles for just the two of us. Why don't you take some?

We meet in the street, show off our catch, exchange stories, and share the plentiful, highly perishable bounty. Roland Audic will even show up on my doorstep a little later with a bowlful of the small, common prawns he catches – perfectly cooked and still warm.

These are delicacies money cannot buy. Their freshness defies anything even the very best fishmonger would be capable of. The sheer quantity of shrimp in Roland's bowl would cost a fortune in a restaurant, if a restaurant even served that many.

And to think that such abundance is accessible to anyone and everyone who heads out into the bay during a *grande marée*! Jean-Patrick Laravine, one of Vincent Cobat's abalone gang, perhaps expressed it best when he took his first bite of an abalone he had caught, sautéed in butter. He looked around the table that was also laden with oysters, crab, clams, and razor clams and said, 'Je ne serai jamais roi, mais je mange comme un prince' ('I will never be a king, but I eat like a prince').[16]

About the Author

Mary Margaret Chappell is an American food writer who lives in Brittany, France.

Notes

1 UNESCO, 'Mont-Saint-Michel and Its Bay' <https://whc.unesco.org/en/list/80/> [accessed 1 May 2025].
2 Emmanuèle Savelli, 'Pourquoi, en Bretagne, les marées sont-elles si différentes de celles du reste du littoral français?', Observatorie de L'Environnement en Bretagne, 17 September 2025 <https://bretagne-environnement.fr/article/bretagne-marees-records-europe > [accessed 26 January 2026].
3 Mairie de Saint Jean le Thomas, 'Les Marées', St Jean le Thomas <https://m.saintjeanlethomas.com/Les-Marees_a12.html> [accessed 1 May 2025].
4 Joseph Pichot-Louvet, *Cancale et sa baie* (Les Éditions du Phare, 1998), p. 5.
5 Roland Audic, interview with the author, 19 March 2025.

6 Manu Grossin, interview with the author, 22 February 2025.
7 Fanny Le Hir, 'Une parenthèse en Bretagne', Retour de Peche <https://atelier-retour-de-peche.com> [accessed 1 May 2025].
8 *Prediction de Maree*, May 2024 <https://refmar.shom.fr/sites/default/files/2025-01/GT-TSH_CatD_Fiche_Prediction_maree.pdf> [accessed 1 May 2025].
9 Jeanne Grossin, interview with the author, 27 April 2025.
10 Manu Grossin, interview with the author, 22 February 2025.
11 Marie-Christine Masson, interview with the author, 7 March 2025.
12 Réseau Littorea, *Etude et diagnostic de l'activité de pêche à pied récréative: Cahier méthodologique et recueil d'expériences* (Pêch à Pied de Loisir, 2021), p. 12.
13 'La Pêche à Pied', Destination Saint-Malo Baie du Mont-Saint-Michel <https://www.saint-malo-tourisme.com/explorer-notre-destination/inspirations-nautiques/la-peche-a-pied/> [accessed 1 May 2025].
14 Pierrick Jegu, author of *Terre de pêche à pied*, interview with the author, 18 March 2025.
15 Various authors, *Cahier Méthodologique Pêche à pied de loisir*, 3rd edn (2021), p. 27.
16 Jean-Patrick Laravine, interview with the author, 2 April 2025.

9
Protein
The History of a Cinderella Superfood

Sarah Dry

This is a story of protein which takes place at both the largest and smallest of scales, from personal obsessions to global challenges. It is a story of unlikely combinations, of nutrition scientists and oil industry executives, of malnourished children and hyperdeveloped bodybuilders, of pragmatism and techno-optimism, of wildly successful business ventures and dramatically failed ones. Beginning in the 1950s, it was hoped that novel engineered forms of protein could prevent an imminent Malthusian crisis of imbalance between the global food supply and population. Protein reigned for twenty years as the queen of nutrition science before it was dramatically dethroned in the mid-1970s, when a series of famines made the quantity rather than the quality of food supply of primary importance. But rumours of protein's death were exaggerated. In important ways, its status as a master element has barely faltered since the 1950s. This story, therefore, is also of the rise and rise of an obsession of those seeking to engineer the perfectly healthy body, an obsession which has, in recent decades, become once again seen as highly relevant to the health of the planet as a whole.

What follows is a tale in three parts. I tell a story of three innovators – Norman Pirie, Nevin Scrimshaw, and Bob Hoffman – who attempted in distinctive ways to introduce palatable engineered protein to a range of consumers in the second half of the twentieth century. Their approach to the challenge demonstrates the ways in which the reduction of food to its elements was tied to development – of the globe, of individual bodies, and of a profitable food system in which industrial processing played an integral but complex role. In their work, we can see ways in which the eating habits and tastes of people all over the planet were reshaped by a new way of seeing food as an element in a system – of global and individual development.

Pioneer of Leaf Protein

British biochemist Norman Pirie (1907–97) began thinking about leaves as food for humans during the early years of World War II, when German attacks on merchant ships threatened British food supplies. Leaves were a commodity Britain had in abundance. If they could be made into food, the practical benefits would be enormous. And Pirie was an extremely practical-minded man. He understood that the challenge was twofold. First, he needed to create a mechanical process for turning fresh leaves (usually 'soft-textured, sappy leaves' derived from local farm crops such as young cereals, mustard, kale, sugar beet, and even river weeds) into a pulp that was high enough in protein to be worth feeding to humans. Equally important was creating new food that humans both could and, just as importantly, would actually want to eat.

In October 1940, Pirie hosted a 'scientific lunch' for the Minister of Food, Lord Woolton, with a menu determined to show that new foods could be readily generated.[1] Such products could be made either from existing but underutilized products usually considered waste, including kidneys, blood, whale meat, and horse beans, or from completely novel sources of food, including 'cream of yeast' soup, a form of Welsh rarebit made out of dried eggs, and – Pirie's special project – leaf protein soup. Served with yeast bread and washed down with a 1923 Volnay Burgundy, the lunch was attended by a brace of Cambridge scientists. Intended to demonstrate that new foods could be made suitable for a Cambridge high table, the meal did not have the effect on the Minister of Food that Pirie had hoped for. In his diary, Woolton recorded that he had needed to have a Turkish bath and massage the next day to 'get rid of the effect on my liver – it was so full of protein!'[2]

Pirie persevered. While his early research on leaf protein had been undertaken in a context of urgent national need, by the mid-1950s, food needs were increasingly being addressed at the global scale. Together, the newly established World Health Organization, Food and Agriculture Organization, and UNICEF aimed to improve the status of the world's children, its health, and its food and agriculture. To do so, they relied on a new science – of nutrition.[3] Beginning in the early decades of the twentieth century, deficiencies in certain essential nutrients – vital amines – had been shown to cause a range of previously poorly understood maladies, many in children, including blindness as a result of a deficiency in vitamin A, and conditions such as goitre and dwarfism could occur if even tiny amounts of essential minerals such as zinc or iodine were missing.[4] During the war, with food supplies under pressure, new dietary guidelines were drawn up in Britain and the United States and by the League of Nations, stipulating the minimum daily caloric intake and the minimum specific nutrients individuals needed. This 'science of single nutrient deficiency' generated evidence to support the widespread fortification of food. In this way, the nutritional requirements of individuals were linked with the nutritional needs of the global body. Researchers continued to search for such missing elements.

In the early 1950s, protein came to be a focus of concern following a series of meetings at which researchers presented evidence that its absence could cause acute illnesses such as kwashiorkor, a childhood wasting disease in Ghana not present in populations with easy access to protein-dense milk. More worrying was the drag on the overall global development of less acute but still clinically significant shortages of protein that children around the world might be experiencing. By 1952, what had initially been a fairly focused investigation of the role of protein deficiency in childhood diseases among children in coastal Ghana had become a broader study of what the committee considered 'a problem of fundamental importance throughout the world'.[5] In 1955, a joint FAO/WHO Expert Committee on Nutrition was established to address a new phenomenon that had been named 'protein malnutrition'.[6]

By 1968, the situation seemed critical. That year, a panel of experts published a report for the UN titled *International Action to Avert the Impending Protein Crisis*, in which they described the 'world protein gap' as a 'most important scientific, technological and public health problem' of both national and international scale. The concept of a protein gap, or deficit, between the nutritional needs of the world's population and its current supply was predicated on the idea that food could and should be considered as the sum of constituent parts, or elements.[7] The quantity of food, or calories, alone was not an adequate measure of global needs. If children lacked protein, their growth would be stunted regardless of the number of calories they were ingesting. The development of both children and the nations in which they lived depended on adequate levels of protein in particular.

Increasing the yield of traditional agriculture and husbandry, a process begun at the inception of the so-called Green Revolution in the 1940s, was one way to solve the problem. But such increases also contributed to the problem by skewing global food production towards cereals and away from protein-rich legume crops. The nature, scale, and urgency of the perceived race to feed the world as its population rapidly grew made it seem that entirely new sources of food would need to be developed. This included soya, fish meal, algae, and new seed oils, as well as engineered proteins.[8]

For his part, Pirie stayed focused on the leaf protein he'd already spent more than fifteen years investigating. He was in a position to quickly respond to the new interest in protein at a global level while at the same time promoting leaf protein as a substance that could be sourced and processed locally, cutting down on cost and rendering it potentially more adaptable to local settings. Though usefully high in protein, it was not easy to convince local populations to eat the cake-y green substance, which turned biscuits an unappetizing mouldy colour. He suggested a slow and steady approach to winning the hearts and minds of local people in 'various undernourished regions of the world', given the innate 'conservatism' of the population. Nevertheless, noted Pirie, the use of 'judicious propaganda' could speed things along. And a 'brief survey of the packaged foods and drinks now widely sold' demonstrated that 'almost regardless of the actual properties or merits of the novelty, acceptance can be won'.[9]

To do so required canny marketing. Pirie suggested that before introducing leaf protein to undernourished communities, it was necessary to entice 'well fed visitors from ministries, embassies and the international organizations'. Recognizing the importance of 'the presentation of leaf protein on the table' in such different settings, Pirie proposed a series of green-coloured leaf-protein 'cocktail snacks', such as leaf paste on a cracker flavoured with curry, fish, or mint and a deep-fried fritter made from leaf protein blended with banana and wrapped in pastry.[10]

Pirie developed a range of recipes using leaf protein, demonstrating how the wet green pulp could be incorporated into everything from appetizers and main courses to soups and desserts (and baked goods). Semolina shortbreads could be made using either wet leaf protein or freeze-dried leaf protein. Banana coconut rolls made with freeze-dried leaf protein could be served with a sweet orange sauce or with a savoury curry sauce. Samosa aloos could be made with equal parts mashed potato and wet leaf protein and served with a pomegranate-mint chutney also loaded with leaf protein. Banana soup was best made with a liquidizer and the addition of redcurrant jelly and evaporated milk alongside leaf protein, with mint or apple added to taste. Leaf protein could also be made into a sauce or salad dressing mixed with vinegar, milk, salt, sugar, and a dash of chili powder. This recipe card includes a note indicating 'acceptability – rather good'.[11] A leaf-protein-extended prune whip, made of prunes, evaporated milk, and leaf, was best served chilled. And sweet oat blocks, normally made with apple alone, could be extended with leaf protein.

Save for a few annotations – 'rather good' – there is little evidence about how people responded to the various preparations that Pirie and his colleagues developed. In part, this is because palatability was only one of the barriers to widespread adoption Pirie faced in his push for leaf protein. In addition to its 'presentation on the table', leaf protein was doomed by a combination of relatively low protein yields, challenges in processing on a scale that made it economical, and the availability of more desirable and less expensive sources of protein.[12]

If the strong colour, bitter flavour, and pulpy texture of the wet protein posed certain limitations on how it could be used, in this, leaf protein was like any other vegetable or fruit. The difference was that the expectations for leaf protein were so high. A food to close the protein gap would need to be, potentially, a food for everyone, everywhere, and in every dish. For that, a superfood might be needed, and such a food might not be found in nature. Instead, it might be necessary to engineer one.

Petroproteins for the People

MIT nutritionist Nevin Scrimshaw (1918–2013), chair of the Protein Advisory Group, was at the centre of efforts to engineer new forms of protein using bacteria and yeast to produce protein-rich food. Such cells could be engineered to feed on a variety of substrates that were the waste by-products of existing industrial processes, introducing what was seen as a desirable circularity and efficiency into the manufacturing process.[13] There was a precedent for such projects, as waste from milk production in the United States was already being used to develop skim milk powder that had been used to supplement infant foods in Africa and Central America.

Other ways to create a virtuous circle linking industrial production in the developed world with hunger in the developing world – for in a sense commercializing the protein gap – were pursued. In the late 1960s, Scrimshaw collaborated with British Petroleum on a project to produce a protein-enriched yeast by fermenting the by-products of the petroleum refining process. The idea rapidly gained traction within the oil industry. BP was one of many companies to launch projects to produce protein from hydrocarbons. Other companies, including Shell, Liquichimica, and ICI, launched their own 'petroprotein' projects.[14]

Hopes ran extremely high that by applying fermentation techniques which had been developed in the pharmaceutical and food and beverage industries to the petrochemical sector, it would be possible to create yeast species that could metabolize the paraffin in petroleum into protein. 'For the first time', explained an internal BP document, 'a purely industrial process, relying neither on arable land nor on farm produce but nevertheless based on a biological *process*, yields a nutrient which can be incorporated into animal feed or food for human consumption. [. . .] Does this not indicate that a third era in man's history is now beginning?' The resulting product would be a 'cream-coloured, free floating powder, odourless and tasteless, high in protein content and constant in quality, a notable feature being its large production of assimilable amino acids, particularly lysine'. Free of the negative cultural or aesthetic baggage that hindered the adoption of leaf protein, such a flavourless, tasteless powder was an ideal – even a wonder – food. Also wonderful was the opportunity to diversify BP, potentially transforming it into 'one of the great food companies of the world'.[15] At an important meeting held to discuss and promote the idea at MIT in 1967, its name was changed from the unappealing petroprotein to the blander, more marketable 'single cell protein'.[16]

Internal BP documents indicate that those involved understood the scale of what was at stake: 'The unacceptable reality of hunger in the world places on the shoulders of the inventors of such a discovery the moral obligation of developing it to the best of their ability within the limits in which this development can be profitable'.[17] For the BP executives who advocated most enthusiastically for the undertaking, profits and ethics were happily entwined. 'Although the goal to be wished for is the credit of removing starvation from the world, and perhaps a Nobel prize', offered Fyfe Gillies, a manager of the BP research development department in London, 'one can contemplate with some confidence that BP will not be out of pocket at the end of the day'.[18]

By 1965, BP had set up three microbiology labs in Lavera (Italy), Grangemouth (Scotland), and Epernon (France). Other proteins were fast on their heels. Standard Oil and Nestle announced a joint research project. Shell, Gulf, Sun Oil, and Mobil were also looking into petroproteins, and scientists in communist countries, including the Soviet Union, the Czech Republic, East Germany, and China, were also at work. In 1967, there was an air of inevitability around petroproteins: 'There is little doubt', concluded a BP study, 'that sooner or later the world will see the development of a protein manufacturing industry'.[19]

If the grand vision of eliminating malnutrition was publicly declared, privately, doubts were expressed within the company that producing marketable BP proteins for human consumption would be easy. A few small forays into testing the product on humans were undertaken, often carefully calibrated to produce positive public relations copy, as in the press release announcing the 1967 conference on single-cell protein held at MIT, reporting that eighty students had been fed single-cell protein for two months with no undue effects, and twenty had gone on to receive their entire protein requirement from the yeast protein.[20] (It was not reported whether the protein was served at the conference dinner.)

Not all trials were so positive. Depending on the manufacturing process, some proteins produced by yeasts contained high levels of certain amino acids that caused severe symptoms, including nausea, vomiting, diarrhoea, and prostration. More alarming still was that such symptoms could emerge after a few days or even weeks of uneventful consumption of the protein. Reports of side effects experienced by those given 150 grams of dried green algae to eat were even more alarming, with petechial haemorrhages, oedema of the face, peeling of the fingers and itching, pain, and oedema of the large toes reported.[21]

Sobering as such results were, the combination of the moral purpose, profit potential, and race for competitive advantage spurred investment. A new subsidiary, BP Proteins, was incorporated in 1970, and by 1971, Toprina, BP's brand name for BP protein, was being sold as a replacement for animal proteins in calf, pig, and poultry feed. But like leaf proteins, petroproteins also never made it to market, where their palatability might test the optimistic projections of BP researchers.[22]

But while the disgust that many felt about eating food made of oil was substantial, and while the technical challenges of making a yeast protein that did not sicken those who ate it remained unresolved, the single-cell proteins BP and others had dreamed would revolutionize their business and save the world in the process failed for market reasons. They never got close enough to being produced or promoted as food for humans to give those humans a chance to accept or reject them. By the late 1970s, a combination of rising oil prices with scientific and manufacturing challenges had scuttled not only BP's protein ambitions but those of most of its competitors. By the early 1980s, the only surviving single-cell protein plants were in communist countries.

The Cinderella of Food

While protein was being extolled, designed, and trialled for the world's poorest, most malnourished people, it was also being sold to some of the world's most developed citizens: American practitioners of the new sport and leisure activity of bodybuilding. For those promoting the body-shaping potential of proteins, profits and nutrition were linked just as tightly as they were in the quest to aid global development.

No one had a bigger appetite for generating profits from the burgeoning subculture than Bob Hoffman (1921–97), an American entrepreneur and founder of the York Barbell company.[23] As early as 1938, Hoffman had articulated in his first book, *How to Be Strong, Healthy and Happy*, the idea that 'a strong and healthy man must consume foods which contain all the necessary elements', including a sufficient amount of 'all of the mineral and vitamins the body requires'.[24] By 1952, Hoffman's rather generic framing of a balanced and complete diet had evolved into a much more precise – and marketable – assertion that, in addition to the right vitamins, a high-protein diet could produce nothing short of miraculous progress in building muscle.

But how could it be that Americans, citizens of the wealthiest country in the world, were undernourished and therefore in need of a supplement to their ordinary diet? The American body had been weakened by a decline in physical exercise and increasing consumption of 'more manufactured, devitalized foods'.[25] Hoffman suggested that part of the blame lay in the damaging effects of intensive agriculture on American soil.[26] The solution was not a return to the simple and wholesome diets of old. For too long, Hoffman declared, protein had been the 'Cinderella' of food, obscured by more 'glamorous' minerals and vitamins. The time had come to recognize that 'protein is the most necessary food element'.[27]

The first 'high-protein food' had been developed by Irvin Johnson in his kitchen. Soon, Hoffman was marketing Hi-Proteen under his own name, having reverse engineered

Johnson's blend by mixing Hershey's chocolate with protein-rich soybean flour, stirring the mass with 'a canoe paddle in a soy bean flour container'.[28] He was more than ready to offer his readers guidance on how best to incorporate the new protein food into their diets. To help consumers introduce 'Hi-Proteen' into everyday diets, he included recipes for broth, milkshakes, and 'hi-pro' cubes made of instant coffee, milk, sugar, and the supplement. To wash the protein down, Hoffman recommended mixing his 'high-protein food' with lemon, tomato, and grape juice. The latter was a particular favourite – a quart jar filled with a pint of grape juice and a cup of protein food made a 'good, palatable, result-producing mix'.[29] Whatever the variation, consuming the product slowly was advised, taking time to ramp up consumption over several weeks and always ingesting it slowly, ideally by sipping over the course of half an hour or more. Certain starchy foods, including potatoes and bread, as well as vegetables, needed to be cut back on so that the protein could be digested more easily. Implicit in such advice were potential digestive side effects that could occur as a result of changing diet too quickly or dramatically.

Photographs illustrated the transformation of a pigeon-chested 13-year-old boy into a muscle-bound young man in just a month, thanks to Hoffman's Hi-Proteen.[30] The small print of the advertisement suggested that it was made of soybean, but no specific information was given on the contents of the miracle food.[31] Results, rather than ingredients, were what really mattered. The powder was designed to be as innocuous as possible – a means to an end.

By the mid-1970s, the idea of a 'complete protein', containing the total complement of amino acids necessary for living organisms to grow new tissue, had been introduced. Since the body could not store these amino acids and not all foods provided them, it was advisable to supplement every meal with a small amount of carefully formulated supplement. 'Hi-Proteens will complete and make stable many times their own weight of incomplete proteins', explained Hoffman. At least a teaspoonful of 'Hi-Proteen with every meal' could make the difference between bodybuilding success and failure.[32] For those seeking variation, vanilla and other flavourings could be added. To meet the need for continuous consumption of supplementary protein, readers of *Strength & Health* could choose from a dizzying array of protein products.[33] Helpfully, Hoffman produced a cookbook to aid housewives in incorporating Hi-Protein into family meals, and a monthly column of reader recipes incorporating Hi-Proteen was published.[34] Adults and children, men and women – there was a protein supplement and an associated benefit for the entire family. Protein supplements would help women achieve a slender physique, men gain muscle weight, children grow up strong, and older people live longer.

Conclusion

Pirie, Scrimshaw, and Hoffman all grappled with the challenge of rendering a novel and deliberately simplified form of relatively 'pure' (or concentrated) protein back into something resembling food that would be appealing to a target audience. While both leaf protein and single-cell protein failed to reach the large numbers of consumers Pirie and Scrimshaw had respectively imagined for them, their fates were also shaped by the dramatic disappearance of the world protein gap. What had once menaced the entire planet had been, by 1974, redefined out of existence.

The gap was in fact a mistake based on a faulty extrapolation of limited evidence from particular settings in sub-Saharan Africa to the entire planet – so declared two influential papers that sought to demonstrate that the world protein gap was a kind of global misunderstanding, or 'fiasco' of scientific hubris, revealing the 'arrogance of supposing that we know the answers'.[35] World hunger itself had not disappeared but intensified, with famines and food shortages widespread in the mid-1970s. When people were starving, worrying about the deficiency of a single nutrient suddenly seemed a dangerous distraction. Malnutrition was usually a problem of insufficient calories, not insufficient protein. 'Protein-calorie malnutrition', not protein deficiency alone, was the new framing for world hunger.

But, as this new formulation suggests, the downgrading of protein did not entail a rejection of nutritionist approaches. The deficiency model held. Vitamin A replaced protein in the 1980s as the missing nutrient, and in the 1990s, it was 'hidden hunger' caused by deficiencies in micronutrients such as iron and iodine.[36] In this sense, protein is just one of many so-called charismatic nutrients, elements which serve to justify the authority of nutrition scientists and policymakers to define the food needs of the global population. The ability to set quantifiable markers of success is part of the logic of these nutrients. But deeper than the quantifying impulse is the drive to produce a technical solution to a problem – hunger – whose social, political, or moral dimensions might otherwise overwhelm an uneasy status quo.[37]

The image of protein as a cure-all – useful both for the planet's poorest and hungriest and its richest, for the weakest, the heaviest, or the thinnest – persists today. 'Healthy' foods are, increasingly, considered to be not only fresh but high in protein. Between 2022 and 2024, the percentage of Americans seeking to consume more protein rose from 59 percent to 71 percent.[38] And the links between novel protein sources and global sustainability asserted by Pirie and Scrimshaw in the 1960s are being reasserted today by a combination of researchers, policymakers, and entrepreneurs who see engineered protein as a replacement for inefficient and harmful animal husbandry. The 'time for a microbial food revolution is ripe', declared the authors of a 2023 review of the field in *Nature Communications*.[39] Individual consumers, too, are more obsessed with protein than ever, and a plethora of products exist (and are being developed) to meet the demand for all that it is hoped that protein can deliver.

Perhaps the power of protein lies in its seeming ability to help us – as individuals and as an imagined global collective – reach some implicit but potent notion of optimality. Protein has become part of an aspirational drive, a desire to be in some unspecified but urgent way better. And the urge seems to be strengthening. Protein is dead. Long live protein.

About the Author

Sarah Dry is a historian of science and writer. Her most recent book, coauthored with Mike Hulme, is *When Temperature Became Global: A Brief History of the World's Most Important Index* (Princeton, 2027).

Notes

1 Norman Pirie papers, Royal Society (hereafter NWP RS), C.149.

2 Oxford, Bodleian Library and Archives, MS Woolton 11, Lord Woolton diary entry 12 October 1940.

3 See, for example, *The Science and Culture of Nutrition, 1840–1940*, ed. by Andrew Cunningham and Harmke Kamminga (Rodopi, 1995). Despite the extensive literature on the history of nutritional science from a medical perspective, there are fewer studies by historians of science on the history of food and cuisine. Exceptions include Julia Adelman and Lisa Haushofer, 'Food as Medicine, Medicine as Food', *Journal of the History of Medicine and Allied Sciences*, 73.2 (April 2018), pp. 127–222; the articles in *Assimilating Knowledge: Food and Nutrition in Early Modern Physiologies*, ed. by E. C. Spary and Barbara Orland, special issue of *Studies in the History and Philosophy of Biological and Biomedical Sciences*, 43 (June 2012); and Emma Spary and Anya Zilberstein, 'On the Virtues of Historical Entomophagy', *Food Matters: Critical Histories of Food and the Sciences, Osiris*, 35 (January 2020), pp. 1–19.

4 Kenneth Carpenter, *Protein and Energy: A Study of Changing Ideas in Nutrition* (Cambridge University Press, 1994), pp. 121–22.

5 Joint FAO/WHO Expert Committee on Nutrition, *Third Report [of a Meeting Held in Fajara, Gambia, from 28 November to 3 December 1952]* (FAO/WHO, 1953), p. 4.

6 Joint FAO/WHO Expert Committee on Nutrition, p. 5. For the role of protein in world hunger, see Josh Ruxin, 'Hunger, Science, and Politics: FAO, WHO, and Unicef Nutrition Policies, 1945–1978' (unpublished dissertation, University College London, 1996); Josh Ruxin, 'The UN Protein Advisory Group', in *Food, Science, Policy and Regulation in the Twentieth Century: International and Comparative Perspectives*, ed. by David Smith and Jim Phillips (Routledge, 2000), pp. 151–66; Richard D. Semba, 'The Rise and Fall of Protein Malnutrition in Global Health', *Annals of Nutrition and Metabolism*, 69.2 (2016), pp. 79–88; Hannah F. LeBlanc, 'The Protein Gap: The Rise and Fall of a Charismatic Nutrient in International Public Health', *Bulletin of the History of Medicine*, 97.4 (2023), pp. 585–613; Kenneth J. Carpenter, *Protein and Energy: A Study of Changing Ideas in Nutrition* (Cambridge University Press, 1994); John Nott, '"No One May Starve in the British Empire": Kwashiorkor, Protein And the Politics of Nutrition between Britain and Africa', *Social History of Medicine*, 34.2 (2021), pp. 553–76; and Mick Worboys, 'The Discovery of Colonial Malnutrition between the Wars', in *Imperial Medicine and Indigenous Societies*, ed. by D. Arnold (Manchester University Press, 1988), pp. 208–55.

7 LeBlanc, 'Protein Gap'; Kenneth J. Carpenter, 'The History of Enthusiasm for Protein', *Journal of Nutrition*, 116.7 (1986), pp. 1364–70.

8 On the history of protein in the context of humanitarian efforts to address world hunger, see Tom Scott-Smith, *On an Empty Stomach: Two Hundred Years of Hunger Relief* (Cornell University Press, 2020), especially chapter 8, 'High Modernism and the Development Decade', pp. 121–36; and Scott-Smith, 'Beyond the "Raw" and the "Cooked": A History of Fortified Blended Foods', *Disasters*, 39.2 (2015), pp. 244–60. On the introduction of multipurpose food in India, see Joanna Simonow, 'The Rise and Demise of Multi-Purpose Food in India: Food Technology, Population Control and Nutritional Development in the Post-War Era, c. 1944–66', *South Asia: Journal of South Asian Studies*, 44.1 (2021), pp. 167–84.

9 Norman Pirie, 'The Present Position and Future Needs of Research on Leaf Protein', in National Research Council, *Progress in Meeting Protein Needs of Infants and Preschool Children: Proceedings of an International Conference* (National Academies Press, 1961), pp. 509–16 (p. 512).

10 Pirie, 'Present Position and Future Needs', p. 512; M. Byers, S. H. Green, and N. W. Pirie, 'The Presentation of Leaf Protein on the Table: II', *Nutrition*, 19 (1965), pp. 63–70; Norman Pirie, 'The Merits of Food Proteins from Novel Sources', *Science Progress*, 54.215 (1966), pp. 401–12; Norman Pirie, 'Orthodox and Unorthodox Methods of Meeting World Food Needs', *Scientific American*, 216.2 (1967), pp. 27–35; Norman Pirie, 'Leaf Protein as a Human Food', *Lancet*, 274.7109 (1959), pp. 961–62.

11 Undated Pirie recipes, NWP RS, C.139–40.

12 See Scott-Smith, *On an Empty Stomach*, pp. 121–36.

13 Nevin S. Scrimshaw, 'Fifty-Five-Year Personal Experience with Human Nutrition Worldwide', *Annual Review of Nutrition*, 27.1 (2007), pp. 1–18.

14 Scott-Smith, *On an Empty Stomach*, pp. 121–36. For a perspective on petroprotein research and development in the Middle East, see Douglas Rogers, 'Petroprotein Dreams: Hydrocarbon Biotechnology and

Microbial Life Worlds in the Middle East', in *Life Worlds of Middle Eastern Oil: Histories and Ethnographies of Black Gold*, ed. by Nelinda Fuccaro and Mandana E. Limbert (Edinburgh University Press, 2023), pp. 198–220. On the context of Cold War planning in which concerns over protein deficiency developed, with particular reference to a correspondence between Carroll Wilson, MIT professor who helped organize the single-cell protein conference at MIT (and coined the term 'single-cell protein') in 1967, and Soviet academician Djermen Gvishiani, see Yakov Feygin, *Building a Ruin: The Cold War Politics of Soviet Economic Reform* (Harvard University Press, 2024).

15 Cited in Rogers, 'Petroprotein Dreams', p. 201, from BP ARC47739, 'Petroleum Microbiology: Reflections on Future Orientations and Recommendations, May 1969' (emphasis in original). The first two stages of history were hunting/gathering and agricultural/livestock raising.

16 R. Mateles and S. Tannenbaum, *Single Cell Protein* (MIT Press, 1968).

17 James Bamberg, *British Petroleum and Global Oil, 1950–1975: The Challenge of Nationalism* (Cambridge University Press, 2000), p. 435.

18 Bamberg, p. 429.

19 Bamberg, p. 432.

20 Office of Public Relations, MIT Press, 18 October 1969 release on single-cell protein.

21 Nevin Scrimshaw, 'Single Cell Protein for Human Consumption – an Overview', report of the 1967 conference on single-cell protein, in Carroll Wilson papers, MIT, Protein MC 29, pp. 13–14.

22 Bamberg, *British Petroleum*, p. 443 n. 99.

23 For more on Hoffman, see John Fair, *Muscletown USA: Bob Hoffman and the Manly Culture of York Barbell* (Penn State Press, 1999); and Daniel Hall and John Fair, 'The Pioneers of Protein', *Iron Game History*, 8.3 (May–June 2004), pp. 23–34.

24 Bob Hoffman, *How to Be Strong, Healthy and Happy* (York, 1938), p. 27.

25 Bob Hoffman, 'Healthful Eating Habits', *Strength & Health*, January 1952, p. 24.

26 Such views were promulgated by Louis Bromfield's *The Farm* (Harper, 1933).

27 Cited in Hall and Fair, 'Pioneers', p. 31.

28 Hall and Fair, 'Pioneers', p. 27. On the relation between Johnson and Hoffman's protein powder, also see Hall and Fair.

29 Hoffman, 'Healthful Eating Habits', p. 31.

30 Hoffman, p. 34.

31 On the production of soy-based versus milk-based protein foods, see 'Rheo H. Blair, Builder of Champions', *Iron Man* 27 (1968), p. 53; Fair, *Muscletown USA*, pp. 178–80.

32 Bob Hoffman, 'Impressions of the World Championships', *Strength & Health*, March 1967, p. 25.

33 See, e.g., *Strength & Health* February–March 1976, p. 2.

34 *Strength & Health*, August 1971. On the history of milk and bodybuilding, see Conor Heffernan, 'Milk, Modernity and Muscles: Raw Dairy and Bodybuilding in 1960s America', *Gender & History*, 34.3 (2022), pp. 881–95.

35 J. C. Waterlow and P. R. Payne, 'The Protein Gap', *Nature*, 258.5531 (1975), pp. 113–17 (p. 117). See also Donald McLaren, 'The Great Protein Fiasco', *Lancet*, 2.7872 (13 July 1974), pp. 93–96.

36 Charlotte Biltekoff and others, 'Interrogating Moral and Quantification Discourses in Nutritional Knowledge', *Gastronomica*, 14.3 (2014), pp. 17–26 (p. 22).

37 See Warren Belasco, *Meals to Come: A History of the Future of Food* (University of California Press, 2006); and Jessica Mudry *Measured Meals: Nutrition in America* (State University of New York Press, 2009).

38 International Food Information Council, '2024 IFIC Food & Health Survey', IFIC, 2024 <https://foodinsight.org/wp-content/uploads/2024/06/2024-IFIC-Food-Health-Survey.pdf> [accessed 26 January 2026].

39 Alicia Graham and Rodrigo Ledesma-Amaro, 'The Microbial Food Revolution', *Nature Communications*, 14.2231 (19 April 2023), pp. 1–10 (p. 1).

10
Food Ink
The Elements of Style in Cookbook Illustration

Rebecca Federman

With its twenty-seven woodcuts of pots, utensils, pans, and a papal procession, Bartolomeo Scappi's 1570 cookbook *Opera* was the first to incorporate didactic illustration. Since then, illustration – defined here as drawings, engravings, or graphics – has remained a key element in cookbooks, used to demonstrate technique, instruct readers, and embellish recipes. But over time, with the explosion of photography and its oversaturation in our lives, glossy photographs have become the standard visual language in cookbooks. Even cookbooks that had once featured only line-drawn illustrations have been reissued to include colour photos to help boost sales and entice new audiences. What is ubiquitous in our lives has become mandatory in our cookbooks. In an era of colour photography and hyperrealism, the number of authors incorporating illustration may be small, but their work endures as a testament to the power of this art form.

A convention that may seem antiquated by modern standards, drawings or infographics remain staples in cookbooks, demonstrating the lasting appeal of illustration within the genre. One need look no further than Samin Nosrat's hugely popular *Salt, Fat, Acid, Heat* to understand the attraction and utility of illustration. The images throughout *Salt*, hand-drawn by artist Wendy MacNaughton, are effective because they make visual what is abstract, with vibrant watercolours and graphics that are as eye-catching and accessible as they are explanatory. Together, Scappi's *Opera* and Nosrat's *Salt* bookend hundreds of years of cookbook titles using illustration to impart the author's own style and culinary knowledge. To review the corpus would be an overwhelming, if not impossible, feat. The goal of this paper, then, is to explore how six titles use illustration to uniquely transmit information and cooking philosophies such as emotion and accessibility. All six of these titles – *The Joy of Cooking*; *Salt, Fat, Acid, Heat*; *Solo*; *Cook as You Are*; *The Vegetarian Epicure*; and the *Moosewood Cookbook* – could have applied the more conventional choice of photography for their recipes (and in some cases were encouraged to by their publishers), but in opting for illustration, the authors demonstrate just how effective, personal, and often emotional the art form can be.

The Element of Instruction

The Joy of Cooking

Let's start with cocktails. That's how Irma Rombauer began *The Joy of Cooking*, one of the most beloved and best-selling American cookbooks of the twentieth century. 'Most cocktails containing liquor are made today with gin and ingenuity', Rombauer wrote in the first edition, privately published in 1931: 'take an ample supply of the former and use your imagination'.[1] Rombauer's daughter, the artist Marion Rombauer Becker, supplied the chapter-heading artwork: two roosters in papercut silhouette strutting across the page, their tails high. Even in silhouette, the roosters suggest whimsy and humour, as do all the chapter headings Becker designed for the book. 'Most cookbooks are as soulless as an empty cupboard. This one is lively and engaging', boasts the jacket copy of the 1936 edition, by then published with Bobbs-Merrill Company.[2] Becker and Rombauer both contributed to making *Joy* so appealing – Rombauer with her breezy and encouraging attitude and her easy-to-follow recipes and Becker with her silhouettes. But Becker's designs communicate joy, not cooking. Rombauer's recipes do both.

With each successive edition of *The Joy of Cooking*, its content increased, and its reputation as a trusted source grew. The Action Method of recipe writing, introduced in the 1936 edition, became a hallmark of the book. Rather than listing the ingredients first, the Action Method incorporates them directly into the instructions of the recipe, in bold, to make the recipe easier to read and execute. Sections explaining wine and tips for cooking in small spaces were also added, as were canning demonstrations and a glossary of culinary terms. Becker's silhouettes no longer reflected the serious reference work *Joy* was becoming. Bobbs-Merrill recommended a redesign, and Becker and Rombauer agreed. But that's about all Bobbs-Merrill and the Rombauers agreed on. Bobbs-Merrill wanted to modernize *Joy* and, much like today, saw colour photography as a draw. As an artist and successor of future editions of *Joy*, Becker was adamant that black-and-white illustrations made the most sense, but Bobbs-Merrill went ahead with their original plan. Based on Becker's response, it wasn't successful: 'Not a single picture is representative of any good cooking', she wrote. An artistic and legal tussle ensued. Rombauer held back the prospective manuscript, hired a lawyer, and sent off a telegram to her publisher: 'Object to illustrations [photographs] on ground that they are damaging to my reputation as a cook and not consistant [*sic*] with my writings as an author'.[3]

Those familiar with the artwork in *The Joy of Cooking* know how this tussle ends. Becker nominated Ginnie Hofmann, her former colleague at the Cincinnati Modern Art Society, as the new artist to communicate the charm and simplicity of *Joy*. Bobbs-Merrill eventually agreed and commissioned Hofmann to mock up the line drawings, which 'constituted one of the most ambitious attempts at illustration in any cookbook to date', according to Rombauer biographer Anne Mendelson.[4] As Becker predicted, Hofmann's drawings perfectly capture both the joy and especially the cooking that was missing from Becker's silhouettes. Instead of chapter headings, which tend to be decorative, Hofmann's didactic drawings appear interstitially throughout the book and alongside recipes. They have an uncanny way of communicating movement: a visual representation of the Action Method. In Hofmann's hands, for example, basting a turkey is both graceful and instructive. There, in a small line drawing, one can learn the proper way to tilt the pan, how to grip the other side so as not to burn yourself, and where to collect the juices with your spoon. You can almost hear the meat sizzling off the page.[5] Illustrated hands appear again and again throughout the book, like cooking manicules, gently drawing the reader in to demonstrate a challenging process: what it looks like to punch down dough, fillet fish, crumble breadcrumbs, or line a trifle mould. Even the lower extremities make an appearance. Turns out it's hard to skin a squirrel without stepping on its tail (serve with polenta). Mendelson called Hofmann's illustrations 'the most genuinely informative artwork to be found in any cookbook of the time'.[6] Today, it's difficult to imagine any edition of *The Joy of Cooking* without its functional, useful, and dynamic illustrations, whether designed by Hofmann or the many *Joy* illustrators who have followed in her squirrel-skinning footsteps.

More recently, in 2019, a new edition of *The Joy of Cooking* was published, this one written by Becker's grandson John Becker and his wife, Megan Scott. This 'new generation of *Joy*' is the largest volume yet, at over one thousand pages. The recipes here reflect an interest in global cuisine with dishes like *fesenjoon* (a Persian duck and pomegranate stew) and *anticucho de corazón* (Peruvian beef heart skewers), along with a sizable section on salting, drying, and fermenting foods. But as modern as this new *Joy* is, it still does not include photographs. Becker and Scott remain committed to the tradition of line-drawn illustration throughout the book for its instructional use, but they also pay tribute to Becker and Rombauer's original design by incorporating papercut silhouettes for each chapter heading. Including cocktails.

Salt, Fat, Acid, Heat

In 2017, Samin Nosrat's *Salt, Fat, Acid, Heat* was touted as introducing a new approach to cooking 'in the tradition of *The Joy of Cooking*', presumably by the way Nosrat's book blends practical instruction, effective illustration, and a heavy dose of charm. In writing *Salt*, Nosrat, a chef and cooking teacher, relied on her own cooking rubric. Used by Nosrat both in the classroom and in professional kitchens, the rubric sets a foundation for applying salt, fat, acid, and heat to food to ensure that dishes taste good. The book is, by Nosrat's own admission, a teaching book. As she was envisioning it, Nosrat found that some concepts were easier to convey visually, so rather than rely on photography to document the process, she reached out to artist Wendy MacNaughton. Together, they collaborated on *Salt* for more than five years, a detailed process of observation and repetition. As MacNaughton described her method, 'I draw from life. I can draw from a photograph, but drawing from life gives a line quality that is very alive. Drawing from a photograph is very flat'.[7] Their collaboration resulted in an instructional cookbook filled with hand-drawn and colourful depictions of foods, techniques, and infographics that are simultaneously playful, referential, informative, and most importantly, extremely effective teaching tools.

Take the 'salt' chapter: 'Understanding how salt works will allow you to make better decisions about how and *when* to use it to improve texture and season food from within'.[8] When a reader is considering the elements of salt, it's helpful to understand osmosis and diffusion, two processes that occur at a microscopic level. In *Salt*, Nosrat describes the process with key concepts bolded, but words alone limit most people's understanding. When MacNaughton translates Nosrat's prose into drawings of salt, water, and a semipermeable membrane, however, suddenly the concept becomes easier to understand. Illustration aids in the understanding of challenging concepts by what it can show that the human eye can't see and in the artistic license to emphasize ideas using colour or texture. In this instance, MacNaughton draws the membrane as a wall in salt water so readers can see and understand how water or salt moves across the membrane depending on the process. In the 'heat' chapter, the cutaway of a frying pan cleverly demonstrates the differences in oil depth for sautéing as well as pan, shallow, and deep frying. These concepts, however simple, could not be communicated through photography or as effectively in text alone.

Nosrat's recipes in the second half of the book are both culinarily and aesthetically simple. This is intentional. 'I only included recipes', Nosrat admits, 'because I knew there was just no way a publisher would publish the book without them'.[9] The recipes for dishes like cabbage slaw, pasta with broccoli, and chocolate pudding are free of any illustrative embellishment. Nosrat's intention with *Salt* is to promote cooking skills without the need to follow a recipe – even her own – so she didn't want to include images that would focus on the picture-perfect results of a professional's finished dish. The aim of MacNaughton's illustrations, interspersed between recipes, is to reinforce the 'philosophy of instinctual cooking' that is Nosrat's goal: frameworks to help prepare the home cook to rely on their senses in the kitchen rather than how a finished dish should look. An interlude on onions, for example, cites recipes a few pages away that would benefit from either blond (corn soup), browned (chicken with lentil rice), or caramelized onions (onion tart), accompanied by a watercolour drawing to illustrate the changes in onion hue for each type of cooking method. Or rather than simply write out the perfect ratio of grain to water in recipes for rice, polenta, and quinoa, MacNaughton devises a visually appealing matrix of nine grain recipes with drawings of filled measuring cups so readers can quickly gauge how many cups of water to use per cup of grain. Just as children's books rely on illustration to help present complex information to commit to memory, Nosrat and MacNaughton use it to create an easy, memorizable, and attractive visual for oft-forgotten calculations. These are the very kinds of appealing rubrics that help someone cook without the aid of a recipe or the constant pressure of perfection.

The Element of Imperfection

Solo

In the 'How to Use This Book' section at the beginning of *Salt, Fat, Acid, Heat*, Nosrat writes that her choice to use illustration was deliberate not because line drawings are more effective in communicating cooking instructions or because illustration aligns with her aesthetics – though both may be true. Instead, Nosrat has her readers' feelings in mind. She writes that the choice to embellish *Salt* with illustration can hopefully 'liberate you from feeling that there's only one perfect version of every dish'.[10] Nosrat's goal of liberation from perfection reads like a response to the modern-day 'compare and despair' syndrome that equally afflicts scrollers on Instagram as it does home cooks comparing their dry, sunken cake to the flawless version photographed in a cookbook. As a counterpoint to that syndrome, Nosrat and others have chosen in some of their books to forgo photography with its unattainable ideals in favour of intentionally imperfect illustration. Chef Anita Lo is one of those authors.

As the first female guest chef to oversee a State Dinner at the White House and as the chef of the Michelin-starred Asian-French restaurant Annisa, one might assume that Anita Lo is the perfect candidate to publish a restaurant-style cookbook with glossy photographs. And you'd be right: Lo's first cookbook, *Cooking without Borders* (2011), is filled with recipes for, and colour photographs of, the global fusion cuisine she was famous for at Annisa. But in 2017, after seventeen years of starred reviews, Lo closed her restaurant. The next year, she released a very different kind of project, the cleverly titled *Solo: A Modern Cookbook for a Party of One*. Lo's focus in *Solo* is on giving home cooks simple and delicious dishes to make when they find themselves alone, whether due to the breakup of a relationship or a restaurant – more fission than fusion. Unlike her geographically expansive first title, *Solo* is personal and contained yet celebrates the imperfections of food (and life) that most home cooks have no choice but to embrace, especially when eating alone.

'I put the "Lo" in "solo", the "A" in "alone" ', Lo writes. 'I've been dumped almost as many times as I've been in relationships [. . .] that's a lot of solo meals'.[11] So begins *Solo*, the rare cookbook that lists relationship failures as frequently as pantry staples, with a humour and vulnerability that few chefs discuss in their work. It's no wonder, then, that Lo's editor chose artist Julia Rothman to illustrate the book rather than having it photographed. Rothman's work is distinctive; she's known for her 'wonderfully messy' aesthetic that captures Mother Goose as accurately as the Brooklyn bar scene. For *Solo*, Rothman's uncomplicated drawings complement Lo's deadpan delivery: a couple of button mushrooms, a cluster of red grapes, a lobster. Nothing is ornate or cutesy. Occasionally, Rothman inserts a drawing of a plated meal, such as a few tacos or a broken soft-cooked egg on a plate of wilted greens. When a recipe calls for slightly more involved instructions, Rothman's illustrations easily segue into the didactic, such as with a series showing how to properly segment a grapefruit or how to assemble pork and chive dumplings: 'Shape dumplings and your life – both can be delicious'. In subsequent interviews for the book, Lo was asked why she chose illustrations over photography. For Lo and her editor, the decision was about avoiding the pitfalls of perfection that photography suggests: 'The fact that it's a drawing makes it a little bit more approachable. It's not like it's some perfect thing that you have to make. I think it gives the reader options just to make it and put it on the plate'.[12]

Cook as You Are

That sense of approachability is also part of the foundation on which UK-based food writer and former *Great British Bake Off* contestant Ruby Tandoh built her 2022 *Cook as You Are: Recipes for Real Life, Hungry Cooks, and Messy Kitchens*. With its open-minded approach and inclusive outlook, *Cook as You Are* seeks to reimagine what a traditional cookbook can be. Tandoh's aim is to meet readers where they are and steer them toward the kitchen, regardless of the ingredients they can afford or the limitations they may have – physically, emotionally, economically, or temporally.

The design and even the organization of the book reflect Tandoh's philosophy of inclusivity as much as the recipes do. Like Nosrat, Tandoh's focus is on the reader: 'Food needs cooks and eaters; without those human elements, it's just *stuff*', she writes.[13] Here the chapters are organized by the needs of the cook, not the elements of the food or the parts of the meal. The chapter titled 'More Food, Less Work' includes recipes that require less chopping or stirring and therefore fewer physical demands on the body for when one's energy is low or mobility is compromised. The chapter called 'Hidden in Plain Sight' encourages the use of pantry remnants to avoid food waste. The philosophy of diversity applies to the book's images, as well. Illustrator Sinae Park – who also contributed to Tandoh's previous title *Eat Up* – was tapped to feature different kinds of lifestyles and people. Park's work and Tandoh's mission align. As a Korean artist living in London, Park describes her work as exploring themes of food, care, and togetherness. Unlike Rothman's images in *Solo*, Park's illustrations are less interested in food on its own and more so in how food and cooking fit into the lives of different kinds of people. 'By photographing a cookbook in one kitchen, with one cook, I'd be capturing only one very narrow vision of what cooking looks like and who these recipes are for', Tandoh argues.[14] Illustration offers Tandoh freedom to represent the fullest range of kitchen experiences. Park's vignettes celebrate eaters and cooks who are gay, Black, mixed race, Asian, adopted. They leave work IDs on the dining room table, are covered in cat hair, and don't hide their power strips. They are familiar. Illustration both anonymizes and every-mans the person depicted.

Like in *Salt*, Tandoh's recipes are written with an attention to sensory detail that is highly specific, in part to avoid having to rely on overly didactic images or photography. A good dish isn't just about what something looks like, but how it smells and sounds and feels. For Tandoh, teaching reluctant cooks to trust their own senses is far more empowering than comparing their work to an image: 'A photo might speak a thousand words, but it can't tell you what something should smell like, or explain the texture of kneaded bread dough, or reassure you that it's all gonna work out fine'.[15]

The Element of Emotion

The Vegetarian Epicure

In the early 1970s, UCLA film student Anna Thomas put together a proposal at the urging of her friends for a cookbook featuring recipes for the comforting vegetarian dishes that were so beloved at her dinner parties. Thomas was in her early twenties when she submitted the proposal and had only recently gained confidence in the kitchen. Though she had been raised in a large Polish family with lots of food at the table, she was relatively new to preparing vegetarian dishes when she started school. Thomas taught herself to cook by experimenting with meat-free recipes from cookbooks like Julia Child's *Mastering the Art of French Cooking* and hosting get-togethers for friends. Her book proposal eventually made its way to Judith Jones at Knopf – Julia's editor – who in Thomas recognized a timely, fresh voice whose full-flavoured dishes could appeal to a wider, vegetarian-curious audience. Jones was especially drawn to a 'certain joy' that emanated in Thomas's writing and wanted to surface that joyful energy above any hint of preachiness often associated with a vegetarian lifestyle. Thomas agreed but remained 'adamant that the book position vegetarianism as a lifestyle of abundance rather than focusing on what the diet lacked'.[16] *The Vegetarian Epicure* was published by Random House in 1972 with joy and abundance at the forefront both in the text – 'this book is about joy, not pollution' – and especially in the recipes and design.

The Vegetarian Epicure is one among a wave of natural food and vegetarian cookbooks released by both small and mainstream publishers in the 1970s. Books like *Laurel's Kitchen*, *Tassajara Cooking*, and especially Frances Moore Lappé's *Diet for a Small Planet* saw the consumption of food as a political act and 'translated the sixties movement outlook into vegetarian dissent'.[17] *Diet for a Small Planet*, published in 1971, laid the foundation for the books to follow by exploring the devastating environmental impact a meat-based diet has on the planet. In its recipes, *Diet* promotes the consumption of plant-based protein combinations, designed to achieve optimal health for both our bodies and our planet. Lappé's political message resonated widely. Her recipes – not so much.

The recipes in *Vegetarian Epicure*, however, appealed to those whose interest in a vegetarian lifestyle didn't mean having to forgo delicious food. Thomas's menu is heavily Eurocentric, with recipes for *dolmades*, *pastitsio*, zucchini quiche, cheese soufflé, and pasta with tomato and wine sauce. Her recipe for French bread, or *pain ordinaire*, is the first recipe in the book: 'Broken or sliced, light as baby's breath inside, only a sliver of butter is needed and you have a masterpiece in purity of taste'.[18] But despite *Epicure*'s European sensibilities, Thomas's inclusion and promotion of whole grains, natural foods, soybean croquettes, and the enjoyment of after-dinner marijuana 'correlated her more with the counterculture than with James Beard blue bloods'.[19]

This is especially true of the design. The artwork, done by artist Julie Maas, dispels with the didactic – 'this is not really a book for those who have never seen a kitchen before' – and instead appeals to the home cook on an emotional level by translating into images a lifestyle of vegetarian abundance. That lifestyle is rendered through images that suggest texture, warmth, and informality: a water pitcher on a counter, plants in a kitchen window, branches of herbs, table linens and loose clothes, patterned bowls and tiles. In one drawing, a woman, dressed in what appears to be pyjamas, prepares a dinner menu while a cat sleeps on her lap – an accurate depiction of what most menu planning looks like. In another, perhaps less realistic but far more idyllic vignette, a man stretches bread dough on his wooden countertop while freshly baked loaves cool around his kitchen. Many images in *Epicure* emphasize socializing and merriment– a nod to Thomas's famous dinner parties. In the 'entertaining chapter' at the beginning of the book, a man and woman greet each reader to the *Epicure* party, arms around each other, with a wine bottle in one hand, a glass in another. A few chapters later, and we're at the table, toasting with drinks around a table overflowing with food and revelry.

The Moosewood Cookbook

As influential as *Epicure* was among the wave of vegetarian cookbooks that emerged in the early 1970s, if asked to conjure up the platonic ideal of a vegetarian cookbook, many people would call forth the *Moosewood Cookbook*. It remains one of the best-selling cookbooks of all time, and it continues to resonate and inspire readers today, both with its recipes and its design.

As widely known as it is today, the cookbook started small. In 1974, the collective that formed the Moosewood Restaurant in Ithaca, New York, self-published a spiral-bound compilation of their most-asked-for recipes. It was a short volume with a long name: *The Moosewood Cookbook Recipes from the Moosewood Restaurant in the Dewitt Mall, Ithaca*. Mollie Katzen was one of the original founders of the collective (along with her brother Josh) in 1973. As members of the collective, workers received the same hourly wage and shared the responsibility of running the restaurant that served the Ithaca community in the Finger Lakes region of upstate New York. Ithaca, known today for its excellent wine and its rich growing conditions, is home to several colleges and universities, including Cornell – a land-grant Ivy League with an excellent hospitality programme. In 1973, this meant the restaurant provided for a community of students, faculty, and Ithaca residents with an increasing familiarity with, and affinity for, natural and deliciously prepared food. Almost immediately, the restaurant became inundated with patrons asking for recipes, so Mollie Katzen decided to sell a cookbook. She designed, illustrated, and handwrote the recipes and text, and she sold it at a local bookstore. The 800 copies of the first printing sold out within the week. So she started taking mail orders as word-of-mouth grew to other college towns in the area. In 1977, Ten Speed Press – at the time a burgeoning publishing house based in Berkeley, California– released the first professionally produced *Moosewood*, and by 1981, it had sold 250,000 copies. Today, it's estimated that *Moosewood* has sold nearly four million copies.

So what's the draw? The recipes, for one. Katzen's recipes for Gypsy soup, Hungarian mushroom soup, Indonesian rice salad, and mushroom moussaka go beyond beloved – they are evangelized. (When I consulted the original 1974 spiral-bound *Moosewood* in a library reading room, the reading room attendant who served me the book couldn't help but share her favourite recipes.) Based on recollections by *Moosewood*'s legions of fans, some of the affection is based in nostalgia. Many cite *Moosewood* as their first introduction to global flavours that are applied so effortlessly in Katzen's recipes, and often fans reminisce about the pivotal role *Moosewood* played in cooking for friends, usually in college. 'The most-loved cookbook, the one where the cover had fallen entirely off and the pages were full of cook's notes and food stains? That would be the original 1977 edition of *Moosewood Cookbook*', wrote a Cornell alum reminiscing about her cooperative meal planning days.[20] 'People have an emotional attachment to it', Katzen agreed.[21] Part of that attachment may stem from the book's origins. Katzen designed the *Moosewood Cookbook* with her local restaurant patrons in mind. The hand-lettering, informal tone, and delightful illustrations are intimate and personal, suggesting a notebook you might share with friends or a menu printed at the restaurant. Katzen's illustrations are not used to communicate information – they're used to communicate emotion, through both calligrams and the drawings customized to each recipe. Sometimes those drawings resemble doodles, like the tiny stars clustered within the Brazilian black bean soup recipe (it's a five-star soup) or the dozens of hearts that dot the recipe for crepes. Other images link more directly to a recipe's origins or ingredients. Gazpacho is announced in the shape of Spain, while a giant eggplant takes up most of the real estate in the recipe for stuffed eggplant (the instructional text encircles it). The 'Perfect Protein Salad', meanwhile, is visualized as just so perfect that it could only have been made by the hands of God, which is why the recipe looks like it's descending from the heavens by cherubs. And a subtle Hebrew shalom – a nod to Katzen's upbringing – is baked into one of the noodle kugel designs. Katzen's illustrations are more than charming; they're intentional, endearing, and delightful, and millions of home cooks agree. Who could resist?

The Element of Joy

The six cookbooks referenced here demonstrate how, as a visual language, illustration offers readers an invitation to engage differently with the text. Hand-drawn images suggest an intention and a focus on detail that make it the perfect medium for cookery books. Illustration also provides a latitude that allows readers to meet the recipes where they are, without forcing a platonic ideal of what a dish or lifestyle should look like. And as in children's literature, illustration invites readers to tap into their own creativity, imagination, playfulness, and yes, joy of cooking. The imperfect handwriting, the subtle gradations of watercolour, the exaggerated sketches – these elements often reveal more about the process of cooking than a staged photograph ever could. Each drawing celebrates the messy but joyful reality of cooking beyond a perfect display and precise measurements. Because, in these books, cooking is not just a science – it's an art.

About the Author

Rebecca Federman is the Assistant Director of the General Research Division and the Vartan Gregorian Center at the Stephen A. Schwarzman Building. She is also involved with the library's culinary collection.

Notes

1 Irma S. Rombauer, *The Joy of Cooking* (A. & C. Boni, 1931).
2 Irma S. Rombauer, *The Joy of Cooking: A Compilation of Reliable Recipes, with a Casual Culinary Chat*, illus. by Marion Rombauer Becker (Bobbs-Merrill Company, *c.* 1936), jacket copy.
3 Anne Mendelson, *Stand Facing the Stove: The Story of the Women Who Gave America* The Joy of Cooking (Henry Holt, 1996), pp. 205, 207.
4 Mendelson, p. 215.
5 Mendelson, p. 386.
6 Mendelson, p. 269.
7 Erin Scott, 'Drawing the Art of Cooking: How Samin Nosrat and Wendy MacNaughton Made *Salt Fat Acid Heat*', *Saveur*, 4 October 2017.
8 Samin Nosrat, *Salt, Fat, Acid, Heat: Mastering the Elements of Good Cooking*, illus. by Wendy MacNaughton (Simon & Schuster, 2017), p. 29.
9 Samin Nosrat, interview by Francis Lam, *Splendid Table*, 5 May 2017 <https://www.splendidtable.org/story/2017/05/05/samin-nosrat-on-mastering-salt-fat-acid-and-heat> [accessed 11 May 2025].

10 Nosrat, *Salt, Fat, Acid, Heat,* p. 12.
11 Anita Lo, *Solo: A Modern Cookbook for a Party of One* (Knopf, 2018), p. vi.
12 Anita Lo, interviewed by Suzy Chase, *Cookery by the Book*, episode 128, 25 February 2019 <https://www.cookerybythebook.com/home/2019/2/25/128-solo> [accessed 11 May 2025].
13 Ruby Tandoh, *Cook as You Are: Recipes for Real Life, Hungry Cooks, and Messy Kitchens* (Knopf, 2022), p. 3.
14 Tandoh, p. 3.
15 Tandoh, p. 3.
16 Sara B. Franklin, *The Editor: How Publishing Legend Judith Jones Shaped Culture in America* (Atria Books, 2024), p. 133.
17 Maria McGrath, *Food for Dissent: Natural Foods and the Consumer Counterculture since the 1960s* (University of Massachusetts Press, 2019), p. 54.
18 Anna Thomas, *The Vegetarian Epicure* (Knopf, 1972), p. 29.
19 McGrath, *Food for Dissent,* p. 81.
20 G. Daniela Galarza, 'Cheesy Broccoli Strudel Is a Taste of the Moosewood Cookbook's Magic', *Washington Post,* 4 May 2023 <https://www.washingtonpost.com/food/2023/05/04/moosewood-cookbook-recipes/> [accessed 11 May 2025].
21 Leslie Kaufman, 'Mollie Katzen Is Still Cooking', *New York Times,* 28 August 2013 <https://www.nytimes.com/2013/08/28/dining/mollie-katzen-is-still-cooking.html> [accessed 11 May 2025].

11

The Sustainability Stones

Or, How Traditional Food Pairings Can Save the Marine Environment in French Polynesia

Russell Fielding and Fiona Gimenez

'Ritual actions', claimed the sociologist G. C. Homans in 1941, 'do not produce a practical result on the external world'.[1] Homans's conclusion has since been roundly refuted through myriad case-study examples of taboos, rituals, and cultural norms that do directly cause real-world effects, particularly in the area of food-related customs and natural resource conservation. In one early example, the anthropologist Roy Rappaport showed in his groundbreaking 1967 study of ritualized warfare among neighbouring cultural groups in New Guinea that the cycle of peacetime prohibitions and wartime feasts served to maintain the ecological stability of both local forests and animal populations.[2]

Later, in 1997, ecologists Johan Colding and Carl Folke identified dozens of cases in which traditional, often ritualized, conservation regimes provided protection for species identified by the International Union for the Conservation of Nature (IUCN) as threatened or endangered.[3] In their systematic analysis, however, Colding and Folke cautioned that while these conservation regimes may be effective at protecting vulnerable species, the very existence of the traditions themselves is endangered through the loss of cultural identity in many parts of the postcolonial world. Later research would echo Colding and Folke's warning that cultural diversity and biological diversity can thrive together or diminish together, but rarely will they take separate paths.[4]

To this end, fisheries scientist Bob Johannes poignantly appended the title of his study of one such traditional conservation regime as 'Traditional Marine Conservation Methods in Oceania *and Their Demise*'.[5] Similarly, given the rapid decline of traditional cultural practices, coupled with the global crisis of biodiversity loss, Colding and Folke understandably urged 'scientists [. . .] to make explicit the potential linkages between social mechanisms, like taboos, and practices of species conservation' while also urging 'conservationists to support traditional peoples that conserve species and ecosystems [. . .] and to conserve and make use of traditional ecological knowledge before it is lost'.[6] Within the Pacific islands, some success has been observed regarding the preservation or restoration of culturally embedded conservation strategies, as evinced by Johannes's 2002 follow-up paper titled '*The Renaissance* of Community-Based Marine Resource Management in Oceania'.[7] Here we consider one such successful – yet challenged – conservation strategy based upon traditional food rituals on the Pacific island of Maupiti.

Study Site and Context

Maupiti is a small island (Figure 1) at the western end of the Society Islands archipelago of French Polynesia (Figure 2), a 'near-atoll', consisting of an eroded and subsided central volcanic peak, surrounded by a shallow lagoon, which itself is bordered by a barrier reef capped with sandy islets known as *motus* (Figure 3).[8] Botanical surveys and archaeological research have indicated that Maupiti was once almost 'completely forested', but its native forest was nearly totally replaced with introduced, nonnative species – mainly fruit trees and other food crops – during the original settlement of the island, which occurred between 1000 and 1200 CE.[9] These fruits, vegetables, and tubers, along with locally caught fish and other marine animals and locally raised swine, comprise the majority of the island's locally produced food supply. Other foods are imported from elsewhere in French Polynesia and beyond.

As throughout the Pacific islands, breadfruit (*Artocarpus altilis*) is a major staple crop on Maupiti (Figure 4). Given to frequent somatic mutations and propagated vegetatively, many distinct, named cultivated varieties ('cultivars') of breadfruit have emerged and have been propagated throughout the region.[10] Local experts disagree about the precise number of breadfruit cultivars that grew on Maupiti in the past, but most cite numbers just on either side of one hundred.

Cultivar diversity results in the availability of breadfruit throughout much of the year, as each cultivar's fruiting occurs at a slightly different time of year than the rest.

Figure 1. The island of Maupiti. Image assembled by F. Gimenez, QGIS.

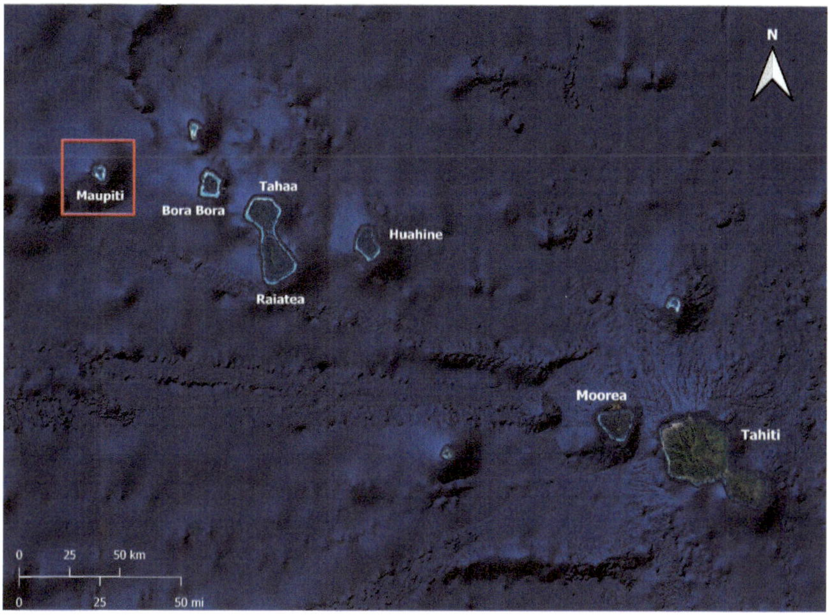

Figure 2. Maupiti's location within the Society Islands archipelago of French Polynesia. Image assembled by F. Gimenez, QGIS.

Figure 3. A partial southward view of Maupiti from Mount Teurafaatiu. The village of Vai'ahe is visible on the left slope of the mountain. Part of Motu Tuanai is visible to the left. Motu Pitiahe and Motu Tiapaa are visible in the distance. Photograph by R. Fielding, May 2024.

Figure 4. A breadfruit tree on the island of Maupiti. Photograph by R. Fielding, May 2024.

Owing to this offset in seasonality, breadfruit of one cultivar or another is generally available year-round, with one cultivar coming into fruit as another goes out of season. 'We are never out of *'uru*', said one local informant, who, using the Tahitian term for breadfruit, explained that with so many cultivars on the island, at least one is always fruiting. According to local tradition, several commonly caught species of fish and other marine organisms are each paired with a particular cultivar of breadfruit. Each of these fish species, then, is traditionally consumed only when accompanied by its corresponding breadfruit cultivar. Since each breadfruit cultivar is available seasonally, this tradition implies a de facto seasonality for the consumption – and therefore the fishing – of each fish species as well.

On the south coast of Maupiti, near Marae Vaiahu – a *marae* being a type of ceremonial structure common throughout the Pacific islands – a monument built in 2009 includes nine tablets of stone representing the island's nine districts (Figure 5). On each tablet, the name of the district is incised across the centre and painted white. Above each district name are pictograms representing the major domesticated plants grown in the district. Below each name are pictograms representing the primary species of fish or other marine animals caught offshore from the district. Most of the images are sufficiently detailed so that a viewer familiar with local agriculture and fisheries can identify the cultivars of the plants and the species of the fauna represented. Each of

Figure 5. A monument near Marae Vaiahu, built in 2009 and featuring nine stones (foreground), representing the nine districts of Maupiti, each with the district's main plant and marine foods pictured. Photograph by F. Gimenez, May 2024.

these 'sustainability stones' (our term) represents a pairing. The marine animal species pictured on each stone is, or are, traditionally eaten together with the pictured plant(s). The monument does not represent an exhaustive list of all fish-plant pairings; rather, it illustrates some of the most common food plants grown and marine species taken and consumed together in each district.

When a particular breadfruit cultivar is unavailable, owing to its being out of season, its corresponding fish species traditionally should neither be caught nor consumed. During each breadfruit cultivar's season of unavailability, then, the population of its corresponding fish species would not face fishing pressure and, instead, would be given time to recover. This traditional system may have represented a case of Indigenous, community-based conservation, if it were strictly practiced and if its resulting seasonal closure for certain fisheries provided sufficient recovery time to each fish population.

The breadfruit-fish pairing traditions in Maupiti, then, may represent a more complex version of a practice observed by the German naturalist Georg von Langsdorff in the early nineteenth century in the Marquesas islands, an archipelago in eastern French Polynesia, by which 'almost all fish are at the time when the bread-fruit is not ripe tabooed, and must not be eaten'.[11] Langsdorff concluded that 'this connexion [*sic*] of fish with the bread-fruit is wholly inexplicable' but offered the conjecture – among other possibilities – that 'it may be that this is their spawning time, and it is intended by such an ordinance to prevent their natural increase being interrupted by their being taken at this period'. Nearly two centuries later, literary scholar and explorer Dennis Kawaharada reflected upon Langsdorff's observation and suggested that the tradition he described 'was perhaps a conservation practice'.[12] Indeed, it is plausible to consider that the closure of certain fisheries – or 'almost all' fisheries, as in the case of Langsdorff's observation in the Marquesas – during certain seasons may be an effective method of conservation of those fish stocks.

Methods

After hearing about the density of breadfruit cultivar diversity on Maupiti and the traditional breadfruit-fish pairings there from informants on other islands in the Society archipelago, we were interested to learn whether this tradition is still adhered to in Maupiti and, if so, whether it might serve to guide human behaviour toward the conservation of fish populations. Lacking historical fish population data, we knew that a definitive conclusion regarding any real conservation effects of the breadfruit-fish pairing traditions would be difficult to make. Instead, we determined to seek examples of local ecological knowledge related to Maupiti's breadfruit-fish pairing traditions and to learn about the relevance of these traditions to people's dietary choices today. With the reasonable assumption that lessening or pausing fishing pressure on certain fish stocks during certain times of the year would positively affect those fish populations, we hypothesized that any evidence of continued adherence to Maupiti's traditional breadfruit-fish pairings would indicate the likely continued existence of an effective conservation strategy. Conversely, significant disruptions to the breadfruit-fish pairing traditions in Maupiti might indicate the demise of this conservation strategy and, in the absence of what had been closed seasons to fishing certain species, the emergence of the risk of overfishing.

We visited Maupiti during May 2024 to observe and learn about these culinary and conservationist traditions. Our work was greatly assisted by the local cultural historian, Jeffrey Tamati, who arranged many of the meetings and interviews that took place (Figure 6). In 2009, Tamati led the construction of the monument pictured here. During our fieldwork, we interviewed sixteen residents of Maupiti, asking questions about adherence to the traditional breadfruit-fish pairings, as well as other traditions related to food and conservation and the changes that have taken place in recent decades impacting food production, food consumption, and environmental conservation.

Data and Analysis

In addition to the district-based plant-fish pairings derived from the monument, according to our local informants, certain traditional island-wide pairings of breadfruit cultivars and fish species exist in Maupiti. Specifically, the camouflage grouper (*Epinephelus polyphekadion*) is traditionally paired with the 'Afara' cultivar of breadfruit throughout the island, 'Hatuma' breadfruit goes with the humpnose big-eye bream (*Monotaxis grandoculis*), 'Puero' breadfruit can be eaten with all the fish ('tous les poissons', according to an informant) but most appropriately with blacktail snapper (*Lutjanus fulvus*), 'Pau' breadfruit is to be eaten with octopus, and the 'Ara ara ha'ari' breadfruit cultivar, a

Figure 6. Jeffrey Tamati, a local Maupiti historian, pointing out features of a breadfruit cultivar that produces small fruits. Photograph by R. Fielding, May 2024.

relatively recent introduction to Maupiti from the neighbouring island of Bora Bora, is typically eaten not with fish but with pork.

In the past, on Maupiti, breadfruit trees themselves were abundant and in no need of conservation. Breadfruit trees grow rapidly, begin producing fruit within just a few years of planting, and reliably produce hundreds of fruits each year. A single breadfruit can weigh up to five kilograms and is a good source of calories, carbohydrates, and starch, as well as some vitamins and minerals (Figure 7).[13] In the linked breadfruit-fish system of food production and consumption practiced traditionally on Maupiti, breadfruit is an abundant resource, available year-round, owing to the differential fruiting seasons of the various cultivars. The association of particular breadfruit cultivars with certain species of fish and other marine organisms, then, provides a mechanism by which conservation of those species could be enacted through adherence to the culinary taboo on consuming certain fish species without breadfruit of their corresponding cultivar. The 'sustainability stones' near Marae Vaiahu indicate a small selection of the pairings of breadfruit cultivars and marine faunal species that would have contributed to this conservation effect.

Maupiti's breadfruit diversity, however, has been declining since the mid-twentieth century. For example, an archived, unpublished report titled 'Ombres & Lumieres sur le "Tumu 'Uru"' ('Shadows and Lights on the "Breadfruit Tree"') was shared with us during

Figure 7. A breadfruit growing on a tree on the island of Maupiti. The white substance on the surface of the fruit is a latex-like sap that naturally exudes from the fruit, leaves, branches, and trunk of the breadfruit tree. Photograph by R. Fielding, May 2024.

our fieldwork. In this report, the anonymous author lists eighty-two named breadfruit cultivars found on Maupiti in a section titled 'Le passé' (the past) and only thirty-two in the next section, 'Le present'.[14]

The common explanation for this decline is that breadfruit diversity has been reduced in stages. At the time of Maupiti's original settlement, breadfruit was planted mainly in the upland valleys of the island's main volcanic peak – more so than on the coastal plain or on the *motus*. Traditionally, the young men of the island would tend these trees, bringing fruit down to the coastal plain where nearly all the island's homes were, and still are, located. Local informants explained to us, however, that when the French government recruited residents of its Pacific colonies during the Second World War, the young men who were preparing to leave the island brought cuttings from some of these upland breadfruit trees down to the coastal plain and planted them there so that the women and children, along with the men who were too old to go to war, might have easier access to the fruit. The parent trees from which these scions were taken eventually died from neglect and competition with faster-growing wild vegetation. The lack of propagation then led to the disappearance of some entire cultivars because only a subset of the one hundred or so upland cultivars were selected for transplantation, resulting in Maupiti's first reduction of breadfruit diversity. A few individual breadfruit trees can still be found in the island's upland valleys, but these are generally not tended, nor is their fruit harvested.

The second reduction of the number of breadfruit cultivars on Maupiti came later in the twentieth century when an island-wide electrical power system was constructed. According to older islanders, the installation of utility poles and power lines that mainly follow the island's circumferential road along the coastal plain required the removal of many breadfruit trees, some of which were the only known individual specimens of their cultivar. Maupiti's rich breadfruit cultivar diversity is precarious. The island's small size and large number of cultivars mean that often just a few trees remain from a cultivar that once was more widespread. Sometimes a cultivar might be represented by a single tree. One local informant told us that his nephew had recently cut down a breadfruit tree to make room to park his newly purchased car; the informant believes that the tree was the last of its variety. Breadfruit trees are resilient, and new growth often emerges from stumps or fallen trunks. Some of the trees cut down to make way for power lines have since regrown, but others have not (Figure 8).

Third, and finally, climate change and coastal land-use change are threatening the diversity of breadfruit cultivars on Maupiti.[15] While breadfruit trees are famously tolerant

Figure 8. A breadfruit tree on Maupiti that has been cut due to its proximity to electricity lines (above) but is showing signs of regrowth. Photograph by R. Fielding, May 2024.

of warming temperatures, they are susceptible to salt spray and belowground saltwater intrusion, hazards closely associated with storms of increased frequency and intensity and with sea-level rise.[16] As the surface level of Maupiti's lagoon rises, the coastline shifts inland, threatening the survival of breadfruit trees planted along the island's coastal plain. Indeed, some of the coastal *marae* have already sunk beneath the lagoon (Figure 9).

Discussion and Conclusions

Most of our interviewees knew the traditional breadfruit-fish pairings but admitted not adhering to them anymore. The most common reasons offered for having given up the traditional pairings included the reduction in the number of breadfruit cultivars grown on Maupiti – making it difficult to acquire a specific fish's breadfruit counterpart – and the ubiquity of imported rice, a starch that has no traditional fish pairing and, thus, might be paired with any fish. For example, one interviewee explained that the blacktail snapper (*Lutjanus fulvus*) is traditionally paired with the 'Puero' breadfruit cultivar and that in the past, he would fish for, and eat, blacktail snapper only when 'Puero' was in season. Now, however, with rice available year-round, the interviewee simply eats blacktail snapper with rice during seasons when 'Puero' is unavailable. This culinary adjustment has almost certainly resulted in increased fishing pressure on the blacktail snapper stock.

Figure 9. A sign marking the location of the Marae Poiheva Hataumee on Maupiti, now submerged by rising sea levels. Photograph by R. Fielding, May 2024.

Another interviewee acknowledged the loss of certain breadfruit cultivars and the rise of rice but reversed the causal direction, telling us that 'God takes away the *'uru* varieties we don't eat'.

The adherence to tradition is strong on Maupiti, and changes to traditional behaviour such as the breadfruit-fish pairings are not taken lightly. In 2005, an international hotel and resort company proposed building an eighty-unit luxury eco-resort on one of the island's *motus*. Wary of the loss of cultural traditions experienced on the neighbouring island of Bora Bora in the wake of extensive resort development there, the residents of Maupiti held a referendum on the question of whether to allow the resort plans to proceed. The vote was 70 percent against the resort, and to this day, the only lodgings available to tourists – or visiting researchers – on Maupiti continue to be small, simple, family-run pensions (Figure 10).[17]

Based upon this demonstrated priority of the island's population to resist international development projects – and thus to preserve their local culture and environment – aided, perhaps, by the island's geographical isolation and remoteness, it may be likely that a traditional conservation strategy on Maupiti based upon the linked seasonality of breadfruit and fisheries can experience a 'renaissance' like that which Johannes identified in other Pacific island contexts.[18] If this were to happen, breadfruit cultivars now

Figure 10. Pension Maupiti Orovaru, one of several small, family-run pensions that serve as the only accommodation option for visitors to Maupiti. Photograph by R. Fielding, May 2024.

extirpated from Maupiti would need to be replaced either agriculturally – that is, with individual trees of the same cultivars still growing on other Pacific islands (if such trees can be found) – or culinarily by creating new pairings of breadfruit cultivars still present on Maupiti with marine faunal species currently associated with extirpated cultivars.[19]

Who would lead such a renaissance? We believe that the most likely scenario is one in which local people on Maupiti – encouraged by the widely held opinion that the convenience associated with the abandonment of food traditions has come at a cost to the culture, economy, environment, and health of the island and its people – seek to shift the balance between innovation and tradition by reasserting or reestablishing the conservation strategies represented by the adherence to traditional breadfruit-fish pairings. Indeed, such a motivation lies behind Jeffrey Tamati's 2009 construction of the monument near Marae Vaiahu. This future, hypothetical renaissance – which might by necessity include a reimagination of Maupiti's specific breadfruit-fish pairing traditions – would fit the pattern recognized by Johannes in which 'communities spread widely throughout the [Pacific islands] region are [. . .] adapting their traditional practices to fit contemporary circumstances'.[20]

Johannes found that the reasons for the reassertion of traditional resource management systems varied across the countries and island groups in his survey area. In some cases, the perceived failure of environmental management efforts established by state or colonial governments to protect marine resources served to motivate local community leaders to reassert their traditional control over the use and conservation of these resources. In others, motivation came from the recognition that tourism revenue can be increased if the biodiversity and ecological health of local reefs, beaches, and islands are maintained. Finally, Johannes attributed the renaissance of traditional conservation strategies in some parts of the Pacific to the reassertion of local culture as a source of pride, especially on islands that – unlike Maupiti – have regained their political independence following eras of colonization.

Despite the fact that several of our informants mentioned the arrival of missionaries in Maupiti as one of the catalysts of the dietary changes that would later ensue, the Christian church today appears to encourage the adherence to certain older food traditions. At the Ètārētia Porotetani Māòhi church one Sunday during our fieldwork, a local teenage girl wearing a fine dress, straw hat, and flower lei stood behind the pulpit above a baseboard decorated with carvings of fish and breadfruit to deliver a prayer of thanksgiving (Figure 11). In it, she thanked God for the abundant fish in the lagoon and the many varieties of *'uru* on the island. The church, like other churches within the Society Islands, produces an almanac that includes daily Bible readings alongside notations indicating the best dates for planting food crops and for fishing for certain species. At Sunday school, children are fed traditional foods like breadfruit, intentionally to counter the influence of imported foods like rice.

Whether through the replanting of scarce or vanished breadfruit cultivars or through the establishment of new breadfruit-fish pairings, the people with whom we spoke in Maupiti seem to have sufficient knowledge of past culinary traditions, nostalgia for past cultural and environmental conditions, and interest in preserving or restoring Maupiti's

Figure 11. Breadfruit and fish carvings along the baseboard at the Ètārētia Porotetani Māòhi church in Maupiti. Photograph by R. Fielding, May 2024.

biodiversity and cultural heritage to initiate a renaissance of their disappearing traditional conservation strategies.

If new breadfruit-fish pairings were to be established, the people of Maupiti might find it fitting to have a new set of stones carved for the monument near Marae Vaiahu. This would be in keeping with the findings of anthropologist Edvard Hviding and geographer Kenneth Ruddle that such traditional resource management systems in the Pacific 'appear to have considerable capacity for handling and adapting to new circumstances thereby becoming potentially important tools in the contemporary management of fisheries and of the coastal zone in general'.[21] If such adaptation were to occur with regard to the breadfruit-fish pairing traditions in Maupiti, these new 'sustainability stones' might carry on the role of Jeffrey Tamati's original 2009 carvings in serving to point the way toward newly effective traditional marine conservation strategies for Maupiti's future.

About the Authors

Russell Fielding is an environmental geographer whose research focuses on sustainable food systems in the coastal and island settings of our changing world. He is the author of two books – *The Wake of the Whale* (Harvard University Press, 2018) and *Breadfruit* (Columbia University Press, 2025) – and is currently an Associate Professor at Coastal Carolina University.

Fiona Gimenez is a water science and technology engineer specializing in environmental issues in Pacific island settings. Currently completing a civic service at the University of French Polynesia, her work focuses on microplastic pollution in lagoon ecosystems and its implications for island environments.

Acknowledgements

The authors wish to thank the symposiasts of the 2025 Oxford Food Symposium whose comments and discussion were instrumental to the revision of this paper for publication in these proceedings. Thanks also to all the local informants, especially Jeffrey Tamati, formerly of UNESCO, who took time to speak with – and in two cases, sing for – us about traditional food practices in Maupiti. Phirmin and Rose Teoroi went far beyond their duties as the hosts of Pension Maupiti Orovaru. Jennifer Kahn of the College of William & Mary advised on fieldwork logistics and important contacts. On Moʻorea and Tahiti, respectively, Hinano Teavai-Murphy and Pevatunoa Levy provided valuable information about Maupiti's diversity of breadfruit cultivars and breadfruit-fish pairing traditions that sparked our original interest in this research. Fieldwork was funded by the Professional Enhancement Grant programme at Coastal Carolina University and a grant from the Alfred P. Sloan Foundation. We are grateful to both funding organizations for their support. This human-subject research was approved by the Institutional Review Board at Coastal Carolina University, protocol #2022.171. An earlier, open-access version of this paper was published in the journal *Geographical Review*. We are grateful to the publisher, Taylor & Francis, for allowing the modification and republication here.

Notes

1 G. C. Homans, 'Anxiety and Ritual: The Theories of Malinowski and Radcliffe-Brown', *American Anthropologist*, 43.2 (1941), pp. 164–72 (p. 172).

2 R. A. Rappaport, 'Ritual Regulation of Environmental Relations among a New Guinea People', *Ethnology*, 6.1 (1967), pp. 17–30, doi:10.2307/3772735.

3 Johan Colding and Carl Folke, 'The Relations among Threatened Species, Their Protection, and Taboos', *Conservation Ecology*, 1.1 (1997).

4 J. Peter Brosius and Sarah L. Hitchner, 'Cultural Diversity and Conservation', *International Social Science Journal*, 61.199 (2010), pp. 141–68, doi:10.1111/j.1468-2451.2010.01753.x; André Frainer and others, 'Cultural and Linguistic Diversities Are Underappreciated Pillars of Biodiversity', *PNAS*, 117.43 (2020), pp. 26539–43, doi:10.1073/pnas.2019469117; Jules Pretty and others, 'The Intersections of Biological Diversity and Cultural Diversity: Towards Integration', *Conservation and Society*, 7.2 (2009), pp. 100–12, doi:10.4103/0972-4923.58642.

5 Robert E. Johannes, 'Traditional Marine Conservation Methods in Oceania and Their Demise', *Annual Review of Ecology and Systematics*, 33 (1978), pp. 317–40, doi:10.1146/annurev.es.09.110178.002025 (emphasis added).

6 Colding and Folke, 'Relations among Threatened Species', p. 6.

7 Robert E. Johannes, 'The Renaissance of Community-Based Marine Resource Management in Oceania', *Annual Review of Ecology and Systematics*, 9 (2002), pp. 349–64, doi:10.1146/annurev.ecolsys.33.010802.150524 (emphasis added).

8 Emilie Dotte-Sarout and Jennifer G. Kahn, 'Ancient Woodlands of Polynesia: A Pilot Anthracological Study on Maupiti Island, French Polynesia', *Quaternary International*, 457 (2017), pp. 6–28 (p. 8), doi:10.1016/j.quaint.2016.10.032.

9 F. Raymond Fosberg and Marie-Helene Sachet, *Flora of Maupiti, Society Islands*, Atoll Research Bulletin No. 294 (Smithsonian Institution, 1987), p. 2.

10 A. Maxwell P. Jones and others, 'Morphological Diversity in Breadfruit (*Artocarpus*, Moraceae): Insights into Domestication, Conservation, and Cultivar Identification', *Genetic Resources and Crop Evolution*, 60 (2013), pp. 175–92, doi:10.1007/s10722-012-9824-8; Diane Ragone, 'Breadfruit: Diversity, Conservation and Potential', *Acta Horticulturae* 757 (2007), pp. 19–29, doi:10.17660/ActaHortic.2007.757.1.

11 Georg H. von Langsdorff, *Voyages and Travels in Various Parts of the World, during the Years 1803, 1804, 1805, 1806, and 1807* (Colburn, 1813), p. 137.

12 Dennis Kawaharada, 'Journey to the Isles of Hiva', *University of Hawai'i-Mānoa*, 1993 <https://www2.hawaii.edu/~dennisk/travel/1993_hiva/hiva1993.html> [accessed 1 October 2025].

13 Brien A. Meilleur and others, *Hawaiian Breadfruit: Ethnobotany, Human Ecology, Agronomy, Nutrition, Modernity* (University of Hawai'i at Mānoa, College of Tropical Agriculture and Human Resilience, 2024).

14 'Ombres & Lumieres sur le "Tumu 'Uru"' (anonymous, undated, and unpublished document archived in the Maupiti library, Vai'ea).

15 Emma Gairin and others, 'Multi-Decadal Coastal Evolution of Remote Pacific Islands: Armouring of Taha'a, Raiatea, Maupiti, and Rangiroa (French Polynesia)', *Regional Environmental Change*, 22.119 (2022), doi:10.1007/s10113-022-01977-1.

16 Lucy Yang and others, 'Potential of Breadfruit Cultivation to Contribute to Climate-Resilient Low Latitude Food Systems', *PLOS Climate*, 1.8 (2022), e0000062, doi:10.1371/journal.pclm.0000062.

17 Philippe Bachimon, 'Tourist Brownfields in French Polynesia – Revealing a Destination Crisis and a Form of Resistance to International Tourism', *Via Tourism Review*, 1 (2012), doi:10.4000/viatourism.1318; D. Lamarque, 'Maupiti: Un Exercice Médical Sous le Vent', *Médicine Tropicale*, 65.6 (2005), pp. 533–36.

18 Johannes, 'Renaissance'.

19 Fortunately, a living repository of breadfruit cultivars exists on the island of Maui in Hawai'i, thanks to the work of pioneering breadfruit scholar and conservationist Diane Ragone (see 'Breadfruit Collections', National Tropical Botanical Garden, 2019 <https://ntbg.org/breadfruit/about-breadfruit/conservation/> [accessed 1 October 2025]).

20 Johannes, 'Renaissance', p. 318.

21 Edvard Hviding and Kenneth Ruddle, *A Regional Assessment of the Potential Role of Customary Marine Tenure (CMT) Systems in Contemporary Fisheries Management in the South Pacific*, Report 91/71 (Forum Fisheries Agency, 1991), p. 12.

12
From Caramelization to *Cousoumeh*
A Case for Transitioning to a New Ideal Pelau Pot

Shrinagar Francis

Re-Membering a Dish

Pelau, a dish rooted in the culinary traditions of Trinidad and Tobago, is made primarily with chicken and pigeon peas, and the dish stands as one of the nation's iconic one-pot meals. Its preparation begins with the use of two ingredients almost symbolically conversing with the country's history and development: sugar – brown sugar in this case – and oil. Equally interesting are the delicious 'rummy' flavour notes developed during the initial chemical process that goes into making pelau, caramelizing sugar. It is the perfectly caramelized sugar that colours and flavours the dish.

Pelau bears a substantial resemblance to the much-contested and celebrated West African one-pot rice dish jollof rice. 'With its seductive aroma, deep-red colour, and spicy flavour, jollof is the undisputed queen of West African kitchens', and echoes of its influence resonate through the Caribbean in Jamaica's rice and peas, Suriname's *moksi alesi* (mixed rice), and many others. Jollof's origins are as equally debated as pelau's. It can be traced back to the 1300s and the Wolof Empire, where rice farming thrived and jollof began as thieboudienne, a preparation of rice, fish, shellfish, and vegetables.[1] Much of the Caribbean's African diaspora can trace their ancestry to the Senegambian region of the Wolof Empire, a thoroughfare of enslaved people who no doubt carried within them the memory of jollof.[2] In pelau, we find strands of a transplanted, dismembered people, re-membered and adapted to Caribbean soil. Because like language, customs, culture, and belief systems, food traditions are growing things; 'chop them up and [. . .] they just sprout [. . .] whole [. . .]. To re-member is to reassemble the limbs of a story, to make it whole again' – a truly unique thing, a thing that sometimes happens on the inside of an ideal cooking vessel.[3]

Cooking as Endothermic Alchemy

Cooking, the transformation or manipulation of earth products through the use and application of fire, air, and/or water, has long been regarded as a sacred act: a form of alchemy that transmutes raw matter into something elevated and nourishing. This transformation is not merely physical but deeply chemical, involving the interaction of heat through such techniques as baking, roasting, frying, or broiling with proteins, carbohydrates, and starches to enhance their palatability and digestibility.[4] It is categorized as 'an endothermic process, where food absorbs externally applied heat or thermal energy', which results in physical changes such as 'melting, evaporation, and sublimation' and undergoes chemical changes.[5]

The endothermic process of cooking causes reactions responsible for the development of complex aromas and tastes, ranging from subtle sweetness to deep umami, that are essential to the sensory experience of a dish. The success of these transformations depends not only on the ingredients and techniques used but also on the tools: in the case of this essay's focus, the cooking vessel. Material culture plays a crucial role in shaping culinary outcomes. The cooking vessel's elemental attributes bear heavily on its ability to effectively conduct the endothermic process it has been tasked with.

This knowledge of the 'right pot for the job' has almost always been intuitive. Our mothers and grandmothers, with quite possibly little to no deep understanding of the scientific intricacies of these materials, have time and again made the most accurate choice of cooking vessel, ones with the capacity to preserve the sensory qualities of cultural dishes like pelau. However, modernity, shifting trends, lifestyle time crunches, and economics have contributed to the contraction of the cook's intuition or ability to lean into that intuition, somewhat perverting traditional cooking practices.

Caramelization: Carbohydrates, Disaccharides, Monosaccharides, and Heat

The caramelization of sugar is the first step in the preparation of pelau. Caramelization, the decomposition of sugar, breaks the ingredient down into its constituent parts. It is a nonenzymatic browning process responsible for producing volatile aromatic compounds such as buttery, nutty, and toasty notes; intermediate nonvolatile compounds; and brown pigments known as melanoidins. Caramelization initiates and sets the tone for the pelau's layered flavour profile and aesthetic appeal.[6] The caramelization of the polar ingredient, one with both a positive and negative charge, sugar, is managed with the use of oil, a nonpolar compound, as the polarity mismatch allows the cook greater control over the depth of colour and, by extension, the flavour achieved. Sugar will not dissolve in the oil; as well, the oil prevents sugar from rehardening during caramelization, resulting in a rich, flavourful, and syrupy starting point of pelau.[7]

Sugar, a ubiquitous ingredient, occurs naturally in many forms and permutations and is required by most life on earth for survival.[8] Its general formulation, CH_2O, is a combination of the three fundamental chemical elements essential for life: carbon, hydrogen, and oxygen. These three chemical elements represent the organic foundation of molecules like carbohydrates, lipids, and proteins.[9] Table sugar, or sucrose, is a simple carbohydrate and a disaccharide composed of two monosaccharides: glucose and fructose.[10] Together, glucose, the body's preferred energy source, and fructose, notably sweeter, found in fruits and root vegetables, are the building blocks of carbohydrates.[11] Sucrose's chemical formulation, $C_{12}H_{22}O_{11}$, reflects its molecular complexity; both glucose and fructose share the formula $C_6H_{12}O_6$, making them isomers.[12]

For the disaccharide sucrose to be caramelized, heat must be added to remove the atoms of water, bringing about a breakdown or chemical change called inversion. Inversion, the first step in the caramelization process, is where sucrose is separated into its component molecules, glucose and fructose. Altogether, the process involves several steps: first, dehydration, the loss of H_2O molecules that bring about inversion, then isomerization, where sucrose splits into its isomers, fructose and glucose. This is followed by polymerization, where smaller molecules and monomers bond to form larger molecules and polymers.[13] These reactions are accompanied by visual cues, by which the stages can be gauged, and include melting, boiling, foaming, and darkening.[14] And they produce complex intermediate nonvolatile compounds and small volatile aroma molecules.[15]

The resultant brown-coloured syrupy mixture produced by caramelization consists of large brown polymers: caramelan ($C_{12}H_{18}O_9$), carmelen ($C_{36}H_{50}O_{25}$), and caramelin ($C_{125}H_{188}O_{80}$), responsible for the caramel's signature brown colour – the proper development of which contributes to the aesthetics of pelau.[16] In addition, small, volatile molecules called furans, maltol, ethyl acetate, and diacetyl are also produced, which infuse the dish with deep, savoury, and sweet flavour notes.[17]

Notably, the requisite temperature that results in the full caramelization of sugars is dependent upon the chemical structure of the starting sugar, as each sugar, be it a monosaccharide or a disaccharide, caramelizes at a different temperature: fructose at 110°C (230°F) and glucose and sucrose at 160°C (320°F).[18] And it is this variability that underscores the importance of factoring in the elemental qualities of the cooking vessel used to produce pelau. The cookware must be capable of sustaining and regulating these variable temperatures evenly to avoid under-caramelization, resulting in a pale dish with a muted flavour, or over-caramelization, which introduces a bitter, burnt flavour. The ideal pelau pot, therefore, must in the first instance adequately support the full spectrum of sugar transformations.

Caramelization, *Cousoumeh*, and Ideal Types

The final phase in the preparation of pelau is called *cousoumeh*. Pronounced *coo-sue-may*, it is an African hybrid word that references the act of allowing cooked food to settle, to rest after the agitation of stirring, whipping, and the heat of cooking. For one-pot dishes like pelau, *cousoumeh* is an essential phase. Direct heat is removed, and the food is left to sit for some time undisturbed. During this time, the pot gradually cools, and in the stillness of this rest, the flavours developed during cooking coalesce, deepen, and harmonize; as such, the attributes of the cooking vessel are just as salient here as they are at the beginning of the cooking process.

It can take anywhere between forty-five minutes and an hour to cook a pot of pelau. The steps involved in cooking pelau begin with caramelizing the sugar into which the chicken is placed, then, once fully coated, covering and leaving it to stew, allowing it to absorb the flavours developed during the caramelization process, as well as impart its own flavours. With the meat protein achieving the desired rich brown colour, water is added, and the meat is left to stew further. Following this, rice, dried pigeon peas, carrots, and a variety of seasonings are incorporated and left to cook fully in coconut milk.[19] Moving from each of these stages of cooking pelau successfully is a matter not simply of ingredients or technique but also of the elemental nature of the pot utilized. Given that the elemental qualities of the cooking vessel directly influence its ability to sustain the precise thermal conditions necessary to facilitate the dish's complex endothermic reactions, the ideal pot, therefore, must incorporate several vital attributes.

First, to support even caramelization, the pot must have a high heat capacity to avoid over-caramelizing component sugars while under-caramelizing or leaving other component sugars uncaramelized. Heat capacity is a combination of factors: first, the mass of the pot, and second, the specific heat of the material; this second factor is an intrinsic property of the material. Taken together, the heat capacity of a particular elemental material measures the 'thermal sluggishness' of that material or the amount of heat or total energy needed to raise an object's temperature per mass unit.[20] The higher the value, the more heat is needed to increase the temperature of that material, and the more ideal the pot is for facilitating proper caramelization.

Additionally, the pot should have the capacity to adequately conduct heat – that is, have a low thermal conductivity rate. Thermal conductivity is another inherent property and an indication of the rate at which energy flows through the material of the pot from the heat source to the substances contained within the pot. A cooking vessel's thermal conductivity indicates its ability to efficiently and evenly distribute heat through the pot.[21] This attribute is vital in an ideal pelau pot regarding its ability to evenly cook the variety of ingredients added to the dish, especially during those steps that call for the ingredients to simmer.

Finally, good heat retention and emissivity are also necessary attributes an ideal pelau pot should have. The ability to retain heat evenly throughout the cooking process, as well as evenly emit surface radiation when cooking has ceased, is critical for *cousoumeh*.[22] Thus, a pelau pot is one with a high heat capacity and a low thermal conductivity rate to successfully facilitate the ideal caramelization of sucrose to avoid under- or over-caramelization, good heat retention for extended cooking time, and a good emissivity profile capable of facilitating gradual cooling and *cousoumeh*.

Weighed in the Balance

Traditionally, the pelau pot is one that is singular in purpose, heavy, medium- to large-sized, and made of sturdy cast iron. Cast iron, of which the main constituent element is iron, is a substance consisting purely of only one type of atom, incapable of being broken down into other substances. It is an excellent conductor of heat and offers a range of health benefits when used as cookware. Iron has an atomic mass of 55.845 amu and a density of 7.874 g/cm^3 at standard temperature and pressure. Now, regardless of the size of the cast-iron pot, the atomic density remains constant, ensuring consistent thermal behaviour across its dimensions. This high density contributes to its significant mass and excellent heat retention, allowing it to heat slowly and evenly, ideal for the gradual caramelization of sugars, slow stewing, even cooking, and facilitating *cousoumeh*.[23]

In contrast, aluminium (Al), one of the most abundantly available metallic elements found within the earth's crust, constituting 8.3 percent of the earth's weight, has a significantly lower atomic mass (26.98 amu) and density (2.70 g/cm^3).[24] Aluminium's abundance and lightweight nature make it an affordable alternative material; as such, it has become a dominant material in modern cookware.[25] Additionally, its low mass and density characteristics make it a lightweight, highly conductive material that allows for rapid heat transfer.[26] This rapid heat transfer and high conductivity make it ideal in those instances where the intense and immediate transfer of heat is required, which is not the case for cooking pelau.

Additionally, the use of aluminium or aluminium-type cookware increases the probability of the over-caramelization of sugars as well as uneven cooking and an inadequate environment for *cousoumeh* – belying its ability to develop the integral flavours of a pelau, as well as preserve the inherent qualities of the dish. Comparatively, cast-iron cookware has a higher heat capacity than aluminium, meaning more energy is required to raise the temperature of cast-iron cookware than aluminium.

Taken together, considering the cooking needs at each point of pelau's preparation, from caramelization to *cousoumeh*, cast iron's general weight, higher heat retention, and ability to store more energy than aluminium cookware make it a superior pelau pot.[27] Also, aluminium's capacity to react with acidic ingredients can distort the complex flavour compounds formed during the pelau's preparation, where caramelized sugar, coconut milk, and seasonings must coalesce harmoniously. Moreover, these leached ions may compromise the buttery, nutty, and toasty notes formed during caramelization and negatively affect the final stage of *cousoumeh*, where flavours are meant to settle and intensify.[28]

Pelau cooked in an aluminium pot may cook well enough, but due to aluminium's elemental comparative disadvantages, the dish may lack the depth of flavour traditionally associated with it. This is not to suggest that aluminium cookware may not be suitable or comparably superior for other types of cooking or in preparation of other dishes; it is, however, a statement on its ability to suitably render dishes that require slow, layered flavour development.

The Case for Clay

Beasts have memory, judgement, and the faculties and passions of our minds to a certain degree, but no beast is a cook. Man is the only animal that cooks.[29] The material culture with which this only undertaking that separates us from other animals often experiences shifts. These shifts are denoted by a change in how our communities and wider society perceive, value, and interact with our material possessions. These shifts represent deep alterations in the cultural meanings of our things, of the value assigned to those objects that have shaped our lives. And they are indicative of the transformation of how we humans relate to the objects that surround us, illustrating the degree to which we have moved away from how we once valued and perceived these items – oftentimes valuing that which is fleeting and often more abundant above that which is durable and long-standing.[30]

The shift from the traditional cast-iron pelau pot to aluminium-based cookware is understood to be a matter of economics, abundance, and the ability to disguise change aesthetically. Modern, lighter, aluminium-based pelau pots are aesthetically similar to the traditional, heavier, iron-based pelau pots, albeit shinier and having the ability to retain their silver look over time. Conversely, traditional pelau pots would, over time, develop the characteristic black smooth external patina – layer upon layer of use, wear, and memories of every meal.

A shift to clay, however, would require a conscious decision, necessitating the reconciliation of aesthetics with abilities. However, clay cookware offers an excellent alternative to the modern aluminium pot. Clay is a naturally occurring material composed primarily of fine-grained minerals, minerals which demonstrate a unique plasticity that bears significance when used in cooking. Clay has a typical density of around 1.76 g/cm^3, and though less dense than cast iron, its unique mineral composition and porous structure allow it to heat gradually and retain warmth effectively.[31] Its plasticity and moldability are a distinct advantage not only when firing the material to create cookware but in its actual use, given that the thermal properties, such as rigidity at high temperature, thermal shock resistance, heat capacity, and thermal conductivity, are particularly relevant to the preparation of pelau.[32]

The similarity of elemental abilities between clay and cast-iron cooking pots extends to the fact that unglazed, soaked, and seasoned clay cooking pots can effectively caramelize sugar.[33] Clay cookware's low reactivity and porous structure support the integrity of the flavour compounds developed during cooking pelau. As well, the cookware type allows for the gradual transfer of heat and ideal retention of moisture, thus preserving the complex chemical reactions that define pelau's sensory profile. Clay cookware's ability to slowly and evenly distribute heat supports the intricate thermal demands of pelau, from the precise caramelization of sugar to the gentle, flavour-deepening rest of *cousoumeh*, positioning it as the ideal pot type, with the capacity to preserve pelau's cultural and sensory integrity.

Overall, given that the selection of cookware is largely dependent upon the vessel's ability to conduct the functions of the preparatory process in a manner that transcends 'well enough' and matches not only the cooking technique but also the demands throughout the production of the dish, the shift to clay cookware is more than substantiated by its abilities. As such, this case for clay cookware is not only founded on thermodynamic principles but fundamentally founded on the cooking vessel's ability to preserve the sensory and cultural qualities and characteristics of pelau. Consequently, though clay cookware may be aesthetically dissimilar to both cast iron and aluminium-based pots now overtaking kitchens, it inherently holds the ability to preserve the taste, texture, and cultural memories embedded in pelau.

Moreover, in evaluating the ability of contemporary and traditional preparation practices to preserve the sensory qualities of a culturally significant dish, assessment encompasses a judgement of method and materials. The cooking vessel, be it clay, metal, or something else, holds histories and infuses subtle qualities into the foods they transform. It is not a passive receptacle: its latent abilities have a direct effect on the outcome, making it an active participant in elevating the raw into refined.

The cooking vessel itself must be considered as part of the recipe. Thus, this suggested shift to clay as the primary culinary vessel in the preparation of pelau comes from a broader reorientation of the philosophies surrounding continuities and change in material culture, away from the aesthetic toward elemental similitudes. Clay's porous quality facilitates a nuanced exchange of moisture and air. It breathes with the dish; its latent thermal properties encourage absorption and the release of moisture, tempering heat, grounding flavours, and allowing gradual, even cooking – gradual, even re-membering.

To shift how we view the cooking vessel requires us to reframe how we view preparation, to appreciate the scientific knowledge hidden beneath traditional practices. It requires judging honestly whether or not preparation has honoured the relationship between food and cooking vessel, where it complements the dish, amplifying its high points and tempering its lows, or where it instead honours a basic adherence to practice.

Too often, efficiency and convenience dictate our material culture choices, reducing cooking to a mechanical process and stripping the sensory qualities of our food memories, the meanings hidden in particular practices, and the tastes of words on our tongue: *cousoumeh*, a part of our culinary culture, our remembrance, our food meanings. Like the traditional iron pot, clay cooking vessels remind us that the preparation of our culturally rich dishes is about patience, filled with depth and reverence.

The call for a shift to clay is neither a rejection of the value of contemporary elements nor in ignorance of contemporary realities. Rather, it is an acknowledgement that the cooking vessel holds a unique place in the balance of preparation. In making a shift to clay, we honour both the dish that is pelau and each and every part of the process of its preparation and preservation.

About the Author

Shrinagar Francis is a food anthropologist, historian, and independent researcher with a BA in African and Asian studies with gender and development and an MA in the social anthropology of food. She is currently pursuing an MSc in food security.

Notes

1 Patti Sloley, 'Jollof Wars: Who Does West Africa's Iconic Rice Dish Best?', BBC, 8 June 2021 <https://www.bbc.com/travel/article/20210607-jollof-wars-who-does-west-africas-iconic-rice-dish-best> [accessed 2 February 2026].

2 Mark Cartwright, 'Wolof Empire', *World History Encyclopedia*, 20 November 2019 <https://www.worldhistory.org/Wolof_Empire/> [accessed 2 February 2026].

3 Nalo Hopkinson, *Whispers from the Cotton Tree Root: Caribbean Fabulist Fiction* (Invisible Cities, 2001), p. 1, qtd in Elizabeth DeLoughrey, 'Yam, Roots, and Rot: Allegories of the Provision Grounds', *Small Axe: A Caribbean Journal of Criticism*, 15.1 (34) (2011), pp. 58–75, doi:10.1215/07990537-1189530.

4 William D. Powrie, Chiu H. Wu, and V. Paul Molund, 'Browning Reaction Systems as Sources of Mutagens and Antimutagens', *Environmental Health Perspectives*, 67 (August 1986), pp. 47–54; Jessica Wittman, 'Thermal Energy, Temperature, and Heat', *Chemistry of Food and Cooking* (Mt. Hood Community College, 2022) <https://mhcc.pressbooks.pub/chemfoodcooking/chapter/heat-and-temperature/> [accessed 2 February 2026].

5 Lindsay Rohland, 'Endothermic and Exothermic Processes', EBSCO Knowledge Advantage, 2024 <https://www.ebsco.com/research-starters/chemistry/endothermic-and-exothermic-processes> [accessed 2 February 2026].

6 Powrie, Wu, and Molund, 'Browning Reaction Systems'; V. Daniels and G. Lohneis, 'Deterioration of Sugar Artifacts', *Studies in Conservation*, 42.1 (1997), pp. 17–26, doi:10.2307/1506572.

7 Heather Mendenhall and Richard W. Hartel, 'Effects of Fat Content and Solid Fat Content on Caramel Texture Attributes', *Journal of the American Oil Chemists' Society*, 93.9 (2016), pp. 1191–99, doi.org:10.1007/s11746-016-2871-0.

8 John R. White, 'Sugar', *Clinical Diabetes: A Publication of the American Diabetes Association*, 36.1 (2018), pp. 74–76, doi:10.2337/cd17-0084.

9 Scott Horowitz and Raymond C. Trievel, ‘Carbon-Oxygen Hydrogen Bonding in Biological Structure and Function’, *Journal of Biological Chemistry*, 287.50 (2012), pp. 41576–82, doi:10.1074/jbc.r112.418574; ‘What Is Sugar?’, Sugar Association, 2026 <https://www.sugar.org/sugar/what-is-sugar/> [accessed 2 February 2026].

10 Melissa Groves, ‘Sucrose vs Glucose vs Fructose: What’s the Difference?’, Healthline, 19 October 2022 <https://www.healthline.com/nutrition/sucrose-glucose-fructose> [accessed 2 February 2026]; Peter Dowse, ‘Disaccharides – a Definition and Examples’, Sugar Nutrition Resource Center, 21 February 2022 <https://www.sugarnutritionresource.org/news-articles/disaccharides-a-definition-and-examples> [accessed 2 February 2026].

11 White, ‘Sugar’; Dowse, ‘Disaccharides’.

12 Colabroy Provost and Kelly Wallert, ‘Browning Reactions’, 2016 <https://home.sandiego.edu> [accessed 2 February 2026]; White, ‘Sugar’.

13 Powrie, Wu, and Molund, ‘Browning Reaction Systems’; Daniels and Lohneis, ‘Deterioration of Sugar Artifacts’; Provost and Wallert, ‘Browning Reactions’.

14 Provost and Wallert, ‘Browning Reactions’; Anne Helmenstine, ‘What Is Caramelization? Why Sugar Browns’, Science Notes, 15 March 2024 <https://sciencenotes.org/carmelization-chemistry-why-sugar-turns-brown/> [accessed 2 February 2026].

15 Powrie, Wu, and Molund, ‘Browning Reaction Systems’; Daniels and Lohneis, ‘Deterioration of Sugar Artifacts’; Provost and Wallert, ‘Browning Reactions’.

16 Helmenstine, ‘What Is Caramelization?’.

17 Provost and Wallert, ‘Browning Reactions’.

18 Provost and Wallert.

19 Traditionally, *roucou* (annatto), which has a sweet-peppery flavour, would be added to give the dish a deeper colour and additional flavour. *Roucou* has in recent times been excluded or replaced by tomato ketchup.

20 Carla Ramsdell, ‘What’s the Most Sustainable Skillet?’, Know Watt’s Cooking <https://www.knowwattscooking.com/single-post/what-s-the-most-sustainable-skillet> [accessed 2 February 2026].

21 D. W. Van Krevelen and K. Te Nijenhuis, ‘Transport of Thermal Energy’, in *Properties of Polymers* (Elsevier, 2009), pp. 645–53, doi:10.1016/b978-0-08-054819-7.00017-0; Manisha Malik, S. K. Bhattacharyya, and Sudhirkumar V. Barai, ‘Thermal and Mechanical Properties of Concrete and Its Constituents at Elevated Temperatures: A Review’, *Construction and Building Materials*, 270 (February2021), article 121398, doi:10.1016/j.conbuildmat.2020.121398.

22 Ramsdell, ‘What’s the Most Sustainable Skillet?’.

23 ‘Metals and Alloys – Densities’, Engineering Toolbox, 2004 <https://www.engineeringtoolbox.com/metal-alloys-densities-d_50.html> [accessed 2 February 2026]; Dave Arnold, ‘Heavy Metal: The Science of Cast Iron Cooking’, *International Culinary Center’s Tech ’N Stuff Blog*, 16 February 2010 <https://cookingissues.com/2010/02/16/heavy-metal-the-science-of-cast-iron-cooking/> [accessed 2 February 2026].

24 Craig B. Lagrone, ‘Aluminum (Al)’, EBSCO Knowledge Advantage, 2023 <https://www.ebsco.com/research-starters/geology/aluminum-al> [accessed 2 February 2026].

25 Yogendra Kumar Gupta, Meenakshi Meenu, and Sharda Shah Peshin, ‘Aluminum Utensils: Is It a Concern?’, *National Medical Journal of India*, 32.1 (2019), pp. 38–40, doi:10.4103/0970-258X.272116; Okunola A. Alabi and Yetunde M. Adeoluwa, ‘Production, Usage and Potential Public Health Effects of Aluminum Cookware: A Review’, *Annals of Science and Technology* 5.1 (2019), pp. 20–30, doi:10.2478/ast-2020-0003.

26 Lagrone, ‘Aluminum (Al)’.

27 Arnold, ‘Heavy Metal’.

28 Alabi and Adeoluwa, ‘Production, Usage’.

29 James Boswell, ‘The Journal of a Tour to the Hebrides’, in *Journey to the Western Islands of Scotland and the Journal of a Tour to the Hebrides*, by Samuel Johnson and James Boswell (Oxford University Press, 2020), pp. 135–432 (p. 155).

30 'Material Culture Shift', Sustainability Directory, 2025 <https://lifestyle.sustainability-directory.com> [accessed 2 February 2026].

31 J. A. Odubiyi, 'The Properties of Clay', *Multidisciplinary Journal of Research Development*, 15.4 (September 2010) <https://www.globalacademicgroup.com/journals/multidisciplinary%20journal%20of%20research%20development/THE%20PROPERTIES%20OF%20CLAY.pdf> [accessed 21 May 2026]; Robert B. Heimann, 'Ancient and Historical Cooking Pots and Food: An Eternal Communion. A Topical Review', *Archaeometry*, 67.1 (February 2025), pp. 219–34, doi:10.1111/arcm.12986.

32 Anno Hein, 'Ceramic Cooking Pots and Their Thermo-Mechanical Performance', *Bulletin de Correspondance Hellénique*, 146.1 (2022), pp. 321–33, doi:10.4000/13g2z.

33 Erin Skahan, 'Clay Pot Cooking at Home & 3 Must-Try Recipes', *Munchery*, 6 September 2020 <https://www.munchery.com/blog/clay-pot-cooking-at-home/> [accessed 2 February 2026].

13
Meat(y) Encounters
The Element of Caste in Late Nineteenth-Century Colonial Western Indian Meat Markets

Aastha Gaur

Introduction

In the late nineteenth century, colonial cities became spaces for experiments in urban governance. As cities grew into densely populated commercial hubs, colonial governments faced challenges in regulating their development. The quest for urbanism coincided with fears about public health as the regulation of the spaces of human-animal interaction became central to the colonial urban project.[1] As a result, food markets and slaughterhouses became crucial spaces of intervention to ensure the sanitary balance of the city. Animals, both living and dead, became essential to the debates around hygiene, spatial organization, and social control. There were active efforts to move animals to urban peripheries by moving slaughterhouses to the outskirts of the cities. Concerns around hygiene and disease outbreaks also led to the policing of meat production, in turn leading to the creation of the market hall structure. In colonial India, these efforts changed both the economic structure of meat supply and the caste-based social relations involved in meat production.

Set within this time of broader discussions on culinary modernity, public health, urban planning, and hygiene, this paper examines changes in meat supply in Poona in the late nineteenth century. Through an analysis of the meat supply chain in the city's cantonment, the paper traces how the element of caste, rooted in occupational divisions and culinary hierarchies, led to a supply chain run mainly by Dalit and ethnic minority workers. New colonial infrastructures like slaughterhouses and market halls, designed to sanitize culinary spaces, led to the wide policing of meat. As the stigma around meat led to upper castes distancing themselves from its production, these spaces brought minority workers into the public domain. The market hall became a site of mutual financial and culinary dependence, respectively, for the minority workers and European consumers.

This paper builds on scholarship around colonial public health, hygiene and modernity, particularly works on India by David Arnold and Pratik Chakravarti and global studies by Ruth Rogasaki, Robert Peckham, and Philippa Levine.[2] While Poona's pre-colonial Peshwa era has been studied extensively, its colonial cantonment has received limited attention. Works by Wayne Mullen, Vaidehi Lavand, and Jayamala Diddee and Samita Gupta are the only works with a focus on the cantonment.[3] Despite this growing body of scholarship on colonial Poona, meat eating has not been studied at all. While Sohoni's work on market halls briefly explores the opening of market halls in the city and their architecture, it is more of a generic study of market halls across the country rather than Poona.[4] Works by Shireen Mirza, Daniel Williamson, Samiparna Samanta, and S. Gunasekaran have examined how meat-eating practices and spaces have shaped social relations in the Presidency cities.[5] While these works are important to contextualize the larger meat modernities of colonial India, they look at aspects of the supply chain rather than the supply chain at large.

This paper fills a crucial gap in the historiography by examining the meat supply chain as an interconnected system rather than a series of isolated economic activities. Unlike the well-studied Presidency cities, Poona's segregated spatiality offers a distinctive lens to examine the relationship between culinary infrastructure and social hierarchies. Focusing on Poona, this paper argues that the colonial market hall embodied a distinctive Indian culinary modernity, where preexisting caste norms shaped and were reinforced by new colonial infrastructures of meat supply. By analyzing the market hall as a space of exchanges between colonial officials and minority workers, the paper examines the restructuring of socio-spatial caste dynamics in the Poona cantonment.

Urbanization and Segregation in the Colonial City

In the late nineteenth century, colonial cities underwent a spatial reconfiguration with the introduction of new colonial technologies and infrastructures. Densely populated cities became sites of zoonotic disease outbreaks, which led colonial authorities to regulate human-animal interactions by separating animals from urban centres and moving slaughterhouses to the outskirts of the city.[6] In attempts to police human-animal interactions, food became central to colonial administration. The 'biomedical gaze', which placed culinary spaces within the confines of health control, was employed to discipline the behaviour of food producers, vendors, and consumers.[7] In cities across the British Empire, efforts to police animals were characterized by a need to create a clear spatial separation between the slaughterhouse and the markets.[8] The attempt to separate slaughterhouses and markets moved living animals to the peripheries and allowed them to enter city centres only as dead meat commodities.

In India, these practices intersected with the local caste system.[9] Under India's unique caste system, vegetarianism was the norm among the hegemonical upper castes. As a result, not just meat but also meat eaters and butchers were seen as impure and a threat to the culinary hegemony. Consequently, meat-eating Dalit or lower caste and Muslim communities were stigmatized and excluded from dining spaces used by upper castes.[10] An Indian exceptionalism was created, one where not only animals but also people who worked with animals were seen as impure.[11] As a result, the colonial government constructed localized markets adhering to localized caste norms while also catering to the colonial principles of public health. As the British sought to reorganize the city's culinary infrastructures, they built upon preexisting social hierarchies that had long structured the city.

Reay and Connaught: Market Halls in Poona

Poona had been the capital, administrative centre, and political-military hub of the Peshwas in the eighteenth century. With economic and political changes, there were urban changes through the setup of *peths*. These *peths* were administrative divisions and the centre of one economic activity. Most *peths* were organized based on occupational castes, and particular groups clustered in certain *peths*.[12] This first phase of urban organization arranged the city into smaller administrative divisions which coincided with economic and social positions.

With the British takeover of Poona following the Peshwa defeat, the landscape completely changed.[13] After the takeover, the area located east of the city limits was designated as the new primary military cantonment. Poona became the headquarters of the Bombay army and the monsoon capital of the Bombay Presidency. As a result, infrastructure, including numerous administrative buildings, was also set up.[14] The cantonment became an entity separate from the Native City, ushering in the second wave of the city's urbanization.

The spatial construction of the city was altered into a tripartite structure consisting of the cantonment, Civil Lines, and the Native City. The cantonment was a permanent military camp run by the cantonment committees. The committees were also responsible for the neighbouring market district called the Sadar Bazaar and the administrative hub for officials called the Civil Lines. The Civil Lines were a suburban municipality administered by a largely British civilian council and occupied by the British administrative elite. The Sadar Bazaar was a commercial centre which expanded to the southwest of Sachapir Street, merging into the *peths*.[15] The Native City sustained its *peth* structure and remained a separate administrative unit within Poona.[16] With this architectural separation also came social segregation.

The setup of the Poona cantonment created two separate entities – the city and camp. The city, characterized by its narrow lanes, house clusters, and *peths*, contrasted the green, suburban barracks and bungalows of the cantonment and Civil Lines, which were collectively referred to as the camp.[17] The social construction of the city became such that the native Indian Maharashtrians mostly lived in the Native City; elite Indian administrators, Parsis, Jews, and other ethnic minorities stayed in the Civil Lines; and the soldiers lived in the barracks in the cantonment.[18] The camp and city evolved into two isolated spatial and social spheres with minimal interaction and distinctive features.

These physical and social separations also extended to urban planning, including culinary infrastructures like markets. When the city's municipality was formed in 1858, it decided to follow the prevailing practice of designing separate marketplaces to cater to different populations.[19] The market hall, a rather standardized structure in England, was adapted to the local needs and climate in India.[20] In Poona, the efforts to build these markets began as a response to sanitary concerns around butchering and open markets.

Colonial concerns around unhygienic butchering practices were articulated in A. H. Leith's 1864 *Report on the Sanitary State of the City of Poona*. The report exposed the 'evil' of cattle slaughtering on butcher shop premises. Leith recommended the building of a slaughter yard and shed in the city.[21] In addition to butchering reforms, Leith also suggested the building of a regular marketplace to replace the open markets of the city, which were held in the police barracks.[22] While Leith suggested altering the structure of the police barracks to create organized market sheds as early as 1864, it was only in the late 1880s that organized market sheds were built in the market halls.

The Reay Market was completed in 1886 to ensure the sanitary and regulated sale of foodstuffs.[23] The market was named after Lord Reay, the Governor of Bombay, and built for the locals by a local architect, Vasudev Bapuji Kanitkar. The market hall housed 474 shops and municipal offices.[24] The residence of municipal officers allowed for the surveillance of the local vendors, who were often seen as unhygienic. The call to build the market was met with a lot of local resistance, as it meant a reconfiguration of the various scattered marketplaces around the city.[25]

The building of the Reay Market changed the culinary supply chain of the city significantly. While earlier the *peths* had their own small open markets, building the Reay Market altered this traditional structure by imposing the colonial market hall structure, which now required the stall owners to pay. However, the market was also forced to cater to local tastes. Resulting from resistance, the Reay Market only sold vegetables. It was forced to adapt to the localized caste hierarchy, which resisted the sale of meat in public spaces.

To cater to largely meat-eating European and ethnic minority populations, another market was commissioned in the cantonment.[26] Racial concerns around the unhygienic native processing of the food products led to efforts to contain the sale of food products within the market. The new cantonment market was to be controlled and cleaned by the cantonment municipality, thereby allowing the British to oversee food production. Established in Sadar Bazaar, the new market was designed to be the one-stop shop for fresh produce, meat, and poultry, with separate sections for each.[27] Named after the Duke of Connaught, the Connaught Market was intended to replace the chaotic and unhygienic shops in the Sadar Bazaar.[28]

The construction of separate markets at the same time, each catering to the city's two distinctive conclaves, depicts the contrasting hegemonies of the city and the cantonment and reveals the complex ways in which power penetrated culinary spaces. While the politically dominant British were able to impose their ideas of public health and hygiene through the building of these markets, they were forced to adhere to the socially powerful caste norms in the Native City. The upper-caste resistance to introducing meat to public spaces forced the British to create a market which appealed to their tastes. The previously open vegetable, fruit, and grain market was reformed into the monopolized Reay Market. For the cantonment, the unorganized Sadar Bazaar butchers and grocery stores were assembled into the Connaught Market. Contrary to the previously scattered butcher shops, meat came under a contained space. Colonial officials monopolized and strictly regulated the meat supply.

From Carcass to Meat: Food Supply and Meat in the Cantonment

With the building of the markets, food supply to the cantonment was changed significantly. Drawing on primary material such as gazetteers, sanitary reports, General Department correspondences, and colonial primary accounts, this section maps the trajectory of meat right from the farm to the slaughterhouse, market, and colonial kitchen.[29]

The supply chain began by procuring the meat which was to be slaughtered. The meat for slaughtering was often sourced from *dhangar* shepherds, who reared animals on the outskirts of the city.[30] The *dhangars* brought sheep and goats to the butchers. While this system had been in place since before the arrival of the British, by the early twentieth century, animals being bought began to undergo quality checks, stemming from suspicions around the provenance of the animals procured.[31] The imposition of quality checks was not a smooth process and was met with strikes from the shepherds, highlighting the contentious nature of colonial policing and the constant surveillance of meat.

Ultimately, however, the shepherds were forced to adhere to the rules of the empire and follow the quality checks. The inspected animals from the shepherds were brought to the city for slaughtering. Due to the absence of a central slaughterhouse, slaughtering was often scattered. The meat was mostly slaughtered within households or in public places in areas such as Kagadipura, Bukharwadi, and Barragpura.[32] The cantonment had slaughter yards which were seen as more hygienic than home slaughtering. Until the 1890s, the Native City butchers also used it, but in 1894, there was a call to build a slaughterhouse in the Janaicha Mela area in the Native City. This effort sought to prevent city butchers from using cantonment yards. Despite resistance from the upper-caste Hindus of Poit Kasba, the government allowed land to be granted for the building of a slaughterhouse in the Native City in 1895.[33] The call to separate slaughtering spaces for cantonment and city residents highlights the active efforts to reinforce the spatial boundaries of the city and camp.

Meat was primarily slaughtered by Muslim butchers and some Dalit communities.[34] While the location of the slaughter yards within the cantonment remains unclear, numerous butcher shops were in or near the Connaught Market. These spaces, likely positioned close to each other, were highly contested and subject to constant colonial surveillance. Apart from the enduring presence of municipal public health officials and regular inspections, butchering sites were also affected by the sanitary needs of the cantonment. As worries about poor hygiene and sanitation converged with fears of disease in these markets, both animals and butchers were scrutinized and viewed as potential carriers of infection.

The concerns around butchers as disease carriers became particularly amplified after the devastating 1896 plague in Poona. In the aftermath of the plague, all beef butchers and their families were ordered to be inoculated before selling beef at the Connaught Market. In September 1912, some of them were sent back with their cattle from the slaughterhouse, as they were not inoculated. Interestingly, this order was not extended to mutton butchers.[35] The choice of only inoculating beef butchers is noteworthy, as beef was mainly sold only within the cantonment and consumed largely by the non-natives.

The contested sale and consumption of beef was central to the modern meat supply chain in Poona. In a hegemonic upper-caste society, beef was often consumed precariously and in isolation. Before British rule, beef was primarily consumed by Muslims and Dalits, while upper castes considered the cow sacred and viewed beef eaters as impure.[36]

The beef-eating habits of the British were similarly seen as impure by the upper-caste Hindus. Due to their position as powerful heads of the empire, they were not openly shunned for their beef eating. Nevertheless, upper castes usually refrained from working in colonial kitchens, as they were unwilling to cook beef. The consumption of beef by Dalit communities made them suitable to work in colonial kitchens. The willingness to handle and cook beef became a criterion for employment in colonial bungalows, which in turn economically benefitted the Dalit communities.[37]

As a result, most colonial servants were either Dalits or converted Christians. Within Poona, there were numerous Christian Goanese servants and cooks.[38] These local Christians mostly came from Bombay or other parts of India and were most likely Dalit individuals who had converted to Christianity.[39] There were also some Hindu servants who did other errands for the British, but the culinary servants were mainly local Christian boys.[40] The servants were recruited through referrals from British friends returning to Britain or via testimonials written by former employees.[41]

Local servants were the backbone of the colonial meat supply to the cantonment. They usually visited the markets to buy meats. The meats were then brought to colonial bungalow cookhouses to store and cook. The cookhouses were located close to the servant establishments in detached subsidiary structures, and only the butlers usually entered the houses to serve the food.[42] The kitchen, servant quarters, and horse stables were often located at the windward side of the bungalows to prevent cooking odours from reaching the bungalows.[43] The architecture of the bungalow ensured a racial segregation between the servants and the primary British residents. This segregation stemmed from not only ideas around racial impurity but also olfactory sensations, as the separation of the cookhouse ensured that the odours of raw and cooking meat did not reach the British inhabitants.

The spatial and sensory separation within colonial bungalows mirrored the broader socio-spatial hierarchies that governed the city's meat supply. The servants procuring, handling, and cooking the meat, much like those who butchered and sold it, were ethnic minorities such as Muslims or converted local Christians and Dalits. As the urban landscape of the city changed with the introduction of slaughterhouses and colonial market halls, meat entered a more public and visible domain. The Dalit communities, who earlier ate meat secretly, now saw meats being sold in markets to cater to colonial needs. Although colonial eaters continued to view meat as dirty and unhygienic, their Western diet demanded it. To cater to their demands, a colonial meat supply chain was set up in Poona, which was run almost entirely by the ethnic minority and Dalit communities.

Fleshing Out the Changes: Boundary-Breaking Meat

In the late nineteenth century, Poona's culinary landscape underwent a transformation that produced a distinctive culinary modernity. While the colonial government designed infrastructures like market halls and slaughterhouses to police and regulate local meat production, they were inevitably shaped by preexisting local caste hierarchies. Caste hierarchy deeply impacted who procured, butchered, sold, and cooked meat. Rather than replacing preexisting structures, the colonial meat supply chain adapted to them, reinforcing traditional social orders within the new, regulated space of the market hall.

Previously, meat was mostly slaughtered within the confines of Dalit and ethnic minority homes. The shift from household slaughtering to public abattoirs brought meat into a more visible domain, challenging preexisting social norms. While this transition challenged the localized stigmatization of meat by upper-caste Hindus, it rearticulated the boundaries of caste within the colonial frameworks of hygiene and order. Market halls such as the Connaught Market, intended to sanitize and compartmentalize meat production, became an embodiment of colonial sanitation with constant surveillance and regulations around native, supposedly unhygienic meat production methods.

While the shift from bazaars to market halls signalled the triumph of colonial culinary policing, it also created a third space for colonial-native interaction in the form of the market hall. Market halls were hierarchical spaces which housed municipal officers above market stalls to oversee sanitation. Yet at the same time, the market hall was also a rare space where butchers, stall owners, colonial servants, and colonial officials socialized in the same vicinity. Market halls provided a space where elites and minority workers, who typically did not interact, were able to and even forced to meet. Hence, food markets became unique spaces of interaction between the powerful and minority communities of the modern city, thereby challenging the well-established social boundaries of Poona.

Poona's hybrid meat modernity – combining the colonial and local – took a tangible form in spaces where meat was sold and cooked. The introduction of colonial culinary infrastructures reconfigured social relationships by fostering a relationship of mutual dependence between colonial consumers and minority workers in the meat supply chain based on shared beef-eating habits. This relationship of mutual dependence, between the financial needs of the butchers and servants and the meat consumption habits of the colonizers, created a distinctive meat supply chain based around caste dynamics.[44] Minority butchers, servants, and stall owners became an indispensable part of the food supply chain, which, while deeply policed, allowed for indirect interactions between the marginalized and the elites.

As the meat-cooking servants came from the outskirts of the Native City all the way into colonial domestic interiors, they also broke the spatial divisions of Poona's urban landscape. The tripartite Native City-Civil Lines-cantonment structure ensured little to no interaction between the city and the cantonment. Meat workers broke these boundaries and travelled all the way from the city's outskirts to the very interiors of the colonial home, becoming rare intermediaries between the city and the cantonment. While colonial regulations intended to control and segregate the processes of meat production and consumption, the very nature of the new supply chain created points of interaction between communities which were hitherto kept apart. Apart from traversing the geographical confines imposed on Poona's spaces, meat also questioned the social order.

By challenging preexisting caste norms and bringing meat into the public sphere, the new supply chain also contributed to the normalization of meat consumption. In a society where meat was previously confined to domestic spaces, the construction of public slaughterhouses and market halls selling meat questioned the stigma around meat eating. While meat and animals were increasingly subjected to colonial sanitary regulations, they simultaneously became a part of broader public debates around health and society. The shift in the dominant political power – from the upper-caste elites to British colonial administrators – was reflected in the transition of the city's foodscape. What was once a city of caste-bound meat invisibility evolved into one with public infrastructures dedicated to meat supply.

Although spaces of upper-caste hegemony such as the Reay Market remained, the construction of slaughter yards and market halls certainly ushered in a new urban era where meat was more visible and publicly accepted. A study of meat supply in colonial Poona thus becomes a window into wider social, spatial, and urban transformations. In tracing the paths of meat and meat workers, this paper has uncovered how culinary infrastructure unsettled the powerful boundaries of caste and colonial sanitation.

Conclusion

Through a contextualization of Poona's urban history, this essay has traced the changes in meat infrastructures in the late nineteenth to early twentieth century. The shift from home butcheries and scattered markets to organized slaughterhouses and market halls brought meat into a more visible public domain. The dual impurity of meat made it a highly contested and policed commodity within the Poona city and cantonment. A supply chain run by local Dalit and ethnic minority intermediaries emerged to keep meat, seen as socially impure by upper-caste Hindus and as a disease carrier by the British, away from the privileged. Hence, albeit indirectly, meat became a conduit for lower-caste workers to engage with colonial officials, forging an unexpected link between the marginalized and the elite through the networks of production and consumption. As meat found its way into the domestic interiors of the powerful, it became an ironic agent of connection in the deeply segregated urban landscape.

Poona's distinctive spatial division of the city and the cantonment, which created a clear architectural separation between the locals and the British, was also broken through the new meat supply chain. As meat workers travelled from the outskirts of the city to the interiors of the cantonment, they challenged the casteist spatial divisions of the city. By bringing minority workers into public domains they had never explored before, the colonial regulation of meat inadvertently altered Poona's cityscape by bringing meat to public domains and bringing minority meat workers to publicly accessible markets. The minority butchers, servants, and cooks developed a mutual dependence with the colonial eaters, stemming from a shared heritage of beef consumption, challenging upper-caste vegetarian hegemony.

The element of caste, unique to the Indian context, produced a distinctive Indian culinary modernity in which colonial sanitation and caste hierarchies coexisted, clashed, and reshaped each other. Public meat markets became spaces where caste and colonial authority were simultaneously maintained and challenged, illustrating how culinary infrastructures served as critical sites for negotiating the city's transition into a modern urban order.

By tracing the movement of meat, the changing roles of minority workers, and the importance of the market hall in crafting the Indian meat modernity, this paper has highlighted how food creates, reshapes, and challenges socio-spatial boundaries in a colonial Indian setting.

About the Author

Aastha Gaur is a master's (by research) in history student at the National University of Singapore. Her interests include culinary heritage, urban environments, and social history.

Notes

1 David Arnold, 'Pollution, Toxicity and Public Health in Metropolitan India, 1850–1939', *Journal of Historical Geography*, 42 (2013), pp. 125–27, doi:10.1016/j.jhg.2013.07.002.

2 David Arnold, *Colonizing the Body* (University of California Press, 1993); Pratik Chakrabarti, *Western Science in Modern India: Metropolitan Methods, Colonial Practices* (Orient Blackswan, 2004); Ruth Rogaski, *Hygienic Modernity: Meanings of Health and Disease in Treaty-Port China* (University of California Press, 2004); Robert Peckham, 'Bad Meat: Germs, Animal Disease, and Food Safety in Hong Kong', in *Moral Foods: The Construction of Nutrition and Health in Modern Asia*, ed. by Angela Ki Che Leung and Melissa L. Caldwell (University of Hawaii Press, 2019), pp. 173–98; Philippa Levine, 'Venereal Disease, Prostitution, and the Politics of Empire: The Case of British India', *Journal of the History of Sexuality*, 4.4 (1994), pp. 579–602, doi:10.4324/9780203881941.

3 Wayne Thomas Mullen, *Deccan Queen: A Spatial Analysis of Poona in the Nineteenth and Early Twentieth Centuries* (unpublished doctoral thesis, University of Sydney, 2001) <https://hdl.handle.net/2123/495> [accessed 14 May 2026]; Vaidehi Lavand, *Architectural Character of Poona in Late Nineteenth Century* (unpublished doctoral thesis, Università degli Studi di Palermo, 2017) <https://iris.unipa.it/handle/10447/221004> [accessed 14 May 2026]; Jayamala Diddee and Samita Gupta, *Pune: Queen of the Deccan* (Elephant Design, 2000). Beyond these, other theses such as those by Meera Kosambi and S. B. Sawant remain inaccessible outside India. Online articles on various aspects of urbanism, the plague, and politics within the city such as those by Meera Bapat, I. J. Catanach, Surinder K. Mehta, Kishor D. Gaikwad, and Martha Kaplan are available online and useful in understanding the larger historical context of the period.

4 Pushkar Sohoni, *Taming the Oriental Bazaar: Architecture of the Market-Halls of Colonial India* (Routledge India, 2023).

5 Shireen Mirza, 'Cow Politics: Spatial Shifts in the Location of Slaughterhouses in Mumbai City', *South Asia: Journal of South Asian Studies*, 42.5 (2019), pp. 861–79, doi:10.1080/00856401.2019.1644766; Daniel Williamson, 'The Crawford Market: Sanitary Problems, Engineered Solutions, and Symbolic Gestures in Late Nineteenth-Century Bombay', in *Mobs and Microbes: Global Perspectives on Market Halls, Civic Order and Public Health*, ed. by Leila Marie Farah and Samantha Martin (Leuven University Press, 2023), pp. 45–65; Samiparna Samanta, *Meat, Mercy, Morality: Animals and Humanitarianism in Colonial Bengal, 1850–1920* (Oxford University Press, 2021); S. Gunasekaran, 'Caste, Food and Colonialism: "Outcaste" Domestics in the European Houses of Madras Presidency', *Indian Economic and Social History Review*, 61.1 (2024), pp. 97–131, doi:10.1177/00194646231222641.

6 Peckham, 'Bad Meat', p. 176.

7 Tatsuya Mitsuda, 'Snacking, Health, Modernity: Moralizing Confections in Japan, 1890–1930', in *Moral Foods: The Construction of Nutrition and Health in Modern Asia*, ed. by Angela Ki Che Leung and Melissa L. Caldwell (University of Hawaii Press, 2019), p. 151.

8 Arnold, 'Pollution, Toxicity and Public Health', pp. 126–27.

9 The caste system is highly varied throughout the country and does not apply to the whole country in a uniform way. For the purposes of this essay, caste as seen in the Bombay Presidency is considered. It is also important to note that not all Brahmin communities were vegetarians, and some such as the Saraswat Brahmins ate meat. However, these communities were anomalies in the otherwise largely vegetarian Brahmin community.

10 B. R. Ambedkar, *Beef, Brahmins and Broken Men* (Columbia University Press, 2020), pp. 225–50.

11 Arnold, 'Pollution, Toxicity and Public Health', pp. 128–29.

12 Mullen, *Deccan Queen*, p. 125.

13 M. S. Moledina, *History of the Poona Cantonment, 1818–1953* (Poona and Kirkee Cantonments' Citizens' Association, 1953), p. 1.

14 Mullen, *Deccan Queen*, p. 144.
15 Mullen, p. 156.
16 Mullen, pp. 145–46.
17 Diddee and Gupta, *Pune*, p. 3.
18 *Guide to Poona and Kirkee for the Season 1876* (Treacher and Co. Ltd. Poona, 1876).
19 Sohoni, *Taming*, p. 4.
20 Sohoni, p. 20.
21 A. H. Leith, *Report on the Sanitary State of the City of Poona* (Education Society Press, 1864), p. 34.
22 Leith, pp. 32–33.
23 'The Opening of the Reay Market in Poona', *Times of India*, 7 October 1886, p. 5.
24 Lavand, *Architectural Character of Poona*, p. 159.
25 Lavand, p. 133.
26 Sohoni, *Taming*, p. 57.
27 Lavand, *Architectural Character of Poona*, p. 171.
28 Arati S. Petkar, G. R. Diwan, and P. Tikhe Tushar, 'Urban Renewal of the Sadar Bazaar Area in Pune Cantonment', in *Proceedings of International Conference on Advances in Architecture and Civil Engineering* (2012), p. 614.
29 Based on sources from the 1880s to the 1930s, this study offers a preliminary, deductive exploration of the likely meat supply chain during that period.
30 Arthur Crawford, *Our Troubles in Poona and the Deccan* (Archibald Constable & Co., 1897), pp. 203–04.
31 'Shepherds Strike at Poona', *Times of India*, 23 January 1904, p. 9.
32 Government of Bombay, *General Department Order no. 6534*, Bombay Castle, 19 September 1917.
33 Government of Bombay, *General Department Volume 97, s-27*, 14 March 1891.
34 *Gazetteer of the Bombay Presidency: Poona* (Government Central Press, 1885), p. 294.
35 'Poona Butchers', *Times of India*, 14 September 1912, p. 7.
36 B. R. Ambedkar, *The Untouchables: Who Were They? And Why They Became Untouchable?* (Amrit Book Co., 1948), pp. 1–11.
37 Gunasekaran, 'Caste, Food and Colonialism', p. 98.
38 Dennis Kincaid, *British Social Life in India* (George Routledge and Sons, 1938), p. 250.
39 Surinder K. Mehta, 'Patterns of Residence in Poona, India, by Caste and Religion: 1822–1965', *Demography*, 6 (1969), p. 475.
40 Edward Elwin, *Indian Jottings: From Ten Years' Experience in and around Poona City* (John Murray, 1907), p. 182.
41 Mullen, *Deccan Queen*, p. 186.
42 Mullen, p. 42.
43 Mullen, p. 40.
44 Gunasekaran, 'Caste, Food and Colonialism', p. 117.

14
Milk and Cheese as Incarnations of the Elements
Respectful Coexistence Instead of Aggressive Suppression

Ursula Heinzelmann

Earth, water, air, and fire – that is, the soil, lakes and rivers, sunshine and rain, heat and cold – shape the landscapes that ruminants transform into milk. Preserving that milk in the form of cheese, guiding its natural transformation within cultural borders, again is informed by the elements and the selection process they represent. All cheese classics, from Roquefort to Camembert and all the Alpine types, have not been 'invented' as such but are the result of dealing with a specific situation, making do with what there was, and transferring viable results into replicable, structured procedures: 'taming' the elements. Each cheese incarnates an expression of the situation in which it was born, of the animals, mountains, wind, and soil – what is generally called 'terroir'. Like winemakers, cheesemakers try to transcribe a place and its emotions into a food product. This rootedness has taken several centuries and led to the emergence of emblematic cheeses based on particular regions: Gruyère, Comté, Ossau-Iraty, and Ste Maure de Touraine, to name just some French-speaking classics. Today, however, the scenario tends to be almost the opposite, as cheeses are invented to fill perceived slots in the market by shaping the elements, from artificial environments and air-conditioning to imported fodder. Focusing on the Swiss cheese L'Etivaz, this paper tries to show how much we are risking when losing those roots.

Because, as the French fromager-activist Augustin Denous says in *Fromages Sauvages*, the ground, the soil, is important: life stems from the ground. From humus, from people, from humanity – and humility.[1]

In theory, Appellations d'Origine Protégée (AOP, or PDO, Protected Designation of Origin) protect the old know-how and origins. Nevertheless, there is a worrying trend towards standardization because the regulations are, on one hand, too static and yet, on the other, not strict enough, prone to misuse as a smokescreen by industrial multinationals. Ferments, essential for developing aroma and texture, are increasingly produced in laboratories and standardized. Animal breeds are pushed towards productivity. Grazing is less and less on natural pastures. PR sells imaginary links to the land, but in reality, practices are largely mechanized, and today's cheese production is off-ground, in the literal sense of the word.

In this paper – which could easily be a whole book, because as soon as you start thinking about cheese, real cheese, you are immersing yourself in the entanglements of life as such (animals, including humans, and their complex interactions on multiple levels, in so many different scenarios, shaped by the elements) – I focus on L'Etivaz, a hard cheese from the Pays d'Enhaut in the canton of Vaud in western Switzerland, an hour's drive east of Lake Geneva.

L'Etivaz is a hard cheese, in solid round wheels, weighing anything between ten and a muscle-straining thirty-eight kilos (roughly twenty-two to eighty-four pounds). With a hint of smoke, it dares to take corners, to have edges, where technology and industrialization have long since propagated and enabled so-called perfection. L'Etivaz is a primeval creature of the mountains and the forces of nature that prevail here; it does not entice with sweet smoothness but demands attention. Nevertheless, it has not landed on the economic sidelines, and this, in my eyes, is something rare we should be grateful for.

'Fromage d'Alpage au feu de bois' ('Alpine cheese from a wood fire'), the black logo says. The label shows everything that characterizes L'Etivaz: spruce trees, flowering grasses, cows, and a simple hut with a large cauldron, plus a person stirring beside an axe in a block of wood. Our cheese actually belongs to the Gruyère family and is, to a certain extent, a Grand Cru single-variety cheese from this large region in French-speaking western Switzerland. It can only be produced during the summer, when the farmers take their cows up the mountains to one thousand to two thousand metres. Between mid-May and mid-October, the so-called summering, or *l'estivage* (hence the name L'Etivaz), is all about milk and cheese (and hay for the winter). It's an old tradition, because the only way to survive here in the mountains is by making the most of all the natural resources; agriculture as such is almost impossible. The practice is documented since the twelfth century but undoubtedly goes back much longer (and obviously not just in the Pays d'Enhaut). It is said that the herders once made do without any dwellings but were convinced by the (absentee) landlords to place at least the cheesemaking under a solid roof for better results. Today, the summer huts serve as both dwellings and cheese dairies, and even if electricity and running water are the norm today, everything is kept very simple, 'just like in the old days' – and that is precisely what is extraordinary and forward-looking.

The seventy to eighty families here in the Pays d'Enhaut realized almost a century ago (after the usual historical waves of changing affiliations regarding tithes and other duties) that their cheese was much more than just a foodstuff, that it embodied and secured their culture and way of life – and that it was in danger of being irrevocably changed by excessive regulations, new technologies, and market structures. The centralized marketing of the major Swiss cheese brands, with the ensuing increasing uniformity, did not appeal to these independent minds, and they decided to take their fate into their own hands (which also had to do with old territorial feuds and power struggles, but that's another story). True to the revolutionary, perhaps even stubborn, spirit that lurks in every Swiss canton and valley, in 1932, they founded a cooperative and drew up rules and specifications for production, maturing, and marketing, and in 1934, to cap it all, they built a communal ripening cave. In the long term, that gave them a precious USP, a unique selling point, that sets their cheese apart from the large mass of Gruyère, be it Alpage (that is, from the summer mountain dairies) or not. The 420 tons of L'Etivaz produced in one summer represent a mere drop in the market, compared to Gruyère's thirty thousand-plus tons per year – but their percentage of cheese exported at around 40 percent is significantly higher.[2] The AOC, later AOP, which L'Etivaz obtained in 2000, was Switzerland's first.

Today, it is the grandchildren of these visionaries who move to the huts with their own families and still work in exactly the same way. Which means: open fires, the flames twitching and flickering under the large *Kupferkessi*, so that in some cases, depending on the architecture, the smoke fills the whole hut, settling as black soot on the walls and on the outside of the copper cauldron, making your eyes water before it escapes through a hole in the roof or the open door.

For a long time, this was the norm in Alpine cheese dairies, whereas today, the smoke (more or less tangible in the cheese) is a special feature of L'Etivaz. In other regions, the fireplaces have long been enclosed, walled in, the flames contained and tidied up – which is more comfortable, certainly, but it also diminishes the awareness of the living elements that are at work here and are only guided by humans. In a way, the great cheeses of the Alps are creatures of flames.

This is meaningful. In the beginning, there was fire: we became human beings when we learned to tame it. 'We humans', as anthropologist Richard Wrangham famously wrote, 'are the cooking apes, the creatures of flame', explaining:

> Cooking increases the amount of energy our bodies obtain from our food. The extra energy gave the first cooks biological advantages. They survived and reproduced better than before. Their genes spread. Their bodies responded by biologically adapting to cooked food, shaped by natural selection to take maximum advantage of the new diet. [. . .] Fossil evidence indicates that this dependence arose not just some tens of thousands of years ago, or even a few hundred thousand, but right back at the beginning of our time on earth, at the start of human evolution, by the habiline that became *Homo erectus*.

'Those claims', he goes on, 'say humans are adapted to eating cooked food in the same essential way as cows are adapted to eating grass'.[3]

I quote Wrangham for two reasons. First, I obviously like the fact that he mentions cows' adaptiveness to living on pastures (and he could easily have said ruminants instead of cows; the singling out of one species just makes the comparison more obvious, I think). Second, and more importantly, he reminds us of the extremely important role fire plays in our lives. No matter which elemental philosophy or theory you consider (e.g. the ancient Greeks' four elements, with fire, earth, water, and air connected to the four humours, with fire being associated with yellow bile, hot and dry; the Chinese Wuxing concept, where fire interacts with earth, metal, water, and wood and is associated with passion, intensity, and the tongue; or Ayurvedic teachings, with earth, water, fire, air, and ether, or space) – the flames are always there, symbolizing transforming, liberating energy.

As Jean Bottéro writes about cooking and cuisine in Mesopotamia:

> The vocabulary itself implied an intervention by humans and their culture to make *edible* the *food* that by itself was not. [. . .] This intervention, in Mesopotamia as

elsewhere, involved subjecting food to fire – *cooking*, the primary and essential act in any cuisine.[4]

Fresh milk back then was considered uncultured: in the epic of Gilgamesh, 'only a character such as Enkidu, in his "wild" state, could have had the habit of suckling only the milk of beasts' (though it has to be said that milk also enjoyed luxury status, as it turns so quickly, and was therefore placed on the Divine table at the temple of Anu in Uruk, in an alabaster pitcher). Cheese, however, like bread and beer (and obviously also much more digestible for lactose-intolerant adult humans than fresh milk), was deemed to be cultured. Enlil, the king of gods, sends 'large cheeses, mustard-flavored cheeses, small cheeses' to his future in-laws as part of the wedding meal.[5]

At this point, talking about cheese in connection with fire and heat, it is important to note that the very moment the two meet is essential. Fire can liberate energy, but it can also destroy: heating the fresh milk above its 'natural' temperature – that is, the animals' own, as in pasteurizing – has an entirely different effect. Then, fire is killing most of the milk's life, the flora and endemic enzymes, most of what makes it distinctive and leads us to talk about terroir. Heating the coagulated milk, however, helps preserve it.

It was in more Nordic areas, such as the Alps, that cheesemakers at some point 'discovered' not only that fire helped maintain the milk at 'working' temperature but also that heating the curd – that is, the coagulated milk, finely cut and swimming in liquid whey – made for harder cheese. It is easy to imagine an Alpine herder, centuries ago, shuddering in the cold morning air of the fast-approaching winter months and adding some logs to the fire burning under the cauldron filled with milk, then noting at some point that this particular lot of cheese turned out different, harder, that it was maturing more slowly, keeping longer – and then consequently applying the 'new method' in a deliberate way.

What the heat actually achieves is furthering the contraction of the coagulated casein, the curd, and therefore expelling more liquid – that is, whey – making for a larger percentage of dry matter in the finished cheese. As is the case with most natural phenomena which are then standardized, this extra heating is a fluid process and can range from really just heating the milk of the previous night's milking to body temperature, to a moderate 42°C (108°F) for semi-hard cheeses such as Gouda or Appenzell, to 50°C (122°F) or above for most of the wheels produced in the Alps, and to a maximum of 57°C (135°F) for Gruyère, including L'Etivaz.

It is an intervention that, in the first place, only occurred because there was an abundance of wood to feed the fire, because the remote mountain pastures were surrounded by trees. And it persisted because the seasonal transhumance traditions, taking the cows into the high Alpine meadows during the summer months, rarely had only the herders' own winter survival as an objective. In most cases, the goal was the production of cheeses for landlords or markets, and the further away those were, the more the milk had to be concentrated in volume, made transportable and durable. In preindustrial times, a longer shelf life extended a cheese's marketing opportunities – the harder, the more slowly cheeses mature and change, and the more stable and robust they are, not only due to the lower moisture content but also because of the elimination of undesirable microorganisms due to the higher temperature.

L'Etivaz has long been an important export that was traded as far as the Mediterranean since centuries ago – though it has to be said that it is a fine line between maximum expression and best keeping qualities. Historically, hard cheeses 'started' on the southern slopes of the Alps, in northern Italy, and they were an economic necessity – what were considered the best and a real luxury in the Alpine regions were 'vacherin', creamier cheeses made from full-fat, non-skimmed cow's milk. Those much more delicate, fragile cheeses were so highly valued that the Duchess of Savoy offered them to visiting important noble representatives from Milan in 1366, besides gold-embroidered garments.[6]

So far for fire. What about the other elements – water, wind, and air – which in turn influence earth? Studies have shown again and again how the different natural vegetation at altitude makes for fundamental differences in the milk and resulting cheese. Swiss scientists found:

> The cheeses produced at L'Etivaz [at 1300–2100 metres above sea level] were judged by the FAM's [the Federal State Research station in Liebefeld/Switzerland] accredited tasting panel to be more intense in taste and smell than those produced at Posieux [630 metres]. They were more 'pungent' and more 'animal', as well as slightly more 'salty' and more 'acidic'. These differences in taste and smell can be explained by a number of chemical, biochemical, microbiological and technological factors.

The scientists explain:

> The alpine pastures at L'Etivaz and Montbovon are rich in plant species, much more so than at Posieux [in the valley], containing an average of nearly 50 different species belonging to around fifteen botanical families, whereas the lower laying pastures consist of just a few grasses.

It is worth noting that farmers also observe that the higher biodiversity on pastures is linked to a better capacity to store water and survive drought. Analyzing milk and cheese, the study goes on to note:

> The terpenes and some unbranched aliphatic hydrocarbons present in comparable proportions in mature mountain milks, creams and cheeses, but virtually absent in the corresponding lowland dairy products, can be attributed with very high probability to the botanical composition of the fodder consumed by cows in mountain regions. The alcohols and other compounds (particularly unsaturated) found in alpine cheeses can be attributed to the microbial flora of the raw milk used and to the starter cultures used in alpine cheeses on the base of full-fat whey.[7]

Obviously, there are other contributing factors which are in turn linked to specific scenarios shaped by the elements, such as animal breeds. Nothing is legally defined for

L'Etivaz, but the farmers agree that Simmental are the best and most robust 'mountain hikers', and their milk is particularly rich in protein, so it can also accommodate more fat.

Often overlooked is the fact that these Alpine pastures are cultural spaces. They need to be looked after to stay open. 'As soon as you pasture, you cultivate', say Alpine specialists Barbara Sulzer and Giorgio Hösli.[8] Besides the art of grazing as such, there is forest management. It is the cheesemaking families themselves who are responsible for harvesting sufficient quantities of wood to keep the fire under their cauldrons going, part of their winter tasks and the management of the spruce that grows along the hills between the valley and the highest pastures. To cut the trunks into logs of different sizes suitable to fine-tune the fire is an art in itself (and then again to listen to it and 'hear' the temperature during cheesemaking!). The wood is also used for the boards on which the wheels mature in the cave, as well as many traditional tools (which, exceptionally, are explicitly permitted for the making of L'Etivaz). Wood (though not spruce but linden or pear) is also carved into the traditional *cuillères d'alpages*, the round wooden spoons with beautiful handles depicting groundhogs, chamois, or edelweiss. Everyone has their designated one up at the summer dairy, using it for soup, but also for the luxurious crème double ladled over meringues on exceptional occasions.

If all this sounds a tad idyllic, the good people of the Pays d'Enhaut are far from resting on their laurels. They continue to look both forwards and backwards. Many of them realize how unique and fragile the status quo is and how important their cheese, now as then, is as a philosophy of life, this circular economy actually being a positive form of agriculture, not a burden on the landscape. For some years now, there have been (again, quite revolutionary) plans for a centre where the whole world of L'Etivaz can be experienced alongside the open flames – from the wild herbs and flowers to the cows and the people and their skills because it is important to raise awareness regarding biodiversity and CO_2 sequestering, as well as the fact that professional grazing at those altitudes also makes for a higher proportion of omega-3 fatty acids in the milk. Like our ancestors who tamed fire, these mountain dwellers have tamed nature while keeping in mind that it should never be completely mastered.

So many different factors and influences, not easy at all to capture in one paper – and none of them actually stable and fixed, which puts that big question mark on all attempts, however commendable and well meant, to fix tradition. And we haven't even mentioned the elephant in the room, climate change, an upheaval for ecosystems, as each species, in a fragile balance, is adapting as best it can to its environment. When the environment changes and is no longer suitable, the plant moves, dispersing its seeds in the wind. But each plant will adapt differently, and generally speaking, scientific models show a reduction in the diversity of plant species, an increase in the altitude of plants, and the arrival of species that are currently found in Mediterranean countries. What does this mean for cheese?

In contrast to many midland species, Alpine plants are highly specialized.[9] This means: every pasture produces a different cheese. But it also means that as many Alpine pastures as possible need to be preserved through sustainable management – without ever being completely mastered and tamed, keeping the nature in culture.

There are optimistic voices pointing out that climate change or not, cheesemaking is a skill passed down from generation to generation, always adapting, never standardized. Each day anew, you have to use the right tools, take the weather into account, find the right timing, and graze the cows in the right place. Each choice can drastically change the resulting flavour.

Indeed, the flavours of Alpine cheeses are constantly changing, each wheel is slightly different, and the climate is just one influence. Still, research confirms the link between floral diversity and the intensity of the cheese, both in aroma and flavour. Scientists say there's a risk that the cheese of the future will be less intense, less pungent, and less 'animal'.[10] Other factors need to be changed accordingly to stay close to the elements and maintain the umbilical cord between mountains, nature, soil, our humanity – and cheese.

About the Author

Ursula Heinzelmann is a trained chef and sommelier who writes about wine and food, especially cheese. Her most important book is *Beyond Bratwurst: A History of Food in Germany.*

Notes

1 Augustin Denous, *Fromages Sauvages* (Ulmer, 2020), p. 80.

2 Switzerland Cheese Marketing, 'L'Etivaz AOP', Schweizerkaese, 2026 <https://www.schweizerkaese.de/schweizer-kaese/letivaz-aop> [accessed 21 February 2026]; 'Allgemeine Lage', *L'oiseau: Jahresbericht 2023 – Interprofession du Gruyère*, 66 (June 2024), p. 4 <https://www.gruyere.com/fileadmin/documents/gruyere/pdf/L_oiseau/DE/No_66_L_oiseau_d.pdf> [accessed 21 February 2025].

3 Richard Wrangham, *Catching Fire: How Cooking Made Us Human* (Basic Books, 2009), p. 14 (my emphasis).

4 Jean Bottéro, *The Oldest Cuisine in the World: Cooking in Mesopotamia* (University of Chicago Press, 2004), p. 38.

5 Bottéro, pp. 88–89, 101, 115.

6 Arnaud Delerce, *Tout un Fromage: Reblochon, abondance, raclette: la véritable histoiredes fromages des Alpes* (Les Éditions Inverse, 2023), pp. 18, 13, 28–29.

7 J. O. Bosset and others, 'Comparaison de Fromages à Pâte Dure de Type Gruyère Produits en Régions de Montagne et de Plaine', *Revue Suisse d'Agriculture*, 29.1 (1997), pp. 23–34.

8 Barbara Sulzer and Giorgio Hösli, 'Die Alpweide', in *Neues Handbuch Alp* (Mollis, 2007).

9 Manuel Schneider, 'Schweizer Alpweiden unter der Lupe', *Schweizerische Eidgenossenschaft*, 2023 <https://www.agroscope.admin.ch/agroscope/de/home/themen/pflanzenbau/futterbau-grasland/grenzertragslagen-alpwirtschaft/anzahl-tiere-alp.html> [accessed 21 February 2025].

10 Marion de Vevey, 'Le goût du fromage est-il menacé par le réchauffement climatique?', *L'Uniscope*, 5 May 2023 <https://wp.unil.ch/uniscope/le-gout-du-fromage-est-il-menace-par-le-climat/> [accessed 21 February 2025].

15
Salt Legacies on the Costa Brava
The Brine Ladies of L'Escala

Rebecca Ingram and H. Rosi Song

The Costa Brava and Catalonia

The Catalan coast, known as the Costa Brava, achieved local and international fame in the late twentieth century as a gastronomic mecca for food lovers thanks to the reputation of, among others, Ferran Adrià and his now closed elBulli, now transformed into a food museum reflecting on the knowledge, innovation, and history of the iconic restaurant. Even before elBulli, the Costa Brava held special significance in Catalan culture, not only because of its rugged and beautiful landscape, but also because it was a potential site for tourism to help restart the local economy after the waning of prosperity brought by trade with the Americas and the return of wealthy *indianos*. In the nineteenth century, the coast was poor, then hurt by phylloxera and the dying fishing industry, which was slowly emptying the beautiful historic villages that dotted the Catalan coast. Its name was a modern invention, lifted from a poet who used the phrase to describe the geography of Majorca. When the poet and cookbook author Ferran Agulló used the name Costa Brava in his newspaper column, it served as a 'call to tourism'.[1]

Since then, its local and international fame has only grown, but it has also been firmly entrenched in the Catalan imagination as part of its national landscape. Joan Nogué and Joan Vicente have discussed how landscapes became 'a fundamental element in the creation process of a national identity' in Catalonia in the late nineteenth century.[2] While the initial focus was on Catalonia's mountains, that geographical concept evolves into a cultural landscape, 'a social product, the cultural projection of a society on a specific space'.[3] The significance of geographical territories rises from the assumption that people who live in the same space must share commonalities; as their landscapes are imbued with values, they intervene in shaping existing imaginaries.[4] This process of consolidating this imaginary and identity involved excursions where Catalans were encouraged to travel to see their own territory. As Venetia Johannes writes, *excursionisme* was a way of getting to know Catalonia 'through travel, physically experiencing the nation by visiting its different regions'.[5] Even today, as she argues, 'Catalans use specific food maps to travel and eat on their way through Catalonia'.[6] The connection between landscape and food brings us to concepts associated with tradition, heritage, and authenticity.

Efforts to safeguard food-related practices can be linked to our current understanding of concepts like 'heritage' or 'authenticity', which, as Aurélie Gfeller explains, are cultural and social practices 'through which objects, places, or practices rooted in the past are endowed with meaning'.[7] From this perspective, activities connected with food practices and ways of life acquire importance and purpose: they become vehicles to connect with the past. From this perspective, the festivities around salt culture that have been recovered and put into practice in L'Escala, a small coastal town in the Costa Brava, present us with an opportunity to better examine this past and the ways in which memories can be made meaningful in the present. More importantly, as we argue here, the salt culture of L'Escala also offers an opportunity to recover and make visible female labour, as it still endures today in the form of a new generation of workers that also reflects the changing demographics of the region.

Salt Culture

Salt is a basic element for human subsistence. Enabling food preservation and enhancing taste, its use has been documented across cultures throughout history. Salt's chemical term, sodium chloride, NaCl, identifies the reaction of an acid with a base. In this case, it is the contact of chlorine (an acid) with sodium (a base) which takes atoms from the latter to create the crystalized body we recognize as salt. Chlorine is essential for digestion and breathing, and without sodium, which the body cannot manufacture on its own, the transportation of nutrients or oxygen or the transmission of nerve impulses to move muscles, like the heart, would not be possible. Salt is also a basic element of cooking. While salt is one of the most abundant minerals in nature, knowledge and ease of its extraction were not the same for all regions, which propelled its trade. In addition to salt extracted from the sea, the evaporated water also created mineral salts in successive layers covered with sediments and formed the fossil marine salt that we know with the generic name of gem salt or stone salt (halite in mineralogy). Catalonia boasts an excellent example of gem salt in the diapir formed by the Cardona Salt Mountain. Its location in the Pyrenees Mountains was already known by the Romans, who considered it the best salt quarry in Europe.[8]

In the Mediterranean area, salt pits were valued, and they were a significant factor in the development of commercial infrastructure. The *viae salariae*, or Roman salt roads, were used into the eighteenth century, until they were replaced with modern roads and then railways in the nineteenth century. In the medieval economy, salt was a valued commodity, as it was used for salting fish. Along the Catalan coast, from south to north, there were three groups of salt pits, gem salt deposits, and salt springs: the first at the mouth of the Ebro River near Tortosa, Amposta, and Sant Carles de la Ràpita; the second around Tamarit, near Tarragona; and the third around the mouth of the Ter River, near Torroella de Montgrí. Along those routes, the existence of salt warehouses or storage areas has been confirmed – for example, through documents from the Council of Barcelona. In addition to these material records, the importance of salt in everyday life shows in the traces left in the Catalan language. A few notable instances of such words are *salabror*, a combination of the words for salt and flavour (*sabor*) that means saltiness, or the noun *salaó*, which indicates the act of salting a food item, especially meat or fish, or pickling it with salt. The *salí*, at L'Escala, indicates the salt place, the former name of the anchovy factory or deposits of salt. Finally, the *salpás* expressed the Christian custom of blessing homes with water and salt during the Easter season.

At L'Escala, salting anchovies and sardines has been documented since premodern times. The first salting plants and factories were set up around the beach and the port of En Perris. Narratives – and photographic evidence – from the turn of the twentieth century describe the beach brimming with activity when the fishing boats arrived. Fishermen and women remember how, in the mid-twentieth century, salters would buy the anchovies and sardines they needed, and fishermen would transport them to factories, where activity would continue throughout the night so that shipments would be ready by dawn. According to industrial licenses registered in 1847, L'Escala had ten plants engaged in this activity under the name of Fomentadores de Pesca y Salazón. Anchovies were inserted into glass pots or wooden barrels for exporting, and olives were stuffed with anchovies in cans. Sardines were pickled, salted, and pressed or left in brine to be packed in wooden boxes and preserved for two weeks. Salters from this town went on to found salting plants in the Basque Country and Galicia. They also established shops to sell these preserved fish in Llancà, Roses, Palamòs, and Sant Feliu de Guixols. These products were also exported to Italy, especially to Genoa.

In terms of culinary tradition, the process of salt packing and preserving anchovies, as practiced in L'Escala factories, is collected in recipes in the *Corpus del Patrimoni Culinari Català* (2016) as 'Anxoves confitades'. It recommends removing the head and innards of the fish, then putting it in a terracotta basin (the *gibrell*) and covering it with salt. The following day, anchovies should be packed in glass jars, alternating layers of fish and salt. When they are all positioned, they are covered with salt, and the liquid remaining in the *gibrell* is poured into the jar. Finally, after weighting the lid for twenty-four hours, more anchovies are added. The fully confited anchovies would be ready to be enjoyed after six or eight months.[9] The entry for 'Anxoves' in Ferran Agulló's foundational cookbook *Llibre de la cuina catalana* (1928) follows these same steps. In Catalan cooking, anchovies go by the name of *seitons* when cooked – *seitó escabetxat, seitó fregit, seitons amb ceba, seitons salats*. The *antxovada* or *Salsa d'anxoves* appears in Marta Salvia's *Art de Ben Menjar*, the first cookbook in Catalan written by women and published in 1923. Anchovies are also a key ingredient in the *xató*, the popular Catalan salad comprised of escarole, tuna, cod, and anchovies, which has its own gastronomic route along the coastal cities of Calafell, Cunit, El Vendrell, Sitges, Vilanova, and la Geltrú i Vilafranca del Penedès.[10]

Today, anchovy salting continues in L'Escala in eight factories, located outside the city centre, which continue to provide jobs to more than two hundred people, most of whom, it seems important to notice, are women. In addition to the products prepared and sold traditionally, such as salted anchovies and sardines processed in brine (called today 'sardines of the coast'), new products have been added which reflect modern tastes: anchovy fillets preserved in olive oil, tuna and bonito in oil (and also pickled in oil, vinegar, and bay leaves), pâté, olive and cream of anchovies, anchovies in vinegar (the *boqueró*), and also olives stuffed with anchovies.

The salting tradition of L'Escala is, at present, globally recognized with exports to countries like Japan. In 1987, a Quality Designation of Anchovy of L'Escala was established,

Figure 1. 'Group of women preparing boxes of sardines in brine, Solés salting factory, in La Platja, 1950s' ('Grupo de mujeres preparando cajitas de sardinas en las pasadas por el caldo, fábrica de salazones Solés, en La Platja, años 50'). Photo provided by Salaons Solés, Museo de la Anchoa y de la Sal de l'Escala (MASLE).

and while these anchovies have not yet received a Protected Geographical Indication (PGI) from the European Union, producers work with Spain's promotion programme, Foods and Wines from Spain, run by ICEX Spain Trade and Investment.[11] The salters' trade union of L'Escala began celebrating the Anchovy Festival in 1989, where people gather to taste bread with tomato (*pà amb tomàquet*) and anchovies, sing Habanera songs, eat Catalan fish soup, the *suquet*, give an annual Anchovy Speech, and award the Golden Anchovy Medallion to a person with a distinguished trajectory promoting the anchovy from L'Escala.

Women's Work

Salt is the key element in reclaiming the fishing origins of L'Escala. It is part of oral histories, songs and dances, a cuisine built upon a foundation of salt-preserved fish, and efforts to protect as heritage the markets as well as the practices of fishing and salt curing. Curiously, what have not been commemorated or designated in any particular way are the roles of women in this salt-based industry and community identity, even as their labour has long been central to the traditional processes of salt curing anchovies.

Figures of women often show up on the labels of well-known brands of *fleur de sel* – a salt collected by hand from the crusts that remain when seawater evaporates – like Le Paluder de Guérande or the Salinere de Camargue. These images create an association between women and salt in the popular imaginary, which underscores their labour as a foundational element of food economies. In L'Escala, for example, breathing the air perfumed by salt-encrusted rocks was thought to ameliorate lung ailments. There exists an entire mythology in the region about the power of salt to drive away malevolent spirits; for example, women dabbed babies' lips with salt to protect them from danger.

Yet the involvement of women in fishing industries and in the culture of fishing in specific communities has long been overlooked and radically under-researched. Men dominate the roles that involve going out on boats, while women comprise the highest number of workers in fisheries services and in 'post harvest sectors'.[12] The fish processing industry in Spain is dominated by women, who make up 80 percent of workers in this sector.[13] These patterns have existed since preindustrial times, when women took care of the cleaning and processing of fish in *bodegas* while men were out on the water.[14] Despite these figures, women's work and their activities in coastal fishing communities have mostly been understood as that of a 'support' for male family members engaged in extractive fishing.[15] Even statistics gathered by the Spanish state render invisible the labour of women along the Mediterranean coastline by naming it as the work of 'collaborating spouses'.[16] Luisa Muñoz Abeledo discusses the challenges of writing a history of women's labour in the fishing industry: evidence has to be reconstructed from a multitude of sources – balancing state demographic information, where the underrepresentation of women's activity is endemic, with data from specific canning companies and oral histories from former workers – in order to correct the biases present in official counts. In studies carried out by Paloma Herrera-Racionero and her collaborators, women themselves have long minimized the importance of their work, tending to frame their labour in the industry as more compatible with the domestic sphere.[17]

Nonetheless, representations of women anchor fishing heritage projects meant to preserve folk histories. They are also used to attract tourists to areas where the fishing industry can no longer sustain local economies.[18] In Catalonia specifically, the dominant representation is that of a passive *mujer que espera* (a woman awaiting the return of her fishing family member).[19] She is the wife or sister of the fisherman, sometimes surrounded by children, and often is featured gazing out to the sea. This woman shows up in Salvador Dalí's 1925 painting of his sister (*Figura en una finestra*), in literary representations from early twentieth-century Catalan costumbrista writers, and even in Josep Pla's 1947 *Cadaqués*. We also see this representation in the *Monumento a la mujer del pescador* ('Monument to the fisherman's wife') in L'Escala. Sponsored by the Cofradía de Pescadores, Francesc Xavier Vila's sculpture was inaugurated in 1991.[20] Less frequently, mostly only in photographs, do we see representations of women as active participants in the fishing cultures and work of towns like L'Escala.[21] This remains the case even as women continue to labour by hand in the traditional way of salt curing anchovies.

If the actual work of women as canners is an essential element to this salt-cured fish tradition, why do these passive representations predominate? Living women and their work in packing jars by hand, fillet-by-fillet, are featured as part of the Anxoves Solés museum and tour exhibition. But the women themselves are often not celebrated or given primary roles in how this salt-cured fish has become a heritage project for L'Escala.[22] This is also the case elsewhere. Jane Nadel-Klein's ethnographic and historical study of Scottish fishing communities reveals that representations of women help communities 'keep hold' of a particular way of life by using women's images and stories as 'heritage signposts'.[23] These representations are a symbolic resource that allows for the reconstruction of community boundaries and negotiations of collective self-worth that assuage anxieties about the disappearance of past ways of life and about the future of those same communities. In this case, women's roles in the heritage signpost of L'Escala understand their work as rooted in nineteenth-century domestic ideology: their labour must be hidden or go unacknowledged.[24]

Figure 2. 'Maria Frou Junqué putting salt in a jar of anchovies at the Callol Serrats salting factory in Port d'en Perris in the 1940s' ('Maria Frou Junqué posant sal en un pot d'anxoves a la saladora Callol Serrats de Port d'en Perris als anys quaranta'). Photo provided by Martí Callol Pascual, Museo de la Anchoa y de la Sal de l'Escala (MASLE).

La festa de la sal

In L'Escala, the salt warehouse, the Alfolí, a building from 1697, is still standing. When it was built, it operated to control the salt trade and served as a maritime customs house for the goods arriving from the coastal trade (salt and cereals) and products that were exported (salted fish and wine). In 1975, it was in ruins and fated to be demolished, but in 1984, the Generalitat (the Catalan Autonomous Government) declared the Alfolí and the town's cemetery as national historical and/or art monuments and forced its restoration. In January 2006, members from the Centre for L'Escala Studies, L'Escala Fòrum, the former Commission for the Defence of the People's Architecture, working together for the Alfolí platform, agreed to purchase the building and decide what its future would be. Today, it functions as a museum and an event place, holding several archives and organizing cultural exhibits in its different spaces.

Meanwhile, historical festivals were being organized by the Municipal Archives in collaboration with the Centre for L'Escala Studies to commemorate a variety of anniversaries, such as the centenary of the first electrical lighting installation or the two-hundredth anniversary of the construction of the wash basin and fountain, an exemplar of stone hydraulic engineering. The Light Festival and the Water Festival were organized, as well as the Salt Festival and others. What distinguished these festivals was their performance aspect: the entirety of L'Escala would participate in re-creating the historical setting of the moment being commemorated. A form of *tableaux vivant*, the festivals re-created a journey through the past, with locals dressing in historic garments, performing roles, and re-creating scenes older villagers recounted from their memories. In 1997, *The Salt of Life* exhibition was organized at the Alfolí in conjunction with the celebration of the first festival of salt, which was visited by more than twelve thousand people in three weeks. The villagers collectively decided that *La festa de la sal* should be an annual occurrence because it most closely re-created the fishing and salting origins of the town.[25] Since then, every September, the villagers gather to enact the festival, which began almost two decades ago with the following 'plot':

> The fest began with the reading of an article based on the remembrances of a fisherman published in the magazine *Fulls d'Història Local*. As the participants recall their memories, they stage the letters that are read as if they were coming to life. The children would become trainees in the trade that they liked the most, taught by an old proficient person or they simply played. In mid-afternoon the sound of mariners' horn would announce the arrival of the salt ship. Immediately thereafter, the small vessels would dock to load the salt on board and then unload it on the beach. At night, beacons were lit and boats with illuminated sails arrived. Spontaneously, everyone began to sing the famous popular song *Quan jo en tenia quatre anys* (*When I was four years old*) and since then it has happened every year. Afterwards, people performed traditional dances of Catalonia and L'Escala and then celebrated in a dance for all led by a group of traditional musicians.[26]

The festival has served both as an official act of remembrance about the fishing and salting origins of the town and as an homage to the ancestors, relying on oral testimonies and the memories of its oldest villagers, especially the fishermen who still remember when their whole world was associated with fishing in sardine boats using the lateen sail. These recollections have also served to recover and preserve the language connected to this trade, a fishing vocabulary that not only named objects but also included phrases and verbs such as the *barcada* (a sardine and anchovy fishing system in which the fish goes after the net and is trapped in the mesh), the *palanqui* rope (a cable used to raise and lower the vessel using pulleys or tackle blocks), *prima* (casting done just at dusk), and the *qüern* (a unit of measurement applied to sardines and anchovies). One *qüern* equals five sardines, and normally a person would grab in one motion three fish in one hand and two in the other, while one hundred *qüern* would make up the five hundred sardines that would fit inside a vat.

The festival has been evolving since its first iteration in 1997. Starting in 2004, guests were invited to bring salt from their own countries and add it to a large terracotta bowl mounted on a stand called the *Sal de les Cultures* as a symbol of friendship. The large salt stand is normally on display in the Anchovy and Salt Museum alongside old tools of the trade. Currently, the festival takes place over three days in late September. A delegation from the guest country makes the salt offering for a ceremony accompanied by a musical recital. A public presentation narrates the story of L'Escala and welcomes visitors, identifying emblematic locations and participants, such as the port, the taverns, and the dry fruit vendors. The beach and L'Escala's promenade feature a ceremonial departure of sardine boats, children's games, and songs, both traditional ones and others composed specifically for the festival, such as 'Cançó de la sal' ('Song of the Salt'), commissioned in 1997. The historic occupations of a fishing village are on display while awaiting the arrival of the salt vessels: hammer masters, barrel men, cask builders, basket weavers, and coral workers demonstrate their work alongside needlepoint workers and seamstresses. Trawl line fishers cast their round nets (*tirar el rall*) and pull those caught from the net for auction, or the 'singing' of the fish. Fishmongers work alongside the anchovy salters and packers. Attendees and local participants observe how sardines are processed in brine and then see the final steps of patching and repairing nets, spreading them, and folding them. During their visit, they also enjoy the Catalan fish stew, the *suquet*, and grilled sardines, the *sardinada*, during lunchtime. The purpose of the festival is to re-create and repeat past traditions so that younger generations understand, reenact, and ultimately make them their own in an adaptation for the present.

Heritage and Community

In the advertisement for the festival, villagers describe the *Festa* as a journey of rediscovery by the residents of L'Escala who crave community in an increasingly fragmented world. Josep Tero, a resident since 1951, shares his opinion that 'we need this festival to better know where we come from and who we are, and not to forget the first rule of life, the one that those around us followed, the artisans we have lost from so many fields'.[27] In this sense, the L'Escala community is developing its heritage scripts and playbook from recognizable models being deployed elsewhere, as fishing economies with their specific architecture, gear, and practices acquire new importance as cultural and economic resources connected to tourist economies. The focus on the element of salt is one 'differentiating nuance' that distinguishes L'Escala from elsewhere.[28] Just as essential as salt, however, is the labour of women.

Their efforts helped sustain the L'Escala economy throughout the twentieth century as fishing ceded ground to an economy based on tourism. Women's remunerated work was part of this transformation, alongside the cooking labour that has kept emblematic dishes and ingredients as a recognizable and celebrated part of gastronomic culture in L'Escala and in Catalonia more broadly. Anchovies, sardines, and their preservation in salt are essential to the gastronomic legacy in L'Escala, as demonstrated by the texts analyzed here. But how does women's paid labour in L'Escala's factories cohere with the heritage scripts being created and performed there? In 2024, the Museu de L'Escala announced a research project to disseminate the photographic evidence of women's historical roles. Connecting the exhibition to the celebration of International Women's Day, the project seeks to recover and share the 'ancestral wisdom [. . .] passed down from mothers to daughters in

Figure 3. 'Inside the Xillu salting factory on Picasso Street, 2010' ('Interior de la fábrica de salazones Xillu, en la calle Picasso, año 2010'). Photo by Miquel Bataller, Museo de la Anchoa y de la Sal de l'Escala (MASLE).

L'Escala, the town with the longest uninterrupted tradition of salting fish in Catalonia and the western Mediterranean'.[29] Titled *Salaó és nom de dona*, the exhibition is a tribute to the women workers in the salt-curing factories of the community from 'yesterday, today and always'. Featuring forty-seven photographs from the early twentieth century through more contemporary times, the exhibit shows women and their labour treated as protagonists for the very first time in the fishing heritage of the community.[30]

La festa de la sal has been established over the past twenty-five years, but only in early 2025 did women come into focus via this exhibition. The featured images make explicit the association between women and salt in the popular imaginary that underscores their labour as a foundational element of this food economy. Moreover, it opens the door to new kinds of consideration for the cultural meanings of salt brining and women's work.

About the Authors

Rebecca Ingram, PhD, is the Chair of Classical & Modern Languages & Literatures and a Professor of Spanish at Texas Tech University. A specialist in feminist food studies as they relate to Spain, she is the author of *Women's Work: How Culinary Cultures Shaped Modern Spain* (2022) and winner of the Gourmand Cookbook Award for Spain in the category Best Women of the World.

H. Rosi Song is a Professor of Hispanic studies at Durham University (UK) specializing in twentieth- and twenty-first-century Spanish culture and literature. She is the author of *Lost in Transition: Constructing Memory in Contemporary Spain* (Liverpool University

Press, 2016) and coauthor, with Anna Riera, of *A Taste of Barcelona: The History of Catalan Cooking and Eating* (Rowman and Littlefield, 2019), which received the North American Catalan Society Award for Outstanding Work in the Field of Catalan Studies in 2021.

Acknowledgements
The authors would like to thank Jaume Badias Mata and the Museo de la Anchoa y de la Sal de l'Escala for their generous collaboration in researching this project and for sharing photographs included in their exhibition *Salaó és nom de dona* (2025).

Notes

1 Vicenç Pagès Jordà, 'Ferran Agulló, Inventor de La Costa Brava', *Revista de Girona*, 287 (2015), pp. 66–68 (p. 67).

2 Joan Nogué and Joan Vicente, 'Landscape and National Identity in Catalonia', *Political Geography*, 23 (2018), pp. 113–32 (p. 113).

3 Nogué and Vicente, p. 116.

4 Nogué and Vicente, p. 116; Joan Nogué and Stephanie Wilband, 'Landscape Identities in Catalonia', *Landscape Research*, 43.3 (2017), pp. 443–54 (p. 443).

5 Venetia Johannes, 'Catalan Culinary Nationalism: A Contemporary Case Study', in *The Emergence of National Food: The Dynamics of Food and Nationalism*, ed. by Atsuko Ichijo (Bloomsbury Publishing, 2019), pp. 85–95 (p. 92).

6 Johannes, p. 93.

7 Aurélie Élisa Gfeller, 'The Authenticity of Heritage: Global Norm-Making at the Crossroads of Cultures', *American Historical Review*, 122.3 (2017), pp. 758–91 (p. 762).

8 See Yves Malartic's study on marine salt pits, gem salt deposits, and salt springs of Catalonia and Aragon between the thirteenth and fifteenth centuries: 'Le Sel En Catalogne (XIIIe–XVe Siècles)', in *Les Pays de La Méditerranée Occidentale* (C.T.H.S., 1983), pp. 181–200.

9 Institut Català de la Cuina, *Corpus del Patrimoni Culinari Català* (RBA La Magrana, 2016), pp. 57–58.

10 'The Route', Ruta Del Xató <https://rutadelxato.com/en/the-route/> [accessed 30 October 2025].

11 ICEX's 'role is to promote exports of Spanish food products to Spain's priority markets, introducing new products and developing the market for more established ones. ICEX takes a strategic approach to the task, from working with Spanish producers, exporters, exporters associations and foreign importers and distributors' (ICEX, 'About Us', FWS: Foods and Wines from Spain, 2021 <https://www.foodswinesfromspain.com/en/footer/about-us> [accessed 30 October 2025]).

12 Marilyn Porter, 'Why the Coast Matters for Women: A Feminist Approach to Research on Fishing Communities', *Asian Fisheries Science*, 25S (2012), pp. 59–73 (p. 59).

13 Paloma Herrera-Racionero and others, '"The Sea Is Our Life": Woman in the Fishery Sector of the Valencian Community', *Sociologia Ruralis*, 61.1 (2021), pp. 278–300.

14 Luisa Muñoz Abeledo, 'Actividad femenina en industrias pesqueras de España y Portugal (1870–1930)', *Historia Contemporánea*, 44 (2012), pp. 49–71 (p. 59).

15 Herrera-Racionero and others, '"The Sea Is Our Life"', p. 279.

16 Herrera-Racionero and others, pp. 288, 279.

17 Herrera-Racionero and others, p. 295.

18 Jane Nadel-Klein, 'Granny Baited the Lines: Perpetual Crisis and the Changing Role of Women in Scottish Fishing Communities', *Women's Studies International Forum*, 23.3 (2000), pp. 363–72.

19 Eliseu Carbonell, 'La Mujer y El Mar En Cataluña: Realidad y Representación', *Revue d'ethnoécologie*, 13 (2018), pp. 1–21.

20 Carbonell, p. 5.

21 Carbonell's study briefly analyzes photographs of women repairing nets and those who worked on beaches accompanied by their children in contrast to the predominant waiting woman archetype.

22 One woman who contradicts this pattern is Víctor Català, the pen name of Catarina Albert i Paradis, described as the *l'escalenca més universal* (most universal woman from L'Escala) and a foundational figure in twentieth-century Catalan literature ('Victoria Català', Museu de l'Escala <https://museudelescala.cat/victor-catala/> [accessed 30 October 2025]).

23 Nadel-Klein, 'Granny Baited the Lines', p. 9.

24 Raúl Gil and Mada Martínez's *Sobadoras de anchoa: Historias de mujeres de Santoña* (Libros, 2016) is one example of a Cantabrian fishing community's attempt to make visible this important facet of women's labour as part of the community's heritage. See Marcela Garcés, 'In Defense of the Anchovy: Creating New Culinary Memories through Applied Cultural Context', in *Food and Memory: Traces, Trauma and Tradition* (Dublin Gastronomy Symposium, 2024) for an analysis of how their labour and protagonism becomes part of the marketing strategy for tinned fish as an artisanal food (p. 8). Muñoz Abeledo identifies a high point in women's canning/fishery-related labour not being counted as remunerated work in the 1920s and 1930s ('Actividad femenina', p. 52). See also Enriqueta Camps's *La formación del mercado de trabajo industrial en la Cataluña del siglo XIX* (Ministerio de Trabajo y Seguridad Social, 1995) on the existence of fake *amas de casa* (housewives) in the study of women's work in the Catalan textile industry. Ágata Albero's recent publication *Les dones i la cuina marinera* (with Sònia Estévez, Genéric, 2025) also makes women's foodwork visible, along with the workshops she has facilitated that bring racialized and migrant women along the Catalonian coast into broader discussions about contemporary Catalan foodways.

25 For more information about the festival, see Antonio Ortí, 'La fiesta marinera que rinde homenaje a la sal en la Costa Brava', *La Vanguardia*, 17 September 2021 <https://www.lavanguardia.com/magazine/buena-vida/20210917/7709830/sal-viste-fiesta-l-escala.html> [accessed 12 April 2025].

26 Lurdes Boix i Llonch, *L'Alfolí i la Festa de la Sal a l'Escala* (Ajuntament de l'Escala, 2007), p. 196.

27 Boix i Llonch, p. 206.

28 Celeste Jiménez de Madariaga and Juan José García del Hoyo, 'Enhancing of the Cultural Fishing Heritage and the Development of Tourism: A Case Study in Isla Cristina (Spain)', *Ocean and Coastal Management*, 168 (2019), pp. 1–11 (p. 1).

29 'Exposición "Salaó es nombre de mujer" en el marco del 8 de marzo, Día de la Mujer Trabajadora – Museo de l'Escala', Museu de l'Escala, 28 February 2024 <https://www.museudelescala.cat/es/exposicio-salao-es-nom-de-dona-en-el-marc-del-dia-de-la-dona-treballadora/> [accessed 30 October 2025].

30 'Salaó és nom de dona', *Museu de l'Escala* <https://museudelescala.cat/recursos-digitals/exposicions-virtuals/salao-es-nom-de-dona/> [accessed 30 October 2025].

16
Calcium, Cornerstone of Mesoamerican Civilizations

The Importance of *Nixtamalization* in the Creation and Maintenance of Traditional Mexican Food Systems

Diego Emiliano Jaramillo Navarro

At first glance, an alkaline metal such as calcium might seem odd within a list of fundamental ingredients for the perfect corn tortilla. However, its discovery and use as a hydroxide at least three millennia ago determined the way in which Mesoamerican diets and food systems would be articulated, thanks to the food technique known as *nixtamalization*.

From the Nahuatl words *tenextli/nextli* ('lime ashes') and *tamalli* ('wrapped'), *nixtamalization* refers to the cooking of fresh or stored maize grains in an alkaline solution, generating a surge of their nutritional yield, which proved critical for the sustenance of large populations. It is no coincidence that the earliest remnants of nixtamalization are contemporary to the first major settlements in the region; the process was consistently present in the following centuries in depictions, archaeological vestiges, and even tribute accounts from vassal states. The story of the development and transformation of the most important food systems in Mesoamerican societies – throughout the pre-Columbian, colonial, and modern eras – is closely related to the use of calcium hydroxide through nixtamalization techniques.

In the following sections, this analysis will be developed. First, to establish the cultural context in which calcium became a relevant mineral, its use and significance, as well as the production of its oxide and hydroxide, are described. Secondly, we analyze the development, richness, and complexity of nixtamalization as a technical feat. The third section discusses the deep implications of calcium usage to enhance the bioavailability of nutrients contained in corn, affirming that nixtamalization could be considered one of the conditions *sine qua non* for the emergence of complex civilizations in Mesoamerica. The final section calls attention to the most pressing matters regarding nixtamalization in our times: gender roles, trade wars, and privately owned genetically modified seeds within a context of collectively maintained food heritages.

Calcium in the Pre-Hispanic World

Before delving into the relevance of nixtamalization as a food technique, we must consider the cultural and historical context in which the processes took place that gave rise to the specific form in which calcium has been most used for at least the last three millennia: calcium hydroxide, better known as lime. Beyond the food field, lime was one of the main traditional construction materials in the Mesoamerican world.[1] This alone is quite remarkable given the technological complexity required to obtain it from limestone ($CaCO_3$), one of the main forms in which calcium is naturally found in the Mayan area and in some deposits in the Mexican central highlands.[2] Broadly speaking, limestone must be subjected to temperatures of at least 900°C (1650°F) for it to release carbon dioxide (CO_2), which produces calcium oxide (CaO), also called quicklime, a hazardous substance due to its high exothermal reactivity, which can result in chemical burns. When mixed with water, calcium hydroxide ($Ca(OH)_2$) is obtained, which is heavier but far safer to handle.

In the pre-Hispanic cultural context, its elaboration also occurred within a symbolic dimension that gave meaning and related the process to the environment in which it was carried out. The amount of labour, resources, and technical knowledge required to produce it in such quantities shows that the lime industry played a prominent economic and social role in several Mesoamerican societies, particularly the Maya.[3] The symbolic charge of lime related to purity, transformation, and culmination of situations in the past may have motivated its extensive use in architectural structures linked to power.[4]

The technology to produce it was first developed independently in the Maya lowlands around 1200 BCE, the oldest records found so far.[5] Among its main applications, lime was used to create more hygienic flattening, providing adequate sanitary conditions to prevent contagions and allow increasing habitational density in the new urban areas, with Teotihuacán around 300 CE one of the most outstanding examples.[6] Similarly, it was widely used for food preservation as an insect repellent in grain storage tanks and also as a coating for the *chultun*, the water reservoirs characteristic of the Maya area.[7]

Despite the processes of colonization and cultural genocide in recent centuries, several of these elements have survived to the present day through oral heritage and practice in some communities of the Maya lowlands, where traditional lime production is still carried out.[8] Some technical elements, such as the circular pyre of freshly cut wood, continue to be used; however, due to its high fuel costs, it was gradually replaced during the colonial era and the nineteenth century by industrial production. Moreover, the intensive use of wood as fuel and its consequent environmental degradation is considered as one possible reason behind the collapse of several Maya city-states in 700–900 CE.[9]

As for the symbolic dimension, in the traditional Maya process of lime production, female and male roles are strictly mediated, and the burning of lime is envisioned as a birth. Metaphors of birth and fertility are present in all aspects of lime creation, with the furnace or circular pyre being a womb;[10] the resulting calcium oxide is known as *Sak Ch'upal* (young white/pure woman).[11] The types of offerings and remedies placed in or on the furnace and the burning of the kiln vary from one community to another and even among individuals in a given place. Even so, they bring together several recurring themes.[12]

Corn + Calcium: A Stroke of Mesoamerican Genius

Although one of the most evident ways in which the presence and relevance of calcium in the pre-Hispanic world can be physically appreciated is through the remnants of stucco in the ruins of various architectural complexes, its quasi-omnipresence in food was not limited to the pre-Hispanic era; nowadays, it continues to materialize in a variety of presentations thanks to nixtamalization. The virtuous association of calcium with corn is no less than a stroke of genius whose ramifications have been far-reaching.

Essentially, nixtamalization is a food technique of alkaline cooking. Although lime is currently the most commonly used form, in the past, calcium could also come from wood ash, *tequesquite* rocks, or even specific types of crushed shells.[13] There is a wide diversity in terms of cooking times, repetition, and degrees of concentration of the alkaline solution, depending on the variety of corn, whether it is fresh or dry grain, the presentation of the calcium hydroxide to be used, and the purpose/dish for which the resulting dough, also called *nixtamal*, is required. For the sake of simplicity, a standard procedure could be to boil for thirty to ninety minutes a 1 percent solution: ten grams (approximately two teaspoons) of lime powder per kilo of corn. Generally, a higher lime concentration achieves faster cooking but affects the dough's flavour. However, this is also a strategy when using old grain: in this scenario, the cooking time can also be increased. In some communities, white corn varieties are preferred over yellow ones because the yellowish tones of the dough may indicate an excess concentration of the alkaline solution, resulting in a texture that is too sticky. In any case, those who nixtamalize are always looking for specific characteristics in terms of density, colour, plasticity, and flavour; in other words, nixtamalization has not only practical and technical but also aesthetic criteria.

The value of nixtamalization lies in the apparent and underlying impacts it has on corn, which dramatically increases the nutritional contribution of this cereal.[14] On the one hand, alkaline cooking removes the cuticle and parts of the grain pericarp, which both gives plasticity to the dough for preparations such as tortillas and increases its digestibility. On the other hand, at a nutritional level, the removal of the cuticle and pericarp implies the removal of insoluble dietary fibre. At the same time, soluble dietary fibre increases due to starch swelling and increases the calcium content up to three times the original level.[15] Plasticity is due to starch gelatinization, specifically between amylose and amylopectin molecules, which prevents the formation of hydrogen bonds and thus starch retrogradation.

The increase in bioavailable calcium through nixtamalization is valuable given the role of this mineral in bone formation and strengthening, transmission of nerve impulses, regulation of muscle contractions, and blood coagulation, among others. However, the main contribution of nixtamalization is related to the release of niacin (vitamin B3) in the form of nicotinic acid, which is digestible by the human body and prevents diseases such as pellagra, caused by niacin and tryptophan deficiency.[16] In terms of increasing other amino acids, lysine increases 2.8 times, isoleucine to leucine ratios increase 1.8 times, and microbial activity is inhibited with the partial destruction of aflatoxins – linked to liver cancer – in corn contaminated by *Aspergillus flavus*.[17]

As a food technique, the nutritional value of nixtamalization can be reinforced when it is articulated with other elements of Mesoamerican food systems. In particular, combining nixtamal with another niacin-rich element such as beans increases the amount of bioavailable niacin. The relevance of the increased nutritional value provided by this food technique is evident when considering that corn represents approximately 80 percent of the caloric intake in Mexico (45 percent in urban areas), 35 percent of the protein intake, and 50 percent of the calcium intake (up to 80 percent in rural areas).[18]

The development and use of nixtamalization depend on specific culturally contingent knowledge. Even ignoring the nutritional richness it unlocks, at first glance, the objective of removing the cuticle of the pericarp of the kernels and generating a dough with greater plasticity is in a dialectical relationship with the possible types of food preparations. Although corn can be consumed without previous preparation as fresh boiled corncobs, as crushed dry grain flour, or as roasted grain, its nixtamalized forms of consumption range from the solid, such as tortillas and tamales, to the semi-solid, such as pozole or soups with *chochoyote*, to suspensions, such as *atole*, *pozol*, *tejate*, and *tascalate*, to the fermented, such as *tesgüino* and *tepache*.

In the same way that the depositaries and reproducers of the knowledge to elaborate these dishes have been women, nixtamalization as a technique is also a female creation, above all because of the gendered division of labour.[19] As in other regions, in Mesoamerica, this has been a long-standing division. Regarding nixtamalization, considering the causal technical relationship of this process with the subsequent grinding of the cooked grain with instruments such as the *metate* and its elaboration into tortillas by means of the *comal*, from the perspective of the nixtamal/*comal*/tortilla complex, the history of nixtamalization is the history of nixtamal grinding.[20]

However, nixtamalization is also one of the most evident materializations of women's subjugation to domestic work: prior to the invention of nixtamal mills at the end of the nineteenth century, manual grinding with a *metate* meant five to six hours of work per day. The domestic unit was the place where it was usually carried out,[21] thus constituting one of the main factors that practically reinforced the sexual division and social stratification during the emergence of more complex sedentary societies.[22] This phenomenon presents a robust temporal continuity: even in previous instances, such as the production of lime, a sharp sexual division of labour is still observed today. In several Maya communities, it is still common for women to sell lime for domestic use, with them being the exclusive merchants of lime and men its producers;[23] according to descriptions by Fray Bernardino de Sahagún, it is possible that the same division existed in Aztec society.

Nixtamalization as Foundation for Civilizations

Nixtamalization is a food technique and also a means of subsistence, a daily practice, a strategy for relating to the natural environment, and even an agricultural-botanical feature of maize.[24] Regardless of the dimension from which it is approached, to understand its relevance and scope, it is essential to consider that nixtamalization is part of a set of agricultural, milling, preparation, and cooking techniques, with a specific sociohistorical development that some authors identify as the nixtamal/comal/tortilla complex or the corn-tortilla chain.[25]

This technique for food management and nutritional enrichment of maize has been undeniably central in a region where historically diets have been based on this grain. The domestication and development of maize in the *milpa* polyculture system were key factors for the emergence of pre-Hispanic civilizations in Mesoamerica.[26] However, as a step prior to or parallel to domestication, nixtamalization is equally relevant, since it would make no sense to take a species as the cornerstone of a food system without being able to take advantage of all its nutritional yield.[27]

As a result, nixtamalization could be considered one of the *sine qua non* conditions for the development of an agricultural food system capable of sustaining populations of the necessary size and density that gave rise to the complex societies of the pre-Hispanic era in Mesoamerica. Such relevance can be verified by analyzing the degree of mutual influence between nixtamalization and maize domestication, which could even be considered a case of coevolution,[28] as well as by the degree of intentionality reflected in the development and refinement of such a complex food technique.

The relationship between the domestication of maize and nixtamalization can be understood as a correspondence of an agricultural technology with a food technology, where the latter ends up solving what the former does not achieve: making the grain more digestible by increasing the bioavailability of its nutrients.[29] Both techniques seek to generate an input of high nutritional value. On the one hand, domestication implied the selection of native corn according to characteristics related to the storage of a greater amount of amylase, since this protein increased the quality of the food in nutritional terms. On the other hand, it is also responsible for generating more malleable nixtamalized doughs.[30]

These effects were observed and exploited in different regions, producing variants in nixtamalization depending on the tastes in terms of smell and flavour of each population. In turn, in parallel to the increasing importance of maize in Mesoamerican diets, it is possible that nixtamalization began to be generalized and perfected as a practice associated with the crop. As communities dependent on the emerging agriculture began to grow, around 1200 BCE, pottery elements related to food consumption and storage also began to be produced, which already showed the existence of the *milpa* and nixtamalization.[31]

There is currently no definitive explanation for the beginning of the association between lime and maize that gave rise to nixtamalization. However, one of the main hypotheses associates it with the expansion of agriculture: the permanent availability of maize grains required some form of storage, and thus it became exposed to attacks from diverse pests. In this sense, there is evidence of the use of wood ashes, sand, and dust for grain protection in various ancient civilizations;[32] as already mentioned, in the case of Mesoamerican civilizations, lime was used to create hygienic surfaces.[33] As ancient Mesoamericans discovered the utility of powdered lime for protecting stored maize grains, it is very likely that after washing it off, some traces of the mineral remained, and after finishing the cooking process to soften the hardened stored grains, the resulting dough turned out to present desirable characteristics. The proportion of the amount of lime added to the grain for protection against pests during storage (ten to thirteen grams per kilogram) is very close to the proportion needed to prepare nixtamal,[34] quite a strong indicator of the coevolutionary connection between maize domestication and nixtamalization.

Another similar route towards the development of nixtamalization could be that of grinding the hardened stored grains. The resulting flour would have also had traces of lime powder and thus may have resulted in more flexible doughs when hydrated. Once nixtamalization became a generalized process, the demand for lime or other forms of calcium hydroxide must have also become an integral part of local economies: there are suggestions that after 200 CE, several cities such as Teotihuacán, Chichen-Itza, and Tenochtitlan might have used the lime obtained during construction.[35]

Eventually, nixtamalization continued to develop according to the specific tastes and needs of local food systems. By 100 CE, with the peak of the political units of the Classic period, archaeological records and the high frequency of certain genetic components involved in flour quality suggest that the selection of grains for tortilla production had already been established.[36] In that period, maize provided approximately 65 percent of the caloric requirements of the Mesoamerican population.[37] Together with the population increase, the increasing complexity of societies began to generate differentiations between sectors, whereby some became detached from the production and transformation of food, while others began to specialize in that sector. The new production and subsistence strategies had their parallel material expression: the high point of the expansionist states tended to be during periods where *comales* for tortilla production were most commonly found;[38] this food format offered greater portability for consumption outside the family unit.[39] In the case of the sixteenth-century Maya, chronicler Diego de Landa recorded that there were three grinding routines for preparing corn: the one with a single grind was for *pozol* or *keyem*, the one with two grinds produced dough for tortillas, and after the third grind, *atole* could be prepared.

Nixtamalization technology was influenced by agriculture and changes in Mesoamerican societies and vice versa in a process of continuous readaptation through which maize was established as the basis of the diets of these populations. In contrast, this was not possible when corn arrived in Europe after the Spanish conquest, since although it was transported as a food product, the Mesoamerican knowledge of nixtamalization remained in America. Consequently, the predominant consumption of corn in European peasant populations led to malnutrition and outbreaks of pellagra in the seventeenth and eighteenth centuries: without nixtamalization, there was no increase in niacin bioavailability.[40]

This difference demonstrates the importance of the technological development achieved by the inhabitants of Mesoamerica in a historical and collective process of systematized knowledge generation. In addition to the hypothesis of the emergence of nixtamalization from the use of lime to protect stored grain against pests, archaeological records of the presence of lime in milling and cooking implements, such as *metates* and *comales*, reveal a sequence of technogenesis over time, in which the nixtamalization process was refined in terms of energy cost and complexity.[41] Specifically, while the more archaic alkaline cooking techniques were mainly applied to wild maize varieties, the more developed modalities had their corresponding application on domestic varieties.[42]

The result of these processes of experimentation and systematic refinement was a highly complex technique, not because of its degree of mechanical structuring, but because of its functional precision in guaranteeing the supply of sufficient calcium, lysine, tryptophan, and niacin.[43] From the environmental interaction, that meant the domestication of maize and the emergence of agriculture; nixtamalization developed in parallel as a response to the search for a functional incorporation of maize as a fundamental element of Mesoamerican diets. The absence of major crises in Mesoamerica due to pellagra confirms the success of these populations in solving the dilemma of the nutritional bioavailability of maize by developing a mechanically simple but biochemically precise procedure, revealing the degree of rationality behind its development.

Past and Present: Nixtamalization as Cultural Resistance

The previous sections developed an analysis of the nixtamalization process as a food technique that allowed the emergence of complex societies in pre-Hispanic Mesoamerica. As a technology, it developed in coevolution with the domestication of maize and thus can be considered as a causal enhancer of the demographic and social phenomena of population increase and the emergence of complex societies, triggered by the generation of a nutritious diet based on maize.

After the Spanish conquest, nixtamalization's central role in the Mesoamerican diet ensured its survival in the following centuries; however, this was limited to the confines imposed by the dominant sectors of colonial society.[44] The separation of maize as a crop from nixtamalization as associated knowledge was one of the main factors that led the dominant classes to attempt to replace it with wheat. After the outbreaks of pellagra in the Old World, corn began to be considered as an inferior food and wheat as a civilizing good.[45]

However, in the New World, the cultivation and consumption of nixtamalized corn became a material expression of the cultural resistance of subaltern groups, marginalized by a caste system during colonization or by their exclusion from the official narratives of national identity construction during the nineteenth century.[46] The importance of the traditional consumption of corn as a symbol of resistance is perfectly visible today in popular Indigenous and urban sectors. This is visible even in the urban activist sphere, considering that one of the main campaigns in favour of food sovereignty and food safeguarding is entitled 'Sin Maíz No Hay País' ('Without maize there is no country').

At the same time, due to the maintenance of corn consumption and, therefore, the relevance of nixtamalization as a food technique, this process also evolved and became industrialized, moving milling from the domestic space to the industrial with the emergence of the first mills.[47] Instead of the industrialization of nixtamal production liberating women from gendered labour, domination from gender roles was inserted into the labour's domination of the industrial era, through which this type of structural violence only acquired new characteristics.[48]

Currently, the main challenges affecting the survival of nixtamalized corn as a collective and culturally relevant practice tend to be economic. Although 98.6 percent of the Mexican population currently consumes corn daily, with an annual per capita intake of 120 kilograms, the monopolistic composition of the national production of nixtamalized flours in Mexico and the dependence on US corn imports due to economies of scale imply two important and interrelated challenges. First, trade renegotiations in an increasingly hostile and nationalistic international environment threaten a basic food input. Second, the transformation of the corn sector in the coming years will depend on the irruption of genetically modified seeds associated with an industrial intellectual property scheme that favours the privatization of genomic knowledge – and in a country where the biological wealth of this particular crop is the result and collective property of the common work of domestication of Mesoamerican peoples for several millennia.

About the Author

Diego Emiliano Jaramillo Navarro, a cultural anthropology master's degree student at the University of Bologna, received a bachelor's degree in politics and public administration from El Colegio de México and has served as a cultural diplomacy analyst in the Mexican government. His main research interests are food sovereignty, Indigenous food systems, processes of patrimonialization, and narratives of identity and authenticity.

Notes

1 *La cal (historia, propiedades y usos)*, ed. by Luis Barba Pingarrón and Isabel Villaseñor Alonso (UNAM-IIA, 2013).
2 Jorge Luis Méndez Martínez, 'Caos-nixtamal: materia y energía de una técnica alimentaria mesoamericana: Perspectivas de la antropología energética y la historia material' (unpublished master's thesis, CIESAS, 2011).
3 Kenneth E. Seligson, *The Prehispanic Maya Burnt Lime Industry: Socio-Economy and Environmental Resource Management in the Late and Terminal Classic Period Northern Maya Lowlands (650–950 CE)* (University of Wisconsin Press, 2016).
4 Thomas P. Schreiner, 'Aspectos rituales de la producción de cal en Mesoamérica: Evidencias y perspectivas de las Tierras Bajas Mayas', in *XVI Simposio de Investigaciones Arqueológicas en Guatemala*, ed. by J. P. Laporte and others (Museo Nacional de Arqueología y Etnología, 2003), pp. 480–87.
5 Juliette C. Gerhardt, *Preclassic Maya Architecture at Cuello, Belize* (BAR International Series, 1988), p. 140; Norman Hammond and J. C. Gerhardt, 'Early Maya Architectural Innovation at Cuello, Belize', *World Archaeology*, 21.3 (1990), pp. 461–81; Edward Wyllys Andrews and Norman Hammond, 'Redefinition of the Swasey Phase at Cuello, Belize', *American Antiquity*, 55 (1990), pp. 570–84 (p. 571).
6 *La cal.*
7 *La cal.*
8 Schreiner, 'Aspectos rituales'.
9 Seligson, *Prehispanic Maya Burnt Lime Industry.*
10 Janferie Stone and Thomas P. Schreiner, 'Ritual Transformation in Yucatec Lime Kiln Burning', conference delivered at Annual Reunion of the California Folklore Society (University of California at Berkeley, 2000).
11 Diana Magaloni, 'Materiales y técnicas de la pintura mural maya' (unpublished master's thesis, UNAM, 1996), pp. 119-120.
12 Schreiner, 'Aspectos rituales'; John Eric S. Thompson, *Maya History and Religion* (University of Oklahoma Press, 1970); Ralph L. Roys, *Ritual of the Bacabs: A Book of Maya Incantations* (University of Oklahoma Press, 1965).
13 Méndez Martínez, 'Caos-nixtamal'.
14 Méndez Martínez.
15 Elsa Gutiérrez and others, 'La importancia del calcio, fósforo y relación C/P en los productos nixtamalizados', in *Nixtamalización, del maíz a la tortilla: Aspectos nutrimentales y toxicológicos*, ed. by Mario Rodríguez García and others (UAQ-UNAM-CLAF-Fundación Produce, 2009), pp. 81–102 (p. 90).
16 Méndez Martínez, 'Caos-nixtamal'.
17 Carlos Alberto Gómez Aldapa and others, 'Cambios en algunas componentes químicas y nutricionales durante la preparación de tortilla de maíz, elaboradas con harinas instantáneas obtenidas por extrusión continua', *Archivos Latinoamericanos de Nutrición*, 46.4 (1996), pp. 315–19; Bruce F. Benz, 'Maize in the Americas', in *Histories of Maize*, ed. by J. Staller and others (Elsevier, 2006).

18 Teresa Shamah-Levy and others, *Encuesta Nacional de Salud y Nutrición Continua 2023: Resultados nacionales* (Instituto Nacional de Salud Pública, 2024).

19 Arnold J. Bauer, 'Molineros y molenderas: Tecnología, economía familiar y cultura material en Mesoamérica: 3000 a.C.–2000 d.C.', in *Mestizajes tecnológicos y cambios culturales en México*, ed. by Enrique Florescano and Virginia García Acosta (CIESAS-Porrúa, 2004), pp. 167–99; Virginia García Acosta, 'El pan de maíz y el pan de trigo: una lucha por el dominio del panorama alimentario urbano colonial', in *Comida y conquista: consecuencias del encuentro de dos mundos*, ed. by Janet Long (UNAM-IIH, 1996), pp. 265–82; Janet Long, 'Invenciones e innovaciones. La evolución de la tecnología alimentaria mesoamericana', *Investigación y Ciencia*, 46.18 (2010), pp. 4–9.

20 Méndez Martínez, 'Caos-nixtamal'.

21 Bauer, 'Molineros y molenderas'.

22 Long, 'Invenciones e innovaciones'.

23 Schreiner, 'Aspectos rituales'; Charles Wisdom, *The Chorti Indians of Guatemala* (University of Chicago Press, 1940), p. 194.

24 Méndez Martínez, 'Caos-nixtamal'.

25 Patricia Fournier, 'El complejo nixtamal/comal/tortilla en Mesoamérica', *Boletín de Antropología Americana*, 32 (1998), pp. 13–40; Yolanda Hernández Franco, 'El maíz en México: Análisis de la historia reciente de la cadena maíz-tortilla', in *Temas selectos de la cadena maíz-tortilla*, ed. by Ana Paula de Teresa Ochoa and Gustavo Viniegra (UAM-I, 2009), pp. 13–54.

26 Arturo Warman, *La historia de un bastardo: maíz y capitalismo* (FCE, 1988); George Cox and Michael Atkins, *Agricultural Ecology, an Analysis of World Food Production Systems* (W. H. Freeman, 1979).

27 Méndez Martínez, 'Caos-nixtamal'.

28 Méndez Martínez.

29 Paula Liliana Sánchez López and others, 'Nixtamalización como herramienta contra la alienación gastronómica', *Revista Latinoamericana de Food Design*, 4 (2023), pp. 267–79.

30 Ma. Susana Xelhuantzi Lopez and José Luis Alvarado, 'Teozintli actual en el Cerro de la Estrella. Reflexiones y consideraciones en torno a esta planta ancestral', *Desgranando una mazorca. Orígenes y etnografía de los maíces nativos*, ed. by Carmen Morales Valderrama and Catalina Rodríguez Lazcano, special issue of *Diario de Campo*, 52 (2009), pp. 18–27.

31 *The Prehistory of the Tehuacan Valley*, vol. III, *Ceramics*, ed. by Richard S. MacNeish and others (University of Texas Press, 1970).

32 Nagnur Shobha and others, 'Indigenous Grain Structures and Methods of Storage', *Indian Journal of Traditional Knowledge*, 5 (2006), pp. 114–17.

33 Silverio García-Lara and others, *Manual de plagas en granos almacenados y tecnologías alternas para su manejo y control* (CIMMY, 2007); Cesáreo Rodríguez Hernández, 'Cal contra plagas', in *Agricultura sostenible*, ed. by C. Rodríguez Hernández and others (Soc. Mex. Agric. Sostenible CP and ITA Tlaxcala, 2007), pp. 81–101.

34 Angel Lagunes-Tejeda and others, 'La nixtamalización del maíz, eje de la cultura mesoamericana, tuvo su antecedente en el combate de insectos', *Agrociencia*, 55.8 (2021), pp. 711–18 (p. 716).

35 Guillermo Sánchez Rodríguez, *Historia, usos y futuros del mayor invento mesoamericano: el maíz* (UAEM-Nueva Época, 2016).

36 Viviane R. Jaenicke and B. D. Smith, 'Ancient DNA and the Integration of Archaeological and Genetic Approaches to the Study of Maize Domestication', in *Histories of Maize*, ed. by J. Staller and others (Elsevier, 2006), pp. 83–95; Daniel Zizumbo Villarreal and Patricia Colunga García-Marín, 'El origen de la agricultura, la domesticación de plantas y el establecimiento de corredores biológico-culturales en Mesoamérica', *Revista de Geografía Agrícola*, 41 (2008), pp. 85–113.

37 William T. Sanders, 'Ecological adaptation in the Basin of Mexico: 23,000 B.C. to the present', in *Handbook of Middle American Indians*, vol. I, *Archaeology*, ed. by J. A. Sabloff (University of Texas Press, 1981), pp. 147–97.

38 Fournier, 'El complejo nixtamal/comal/tortilla'.

39 Elizabeth Brumfiel, 'Weaving and Cooking: Women's Production in Aztec Mexico', in *Engendering Archaeology: Women and prehistory*, ed. by J. M. Goro and M. W. Conkey (Basil Blackwell, 1991), pp. 224–51.

40 Solomon H. Katz and others, 'Traditional Maize Processing Techniques in the New World', *Science*, 184.4138 (1974), pp. 765–73; Ricardo Bressani, 'Effect of Lime Treatment on In Vitro Availability of Essential Aminoacid and Solubility of Protein Fractions in Corn', *Journal of Agricultural and Food Chemistry*, 6.10 (1958), pp. 774–78.
41 *La cal*; Méndez Martínez, 'Caos-nixtamal'.
42 Daniel Zizumbo Villareal, 'The Use of Ash as a Possible Precursor of Nixtamalization in Western Mesoamerica', *Revista de Geografía Agrícola*, 57 (2016), pp. 157–68.
43 Méndez Martínez, 'Caos-nixtamal', p. 259.
44 Méndez Martínez.
45 Bauer, 'Molineros y molenderas'.
46 Antonio Escobar Ohmstede and R. Falcón, *Los ejes de la disputa: movimientos sociales y actores colectivos en América Latina*, siglo XIX (Iberoamericana Editorial Vervuert, 2002).
47 Jaime Aboites Aguilar, *Breve historia de un invento olvidado: las máquinas tortilladoras en México* (UAM, 1989).
48 Méndez Martínez, 'Caos-nixtamal'.

17

Rygeost

Smoke, Soil, and Tradition

Birgitte Kampmann

Introduction: Food and the Elements

In this paper, I have chosen to explore *rygeost* – a Danish smoked cheese that connects directly to the element of smoke. The name *rygeost* or *røgost* derives from Danish: *ryge* ('to smoke'), *røg* ('smoke'), and *ost* ('cheese'). This paper traces the history of *rygeost* from its earliest mentions in the tenth century to its place in today's culinary landscape. I approach *rygeost* both as a historic artifact of Danish food culture and as a versatile ingredient, suited to both savoury and sweet preparations. Because it is not possible to industrialize production without losing the essential qualities and identity of the cheese, it plays a crucial role in preserving a tradition that dates back centuries.

I outline a historical account of this remarkable cheese from its evolution in Scandinavia before the tenth century to the present and explore how its character has been altered significantly with the changes in production methods through the ages. I describe what I learned from spending a day at Tarup Gamle Mejeri, the only artisanal dairy left that makes *rygeost* the traditional way, ladling the cheese curds by hand. As an avid consumer of this unique and wonderful cheese, I cannot resist sharing with you how to make it at home, so I will include recipes at the end of the paper.

The Cheese

Rygeost is a fresh, non-aged curd cheese made from raw grass-fed cow milk, originally without the addition of animal rennet. The milk is soured using buttermilk and allowed to curdle slowly at low heat. The cheese curds are then ladled into cloth-lined wooden colanders and left to slowly drain off the whey.

The defining step is the smoking. The drained cheese is gently smoked over damp wheat or oat straw, which imparts a subtle smoky aroma and extends its shelf life. This method of preservation has deep roots in rural life, where refrigeration was once limited, and smoking was essential. The use of oat straw avoids the bitter flavours that often arise with smoke from other plants. The smoke only hits the surface and penetrates through the water channels of the cheese. The result is an evenly distributed subtle smoky flavour with floral and lactic notes that vary with the seasons. Today, skim-milk-based versions are typically sprinkled with caraway seeds, while whole-milk versions are left plain.

Rygeost is a rare example of a smoked fresh cheese and the only original Danish cheese. Its identity is uniquely tied to a tiny local terroir in the northern part of the island of Funen, with roots reaching back to the early uses of soured milk in the Viking Age. Its history reflects the development of Danish agriculture from an agrarian society dominated by land-owning elites through freehold farming and the formation of cooperative dairies to the industrialized production and centralized value chains in today's food industry.

Although *rygeost* has deep roots in Danish culinary history, its popularity has declined over the past decades. Historically, it arose from the need to be resourceful, to repurpose leftovers, and to preserve food for storage. The cheese would change its character over the course of the year, from a creamy fresh texture in the summer to a harder consistency in the winter when it was preserved with salt. Today, it is made to please the consumer, who expects a standard product year-round. Having lost its original function, it is no longer a staple in most households. It is rarely sought out by younger generations, who associate it with older folks from Funen. In recent years, however, it appears to have found a renaissance as a new ingredient paired with, for example, raw oysters, fatty fish, and new potatoes on rye bread or even made into an ice cream.

The decline of everyday use has not diminished its cultural value. *Rygeost* remains a symbol of resourceful rural cooking and a testament to Funen's agricultural legacy. It is a rare example of a fresh cheese preserved by smoking – a method both ancient and effective.

Early History

Although *rygeost* has its origins in the Viking Age in the tenth century, cheese production as such is believed to date back much further. Historians believe that dairy products have been part of the Danish diet since the middle of the fourth millennium BCE, but the emergence and roots of cheesemaking in Denmark are not fully known.[1] The Vikings certainly knew how to make fresh cheese from soured cow milk from around the year 1000, though there is significant circumstantial evidence that the Nordic peoples knew about cheesemaking much earlier. For instance, around the year 100 CE, the Roman historian Tacitus describes how the Germanic people used cheese curd as food.[2]

Rygeost Arrives

The custom of smoking the curd cheese emerged on the Danish island of Funen, especially North Funen[3] – hence its frequently used name *fynsk rygeost* from the island's Danish name, *Fyn*. According to the single remaining artisanal maker of the cheese, it is the only truly original Danish cheese. The earliest references to *rygeost* in its current form date from the seventeenth and eighteenth centuries. It was originally a farmhouse cheese made by peasant women, typically using the skimmed milk left over from skimming the cream for churning butter. The skim milk was mixed with whole milk and left to sour naturally near the hearth and curdled slowly over low heat.

The curd was then hand-ladled into cloth-lined wooden tubs or round wooden sieves with holes at the bottom. Curds were added in layers to allow the whey to drain slowly. Once firm, the cheese was pressed by hand and formed into rounds. Smoking was done using a metal grate or iron plate over a small fire fuelled with oat straw. Some households added fresh, moist nettles for extra flavour, though this sometimes gave the cheese a bitter edge and black streaks.

Salt and caraway seeds were then added according to taste. Caraway seeds are readily available in Denmark, as the plant grows in the wild. The seeds have been used in a variety of ways, in baking, pickling, cabbage dishes, and akvavit. They were also used to give flavour to the bland-tasting leftover curds, as cheeses based on skim milk were considered inferior to *rygeost* made from whole milk.[4]

In autumn, before the cows were brought indoors, farmers' wives often made a sturdier version of the cheese for winter use. These cheeses were salted heavily and eaten with lard on rye bread.

We have much of this insight from Christine Reimer, an early twentieth-century folklore writer from North Funen, who documented this process in an entire chapter devoted to *rygeost* in her account of peasant life in Northern Funen.[5] Her five-volume work describes the lifestyle, customs, and language of ordinary people in the rural communities of the region and is considered one of the most important historical records of Danish folk memories and related traditions. Her description offers a rare glimpse into the domestic production of *rygeost* and the daily rhythms of rural life.

From Farmhouse Practices to Cooperative Industry

In preindustrial Denmark, milk was typically consumed directly on the farm or processed into butter using simple methods. Left to stand, the milk naturally separated into a low-fat sour milk and a high-fat cream, the latter being churned into butter using hand-operated equipment.

During the sixteenth and seventeenth centuries, knowledge of cheese production expanded greatly, inspired by the practices in Switzerland, Germany, and the Netherlands and adapted to Danish conditions. Archival materials, including letters from King Christian IV (1568–1648), indicate a sophisticated understanding of dairy production on the royal estates. By the eighteenth century, dedicated dairies had been established on Danish manor houses, with production scaled up to meet the increasing demand from urban markets. The agricultural reforms in the late 1700s ended tenant farming and created a new class of independent freehold farmers able to engage in commercial dairy production, primarily butter, which by the mid-nineteenth century was increasingly directed toward export.[6]

From the 1860s, independent dairies began to emerge. These were typically operated by individuals who purchased milk from surrounding farms, processed it, and retained full profit from butter sales. A regional example from Kauslunde Parish on Funen (1875–83) involved joint milk processing, though not under a formal cooperative framework.

The 1880s marked the beginning of Denmark's cooperative dairy movement. Independent farmers, motivated by a belief that manorial dairies underpaid them, began to invest collectively in shared facilities. These cooperatives were structured to ensure equal ownership among members, with shared profits and responsibilities. The establishment of the world's first cooperative dairy in 1882 in Hjedding, near Varde, set a precedent. Within seven years, over six hundred cooperative dairies were founded across Denmark![7]

The movement's success was closely linked to technological innovation, particularly the invention of the continuous milk centrifuge by Laurits Christian Nielsen in 1878. This device significantly improved cream separation and allowed Danish butter to compete more effectively in international markets.

By 1900, Denmark had over 1000 cooperative dairies. Despite the absence of centralized management, the cooperative model facilitated a rapid and widespread restructuring of dairy production. By 1940, approximately 1400 cooperative dairies were operating, although most remained relatively small. From the mid-twentieth century onwards, economies of scale and the need for reliable and consistent production led to a dramatic consolidation of dairies. By the 1990s, the cooperatives MD Foods and Kløver Mælk had emerged as the dominant actors, and they ultimately merged in 1999 with Swedish Arla into a single company, Arla Foods.

Today, Arla Foods remains Denmark's largest dairy cooperative, maintaining the foundational principles of shared ownership despite its global scale. While private dairies still exist, approximately 97 percent of Danish milk is now delivered to cooperative facilities.

Keeping Tradition Alive

Let me now take you to the northern part of the island of Funen, considered the birthplace of *rygeost*. Funen is Denmark's third-largest island, with an area of 3100 square kilometres, nestled between the Jutland peninsula and Zealand. Known as the 'kitchen garden' of Denmark, the island is home to rich, humus-filled soil, fruit orchards, grain fields, and dairy farms. With a mild climate and fertile soil, it has long been a stronghold for dairy and vegetable farming and a centre for the production of crops like tomatoes, peas, apples, cherries, and potatoes. Historically, the island was densely populated by manor houses – over 120 of them – each serving as a hub for farming and animal husbandry.[8] These estates played a critical role in passing down agricultural knowledge, including the art of dairying and cheesemaking.

I recently paid a visit to Kent Thrane, owner of Tarup Gamle Mejeri in the town of Tarup (Figure 1). His dairy is the only one today that produces *rygeost* in the traditional way, but within a modern setting. Remarkably, this tiny dairy covers almost 70 percent of the Danish market for *rygeost*. Most of the remaining 30 percent is produced by Løgismose, a Danish food producer with a strong high-end food brand. It is thus clear that *rygeost* is a niche product that has survived the industrialization of the Danish dairy sector.

As I discovered during my visit, this is because *rygeost* is a delicate soul whose production cannot be rushed without losing its character and quality. As such, it preserves a unique link to Danish agrarian culture reaching back centuries. At the same time, Tarup Gamle Mejeri represents a fascinating compromise between artisanal tradition and the unavoidable present-day demands for hygiene, consistent quality, and reliable delivery.

Tarup Gamle Mejeri was started by Mr Thrane's great-grandfather together with his wife. The family recipe is over a hundred years old. Over time, production has been modernized, but essential parts of the process remain the same.

Mr Thrane sources raw milk daily from two local farming families, ensuring freshness and traceability. The milk, primarily from Jersey cows, has a higher fat content, which gives the milk a creamier taste. He also smokes the cheese using locally sourced straw so that both the milk and the smoke come from the same soil as it always did in years past.

The milk is run through a centrifuge to remove impurities and to separate the milk into skim milk and cream and then gently pasteurized. Mr Thrane explains how this process allows him to control the fat content in his final product and to secure a clean product. He offers two variations of *rygeost*. The low-fat variety, based on skim milk, has a fat content of 1 percent and contains 12 grams of protein per 100 grams. This variety is always sprinkled with caraway seeds to jazz it up a little bit. The other product, based on whole milk, is very creamy due to a much higher fat content (10 percent) and has 9.9 grams of protein per 100 grams. It has a much richer and more complex taste that is best left to stand on its own.

It is important to remember that *rygeost* is a simple product made exclusively of milk. So, for Mr Thrane, milk is not just an ingredient – it is the very foundation of his product. Yet milk is never static: milk from grass-fed cows naturally shifts in flavour with the seasons – rich and floral in summer, sharper and leaner in early spring and during calving.

To meet the stability modern consumers expect, Mr Thrane ensures a consistent flavour year-round by using starter cultures with specifically designed lab-cultured bacteria developed in cooperation with the world-leading biotech company Novonesis.[9] This is a remarkable marriage between artisanal production and high-tech biotechnology, but it comes at a cost. The seasonal variation in taste is eliminated by standardization, even at the artisanal level. Once again, we are reminded how industrial logic reshapes our expectations. It's a careful balance.

Figure 1. Fresh curds are carefully ladled into metal moulds, where they begin to drain and take on the characteristic shape of the cheese.

The physical act of making *rygeost* at Mr Thrane's dairy is still resolutely artisanal. After pasteurization and inoculation with starter cultures and rennet, the milk is left to coagulate. The cultures convert the milk sugar, changing lactose into lactic acid, initiating the acidification process. A small section of the curd mass is then broken manually into small curds, which are carefully ladled by hand into round, perforated stainless steel moulds. The production takes place under sterile air and controlled temperature conditions to meet strict hygiene standards.

Figure 2. Freshly smoked rygeost *straight from the smoker. This* rygeost *made from skim milk is characterized by a sprinkling of caraway.*

Mimicking the slow draining process of earlier times, the curds are poured in stages – first to allow natural whey drainage, then twice more to build a stable, unpressed form. The result is a cheese with a distinctive, lumpy surface, characteristic of unpressed *rygeost*.

Unlike machine-pumped curds, which will generate some heat and disrupt the milk proteins, causing a metallic off-flavour and mushy textures, Mr Thrane's method ensures that the structure of the curd remains intact, resulting in a delicate, layered flavour and a tender bite.

Drainage continues overnight – typically sixteen to eighteen hours – after which the cheeses are ready for smoking. This is done using wheat straw in an automated smoker, staying faithful to the original flavour profile. The precise temperature that keeps the smoke white, clean, and free from PAH (polycyclic aromatic hydrocarbon) remains a well-kept secret. The operation is automated and temperature-controlled using an internal thermometer to prevent the formation of carcinogenic materials. If the oven temperature begins to rise, straw is added to temper the heat; if it falls, airflow is increased. The daily output of ten thousand to twelve thousand cheeses is supervised by a team of twelve employees.

The visible layers of a handmade *rygeost*, thin lines of fat between the curds, are only revealed when cutting into the warm cheese. They form when the fat settles unevenly during the slow, careful ladling of the curd. *Rygeost* made from whole milk shows these layers less distinctly, but it is no less skilfully made.

Conclusion

Rygeost stands at the intersection of food and the elements – where grass-fed cow milk meets fire, straw, and time. From its farmhouse roots in Northern Funen to its careful preservation at Thranes Old Creamery, this cheese is more than a product. It is a practice, a memory, and a flavour shaped by land, labour, and smoke. As we revisit the elements that shape our food traditions, *rygeost* reminds us of the quiet, enduring value of preservation – not just of food but of methods, landscapes, and meaning. At last, let me share with you how to prepare this wonderful cheese.

Cooking with *Rygeost*

Figure 3. The author and Kent Thrane at Tarup Gamle Mejeri, summer 2025.

A few years back, while living in Boston, I had the opportunity to work at Christopher Kimball's 177 Milk Street kitchen. I decided to make a batch of *rygeost* for the cooking team to taste. The tricky part was the smoking. So, naturally, I turned to the experts: the Boston Fire Department. I was living on Beacon Hill at the time and asked the chief if I could smoke my Danish cheese on their premises. To their credit – and perhaps out of equal parts kindness and concern – they agreed. There I was, standing with a metal bucket, a grate, and a bundle of straw, ready to light up. Excited and happy was I, but judging by the look on their faces as they sized up my homemade contraption, I suspect they thought, 'We'd better help this Danish lady before she burns down the neighbourhood'.

In truth, creating a smoke environment isn't hard at all. I hope you give it a try – and do let me know how it goes. Once smoked, the cheese will keep in a sealed container in the fridge for a couple of weeks. It is best enjoyed at room temperature. The creamy smoky flavour is a delicious addition to dips, as a creamy sauce paired with pickled herring, or as a filling in a quiche.

How to Make Your Own Smoked Cheese (*Rygeost*)

You can make *rygeost* in two ways:

1. **With buttermilk only** (traditional method)
2. **With buttermilk and rennet** (faster coagulation; choose either animal or plant-based rennet)

You may also choose between **whole milk** or **skimmed/low-fat milk**, depending on your preferred fat content and texture.

Option 1: Buttermilk Only

- 2 litres good-quality whole milk
- 200 ml buttermilk
- A pinch of salt

Option 2: Buttermilk and Rennet

- 2 litres skimmed or low-fat milk
- 200 ml buttermilk
- 3 drops rennet
- A pinch of salt

Method

1. **Heat and inoculate**
    - Combine milk and buttermilk in a saucepan.
    - Gently heat to **22–25°C**, stirring occasionally.
    - Take off the heat.
    - If using option 2, add the rennet and stir gently together.
2. **Coagulation**
    - Pour the mixture into a clean bowl.
    - Cover with a cloth and leave it in a warm place for **twenty-four hours**.
    - The milk should now be separated into coagulated milk protein and whey. I recommend that you cut the coagulated milk into larger pieces rather than into smaller curds.
3. **Drain the curds**
    - Line a sieve with cheesecloth.
    - Gently ladle the curds into the sieve in two to three passes to allow slow draining.
    - Sprinkle a little salt between each layer.
4. **Refrigerate**
    - Place the sieve with the curds in the fridge and let it drain for **twenty-four hours**.
    - The whey can be discarded or reserved for baking or drinking.
5. **Unmould and smoke**
    - Turn the drained cheese out onto a plate.
    - Remove the cheesecloth.
    - The cheese is now ready to be smoked.

Smoking Your Cheese

Equipment You'll Need

- **Smoke chamber:** This can be a grill starter, a short metal pipe (similar in diameter to your cheese), or even an old metal bucket with ventilation holes in the bottom and sides. You can drill the holes yourself.
- **Grate:** A grill grate or any heat-safe metal mesh larger than the pipe opening. Avoid galvanized materials like chicken wire, as the zinc coating is toxic when heated.
- **Fuel:** Use clean oat or wheat straw – dry and untreated. Optionally, freshly picked nettles can be added to the straw for aromatic variation, but this is a matter of personal taste.

Smoking Method

1. **Prepare the setup**
   - Place the fresh cheese on a plate.
   - Hold your grill grate above it, then carefully invert the whole setup so the cheese sits directly on the grate.
2. **Start the smoke**
   - Light a small amount of curled straw inside the smoke chamber.
   - Once it catches, **press more straw down** over the flame to smother it – this should create **thick, white, non-flaming smoke**.
   - Continue adding straw slowly to keep the smoke flowing.
3. **Smoke the cheese**
   - Lower the grate with cheese into the smoke.
   - Let it smoke for **thirty seconds to one minute**, or until the surface takes on a **warm golden colour**.
   - Remove from the smoke and allow to cool.

Sommersalat

A classic fresh, creamy topping built around smoked dairy – perfect for toasted dark rye bread.

Ingredients

- 200 g *rygeost* (smoked Danish cheese)
- 2 tbsp crème fraîche
- 3 tbsp mayonnaise
- ¼–½ English cucumber, seeded and finely diced
- 1 bunch red radishes thinly sliced
- 1 bunch chives finely chopped
- Salt and freshly ground black pepper to taste

Method

1. **Combine the base**
   - In a mixing bowl, whip together the *rygeost*, crème fraîche, and mayonnaise until smooth and well blended.
2. **Fold in vegetables**
   - Add the diced cucumber and half of the sliced radishes to the cheese mixture.
   - Season with salt and freshly ground black pepper to taste.
3. **Finish and serve**
   - Transfer the spread to a serving bowl.
   - Top with the remaining radish slices and a generous handful of chopped chives.
   - Serve immediately with toasted dark rye bread.

About the Author

Birgitte Kampmann, a classically trained chef, spent many years in the United States, later becoming the chef at the American Embassy in Copenhagen. She is the founder of Kitchen Lab, a monthly online cooking workshop, associated with the Oxford Food Symposium.

Notes

1 J. Burchardt, 'Ostens danske kulturhistorie' ('The Danish Cultural History of Cheesemaking'), *Gefjon*, 4 (2019), pp. 40–85 (p. 45) <https://www.burchardt.name/ostens-danske-kulturhistorie/> [accessed 27 April 2025].

2 Burchardt.

3 J. M. Buch Kristensen, *Den fynske historie bag 6 danske oste* (*The Funen History behind 6 Danish Cheeses*) (Mælkeritidende, 1994), pp. 528–32.

4 Burchardt, 'Ostens danske kulturhistorie'.

5 C. Reimer, *Nordfynsk Bondeliv i Mands Minde: Folkeminder, øste af Folkemunde i det 19. Aarhundrede* (*Peasant Life in Living Memory*) (Andelsbogtrykkeriet i Odense, 1910), vol. 2, p. 196 <https://slaegtsbibliotek.dk/909907.pdf> [accessed 27 April 2025].

6 Burchardt, 'Ostens danske kulturhistorie'.

7 Ferslev Mejeri, 'Kort om mejeriet', Om billederne, 2006 <https://ferslevmejeri.dk/> [accessed 27 April 2025].

8 S. Ryge and C. Holm, *Smag på Fyn* (Gyldendal A/S, 2024).

9 The company is a merger between Novozymes and Christian Hansen. The use of rennet, a coagulant introduced in standardized form by Danish pharmacist Christian D. A. Hansen in 1874, has revolutionized cheesemaking. Hansen's innovation – isolating enzymes from calf stomachs by modifying the pH of the extract – paved the way for consistent and safe production long before terms like 'enzyme' or 'pH' were even part of scientific vocabulary.

18

Super Saffron and Virtuous Vanilla

Can Terroir Be Synthesized?

Priya Mani

From anticipating the annual flooding of the Nile to cultivating crops in the tranquil lakes of Kashmir, civilizations have long worked with the limitations of ecology and landscape, adapting ingeniously to their environments, shaping the land and managing water to grow their food. In other instances, we have mimicked the still, waterlogged biomes of paddy fields – from terraced paddies in Asia to basin irrigation in Africa, man-made landscapes mimicking natural systems have long shaped agricultural innovation.

The Hanging Gardens of Babylon, built about 600 BCE as towering fields in the ancient world, channelled the Euphrates upwards through 'chain pumps', creating small lakes on the upper floors that could irrigate the gardens. The Chinampas of the Aztecs built floating rafts of dense soil and reeds so root systems could use the river water and nourish biodiversity for an omnivorous diet.

In 1915, American geologist Gilbert Ellis Bailey suggested building vertical farms but by boring deeper into the ground, an idea that was, well, grounded by wars. Hydroponics was employed in greenhouse trials in Armenia in the 1950s to look at new growing mediums and environments. Still, it was not before the space rush of the following decade that research into agriculture in synthetic environments attracted serious investments of time and resources. By the turn of the 1990s, American microbiologist Dickson Despommier began wondering how to feed Manhattan. Would converting the island's roofs provide enough arable land and reduce the carbon footprint? Despommier suggested that tall skyscrapers could house the farms of the future. Vertical farming has since been framed as a transformative solution for sustainable agriculture, with year-round crop production and minimal land use.

In the last two decades, the industry has seen many well-funded start-ups raise intense commercial speculation, including attracting huge capital investments in Europe and the United States. However, in the last five years, a significant number of vertical farming start-ups have faced financial difficulties, shut down operations in certain geographies, or quit entirely.[1] The current vertical farming model is complex, and technically advanced infrastructures need deep investments, while high operational costs keep food costs high. However, market misalignment may be the most obvious pitfall of the vertical farm industry – they have focused their technologies and control algorithms on growing greens, microgreens, and fresh herbs as high-value grocery products, often costing way more than traditionally farmed produce. Consumer willingness to pay a premium has been low, and this approach, since these products only constitute the 'fringe of the plate' and are highly perishable, results in loss and food waste.[2] Given those facts, can a future outlook focus on high-value crops such as spices? As dried botanicals, they have high shelf stability, offering more sustainable choices for the digital farmland.

Within the integrated environmental systems of vertical farms, high sensorization means the environment adjusts to plants instead of modifying plants for climate and geographical conditions, thus changing the models of plant-environment relationships – agricultural dependence on soil, seasonality, and topographical factors. Vertical farms' integrated systems allow real-time monitoring and dynamic adjustment of light spectra; indoor temperature, humidity, and CO_2 concentration; airflow; and nutrient availability at each growth stage. Thus, traditional open-loop agricultural systems are transforming into closed-loop systems with cyber-physical plant cultivation in digital environments.[3]

Let us consider the simulation of light in such controlled environments.

Plants respond to light in ways that affect their growth rate, their root development, how large or colourful their leaves become, and, most importantly in this context, the evolution of their flavours. For instance, blue light encourages root growth, thicker leaves, and shorter, stockier plants. Red light, on the other hand, promotes taller growth, thinner leaves, and reduced root development.

When LED lighting was first introduced to horticulture around two decades ago, researchers focused on red and blue wavelengths, since these are the ones most efficiently absorbed by chlorophyll A and B. Early grow lights, therefore, emitted a pink hue – a mix of red and blue – designed to promote healthy plant growth in greenhouses.

But nature does not waste, and sunlight offers a full spectrum of wavelengths, not just red and blue. Thus to understand the role of middle spectrum light – green, yellow, and orange – scientists have been studying *Arabidopsis*, a model plant with a simple genome shared by many plant species. It turns out that green light plays a subtle but important role, particularly in enhancing flavour.[4]

Interestingly, when sunlight passes through a leaf, the red and blue light are mostly absorbed for photosynthesis. What filters through is far-red and green light. This leftover light isn't wasted – it travels through the canopy to lower leaves, signalling the plant that it's shaded and affecting how it distributes energy and produces secondary compounds, including those responsible for flavour.[5]

But such engineering of 'climate by design' also raises deeper questions about the nature of cultivation. The farm complex becomes a programmable 'terroir', where airflow and photon densities substitute for wind and sun angles, and nutrition formulations are as precise as soil amendments. So if flavour and form are consequences of environmental feedback loops, then such controlled environments are not merely production spaces: they may be cultural and biochemical laboratories generating new plant phenotypes.

Since vertical farms maintain constant environmental conditions, they require new evaluation methods. Unlike field farms, which must consider seasonal variation, micro-climate anomalies, and pest pressures, such sensor-controlled indoor environments promise reproducibility. Our extensive food system contains numerous instances of food production from plant (what produces food) to 'plant' (where food is produced), which people accept without hesitation.[6] Consider the use of wild yeast in traditional home fermentation to inoculate specific strains for consistent outputs in fermentation-forward modern restaurants and the food industry – or the case of home cooking, often individualistic to the cook and his or her family, while cookbook writers apply pattern recognition and sensemaking to write down 'recipes' to replicate results. Restaurants maintain identical results between services, just like Big Macs maintain their consistency across all locations. Thus the emotional and cultural significance of field-based farming need not be threatened by this move indoors or by the use of a 'recipe' to replicate, as there appears to be a time and place for both approaches.

To flip the narrative, I suggest that such a technologically mediated reliability supports high-value, high-sensitivity crops like saffron and vanilla, whose quality is not only inconsistent but also where violence, oppression, and security plague labour across traditional supply chains. As these sensor systems and their computing potentials mature, the discourse around environmental control shifts from risk mitigation toward expressive potential: how might we compose an environment to elicit particular flavours, textures, or even cultural resonances?

This turn toward expressive control – toward the aesthetics of agronomy – is where vertical farming enters into philosophical dialogue with older ideas of cultivation.[7] It challenges the notion that agricultural value is rooted in struggle, risk, or fidelity to place. Instead, it reframes cultivation as optimization, even orchestration; vertical farming repositions the farmer from a caretaker of land to a composer of ecologies, mediated by sensors, actuators, and code.

One Morning Only: Saffron's Golden Hour

Saffron (*Crocus sativus*) grows best in Mediterranean climates where summers are hot and dry and winters are cool and slightly wet. The ideal temperature for growing saffron is between 15°C and 20°C (59°F–68°F). The plant can become stressed if it gets hotter than 30°C (86°F). If it drops below -5°C (23°F), the underground corms may be damaged by frost. Saffron prefers light, well-draining soil – like sandy or loamy types – with a pH between six and eight. If the soil holds too much water, the corms can rot. Gentle air movement is helpful: strong winds can tear the fragile flowers, and air that doesn't move enough can lead to fungal problems. That's why saffron has traditionally been grown in open areas with good airflow.[8]

Saffron is a shy late-autumn bloom, opening only when the season strikes a delicate balance. A sudden chill or unseasonal rain in October can shorten the harvest before it even begins. In a normal year, the crocus flowers in three flushes. They bloom all at once – often in the hush of night – and fade quickly by midday. Farmers watch the fields like hawks during bloom to catch saffron at its finest. Harvest begins before the first light and ends by midmorning. Then, workers with nimble fingers peel the flowers of their petals, and the stigmas are separated and dried the same day before their aroma fades. For the farmers cultivating saffron, harvest time is a family affair and, in many instances, seasonal work for transient labour. Thus, harnessing such a crowd to tackle windfall blooming is a huge logistical issue across all geographies involved in saffron production.

One Morning Only: Vanilla's Golden Hour

Vanilla thrives in warm, humid climates, typically in tropical regions close to the equator – like Madagascar, Mexico, Uganda, or southern India. The ideal environment has temperatures between 20°C and 30°C (68–86°F), high humidity, and dappled sunlight, often provided by growing vanilla vines under shade trees like *Gliricidia* or banana. Rainfall should be regular but not excessive, and the soil must be well-draining.

Vanilla (*Vanilla planifolia*) is an orchid whose flowers are delicate and short-lived. Each vine produces long, climbing stems with flowering spikes growing from the points where the leaves meet the stem. Each spike produces several flowers, but they usually bloom one at a time over a few weeks – and each flower only lasts for a single morning. If it's not pollinated within a few hours, it wilts – and the opportunity to grow a bean is lost.

That's why vanilla farmers walk their plantations daily during the blooming season – usually early mornings between September and December, depending on the region. With trained eyes, they scan the vines to spot freshly opened flowers, gently parting leaves to peek at the pale greenish-yellow blossoms tucked along the vines. In nature, vanilla is pollinated by specific bees, like the *Melipona*, found only in parts of Mexico. Outside its native habitat, these natural pollinators are absent, so vanilla flowers must be pollinated by hand. So, in a plantation, once a fresh flower is found, the farmer carefully lifts the flower's flap (called the rostellum) using a small toothpick and presses the male and female parts together – something that generally wouldn't happen on its own. The entire motion takes seconds but requires great precision and a gentle touch. One wrong move can damage the flower.

This work is done by hand, flower by flower, sometimes pollinating hundreds in a single morning. Each pollinated flower, if successful, becomes a green vanilla bean in about six to nine months – though it will still need to be picked, fermented, and cured before it smells anything like the vanilla we know.

While the rise (and fall) of vertical farms has been a subject of intense scholarly study and commercial speculation, their effects on crop identity and sensory quality remain poorly understood.

Plant Reactions to Stress: From Survival to Taste

In 2020, Harold McGee gave a keynote at the Oxford Food Symposium, where he shared research from his book *Nose Dive: A Field Guide to the World of Smells*. He explained that while nature and growing conditions shape plants, their smell often comes from the stress or damage they experience – like being grazed by animals. For example, the strong scent of rosemary isn't just for our pleasure – it is the plant's way of defending itself. The MIT Media Lab's OpenAg project, too, focused on the role of stress in flavour, although the project has since closed.

Thus, the flavour of spices is not merely a product of species or genotype – it is the echo of environmental change. When plants are under stress, their response, like humans, is aimed at survival and adaptation rather than growth. This stress activates changes in the plants' metabolism and, in turn, their defence mechanisms. Special plant hormones – like jasmonic acid, salicylic acid, and abscisic acid – send signals that turn on stress-related genes. As plants respond to their environments, these genes help them produce certain compounds called secondary metabolites, such as terpenoids, alkaloids, flavonoids, and phenylpropanoids – the key to the development of flavour, aroma, and pungency that have come to define the human experience of spices. These compounds are part of the plants' defence system, as they deter herbivores, attract pollinators, or mitigate oxidative stress in a poly-faceted relationship between plant physiology and flavour chemistry.

In saffron, the main flavour and colour compounds – crocin (for colour), picrocrocin (for bitterness), and safranal (for aroma) – are made in response to light, temperature, and water stress.[9] These compounds don't just appear independently; they develop when the plant experiences specific growing conditions. For example, intense sunlight, cold nights in autumn, and dry weeks just before the flowers bloom all help trigger these changes. Farmers in geographies as diverse as Spain, Morocco, Iran, India, and Sweden agree that these environmental cues are key to getting the crocus to bloom and produce high-quality saffron. These changes affect how plant genes work, and certain enzymes act inside the flower, especially in the stigmas – the part of the flower that becomes saffron. How saffron is dried and handled after harvest also plays a vital role in its final flavour.

Vanilla is a complex crop in terms of flavour. Its primary flavour compound, vanillin, doesn't develop fully until after harvest during a long curing and fermentation process.[10] But the building blocks for vanillin – special sugar-linked compounds stored in the vanilla pod – are made while the plant grows. These precursors are strongly affected by environmental conditions like humidity, root temperature, fungal activity, and even wind, meaning that the potential for good flavour starts on the vine and depends significantly on how environments can stress a plant during its growth. Mild stress – like heat, limited water, or controlled exposure to microbes – can help set the stage for higher vanillin levels later.

Green pepper (the unripe fruit of *Piper nigrum*) gets its sharp, spicy taste from a compound called piperine. But the amount of piperine isn't always the same – it changes depending on how mature the pepper is, how it's dried, and what kind of soil it grows in.[11] Studies show that giving the plant a little water stress during fruiting can increase the amount of piperine. By contrast, using too much fertilizer can reduce its spiciness because the plant focuses more on growing leaves than making flavour compounds. The minerals in the soil, especially potassium and calcium, also play a role by helping the plant's enzymes make piperine more effectively.

When faced with challenges like drought, intense light, or nutrient scarcity, plants shift gears, changing from growing to a survival strategy. A precise botanical response – an internal alarm bell – plays a key role in producing the secondary metabolites responsible for the distinct flavours of spices. Such stress-induced physiological reorientations provide a significant opportunity for vertical farming systems to evoke their best flavours. It is not about pushing the plant to the edge but guiding it to produce its full aromatic potential.

Let me illustrate this with a few examples from current industry practices.

Air quality has a sufficiently important role, and the levels of CO_2 are a huge factor in affecting photosynthesis. For a plant like cannabis, where the leaf is the primary product, speedy growth can be a valuable asset. Maintaining high levels of CO_2 promotes high leaf growth in cannabis plants. So, while atmospheric CO_2 is approximately four hundred parts per million, the levels can be many times higher in environment-controlled indoor cannabis farms: about twelve thousand parts per million for high-yielding cannabis. In the case of saffron, a dark dormant period of at least 1100 hours (forty-five days) is vital to trigger blooming. To do this, in vertical farms, batches of saffron corms are deprived of light for the required period and then exposed to the required light levels. Vertical farms can modulate this phenomenon in batches, which in real farms can be an unpredictable situation, creating peak demands of labour with little notice.

One cannot consider natural elements alone, as they are always synergetic. For example, in the case of vanilla, the need for CO_2 is higher at night to mimic the plant's photosynthesis cycle in its natural habitat. Vanilla plants take in most of their carbon dioxide (CO_2) at night because they use a special water-saving process called CAM metabolism. During the night, when temperatures are cooler and there's less risk of losing water, the plants open their pores and absorb CO_2 efficiently – about 81 percent of their total daily intake happens during this time. As the sun rises around 6 a.m., CO_2 uptake starts to drop sharply, and by 10 a.m., under strong sunlight, the plants stop absorbing CO_2 altogether and may even lose some. Although CO_2 levels rise again after 4 p.m., the plants usually don't start taking in CO_2 again until after 6 p.m., when it's already dark. This pattern confirms that vanilla is a vigorous CAM plant that is well adapted to hot, humid environments where conserving water is key.[12]

Some vertical farms even programme terroir by integrating local air, water, and microbial traces into controlled systems. However, their flavour is the most critical parameter when speaking of spices. How do technologies such as LED lighting, hydroponics, and machine learning optimize the biosynthesis of flavour compounds?

Flavour by Biomimicry

In the controlled farming environment of vertical farms, engineers can 'design stress', aiming to create specific stress that improves flavour, with a degree of precision that is impossible in traditional agriculture. Using controlled-environment agriculture (CEA), vertical farms can precisely adjust light, temperature, humidity, and nutrients to trigger helpful changes in the plant's metabolism. With new tools like metabolomics (which looks at the small molecules – metabolites – inside a plant) and transcriptomics (which studies how plants express or suppress genes), scientists are trying to understand how plants change at the molecular level when stressed – and how that stress affects flavour.

A method called stress priming is employed, giving the plant a mild, harmless dose of stress to encourage it to make more of the compounds that create flavour. For example, exposing plants to small amounts of UVB light (a kind of ultraviolet light) helps them produce more antioxidants like flavonoids and phenolics, which improve taste and health benefits. Another trigger is holding back nutrients – like nitrogen or phosphorus – which can provoke the plant to make more flavourful compounds, as seen in spices like black pepper and oregano. Changing the temperature around the roots is another way to stimulate these changes, which has been tested successfully in crops like tobacco and tomatoes.

When these stress techniques are timed carefully to match the plant's growth stages, vertical farms can mimic natural challenges in a predictable and repeatable way, giving them a special advantage: the ability to design flavour through controlled stress.

Therefore, the flavour of a spice becomes not a static essence but a dynamic outcome – a tasteable record of its environmental story. While the traditional notion of 'terroir' is grounded in a deep connection to the land and natural elements, vertical farming can create a 'synthetic terroir', constructing the elements that precisely influence flavour. So a plant's flavour is not simply a product of soil and sun but a controlled manipulation of environmental stressors. Thus I propose that since spices are not a singular aroma or flavour molecule but instead a bouquet with highly variable notes, there is also a potential to incite a new spectrum of spice flavours – making flavour a product of deliberate design.

In the human history of spices, these dried botanicals have held great value for their capacity to transform cuisine. Trade and transhumance took spices around the globe through sea and land, but in spice's hyper-regionality lies its actual value. Nutmeg came from the Banda Islands, cloves came from Ternate, white cardamom was unique to the dense forests of Indochina, and so on. Driven by its commercial value, colonial powers launched a global hunt for microclimates to acclimate spice plants. So cardamom from India reached Guatemala, and vanilla from Mexico found a new home in the Reunion Islands, setting off a keen study of the natural biomes of these spice species and how such ideal conditions might be replicated (albeit by finding another natural terroir) so the new transplants could be commercially viable.

Vertical farming and environment-controlled hydroponics of spices are extensions of this idea. So with timed stressors and monitored plant responses, spices with consistent and potentially intensified flavour profiles can be produced regardless of their geographical location, making precision farming a paradigm shift for the spice industry.[13] If so, can vertical farming offer a more sustainable and efficient system of spice production?

Artificial intelligence (AI) and machine learning help make these systems smarter by learning from data collected during each growing cycle from various systems that automatically deliver nutrients, special LED lights that mimic sunlight, and climate sensors that monitor temperature and humidity.

For example, machine learning can find patterns between small environmental changes and how they affect a plant's flavour, allowing farms to fine-tune growing conditions for the best possible taste. In saffron farming, AI can adjust light and water levels depending on the growth stage and the desired flavour compounds. In some farms in Sweden, robots even handle harvesting – gently picking saffron flowers without human hands, making the process clean, fast, and consistent.

Together, these digital tools turn farming into a high-tech, data-driven process. The implementation of precision techniques through technology enables efficiency but raises profound questions about what it means to grow something naturally in an age of algorithms and machines. I must hasten to reiterate that such technology is not inanimate; it is also a labour of the human mind and the power of its imagination.

What are the cultural and philosophical implications of growing culturally rooted crops in nontraditional, artificial environments? By shifting tasks from manual agriculture to automated supervision, vertical farming reconfigures both labour dynamics and farmers' dignity. Traditional spice farming is physically demanding and precarious, often involving seasonal or gendered labour. It percolates and perpetuates bucolic ideals that maintain social hierarchy. Indoor farming promises safer, year-round, gender-neutral employment with technical skill development.

Vertical farming also offers sustainability advantages regarding land and water use – but also faces scrutiny regarding energy consumption. Due to their economic margin, high-value spices are more suited for vertical farming than staple crops, as market prices offset the energy costs. As for the carbon footprint of such energy, the technologies to integrate solar panels and energy-efficient LED systems, as well as exploring geothermal and battery-backed grid systems, are becoming more commercially viable, as the energy demands of vertical farms for lighting, HVAC (heating, ventilation, and air-conditioning), and automation are better met.

Water recycling systems, closed-loop nutrient delivery, and drastically reduced pesticide use are all strengths of vertical farms. Saffron cultivation in India's indoor farms has shown up to 90 percent water savings over field cultivation. As climate change impacts rainfall patterns, the tropical world has become the most vulnerable to depleting resin-fed freshwater resources.[14] The long-term sustainability of spice vertical farms will hinge on innovations in renewable energy, climate-neutral building design, and efficient energy use across the value chain.[15]

However, these benefits depend on access to training, technological equity, and the ability to localize systems in diverse contexts. Robotic systems and software-heavy operations can centralize power in the hands of agri-tech companies, risking new forms of dependency. If the vertical farming industry must reinvent itself, it can consider a decentralized model – open-source climate algorithms, local and modular farms, and community cooperatives – as this could allow vertical farming to uplift rural economies while bringing spice cultivation closer to consumers.

The most profound question, however, is, When saffron can be grown in a warehouse in Stockholm with the same or enhanced aroma as Kashmir's harvest, what becomes of origin, authenticity, and cultural value? The 'synthetic terroir' notion challenges romantic ideas of place-based flavour. Yet it also democratizes access to rare crops. As vertical farming reshapes terroir from a passive inheritance to an active construction, raising essential questions about taste memory, identity, and innovation in food is critical.

About the Author

Priya Mani is the author of *Spiced: Recipes Celebrating Spices from around the World* (Phaidon Press, 2026) and the creator of the first Visual Encyclopedia of Indian Foods, a vast artistic exploration of India's culinary landscape.

Acknowledgements

With thanks to Mr Stephen Edwards, Chief Executive Officer, Vertically Pure Ltd, and Ann Sofie at Dansk Vanilje for their kind inputs.

Notes

1 Freya Pratty, 'The Vertical Farming Boom Is Over – for Now. What Went Wrong?', Sifted, 18 October 2023 <https://sifted.eu/articles/vertical-farming-boom-over-for-now-what-went-wrong> [accessed 17 May 2025].

2 Sidney W. Mintz and Daniela Schlettwein-Gsell, 'Food Patterns in Agrarian Societies: The "Core-Fringe-Legume Hypothesis": A Dialogue', *Social Science Information*, 19.6 (1980), pp. 937–53.

3 See, e.g., Rabia Khan and others, 'Internet of Things (IoT) Based Saffron Cultivation System in Greenhouse', *Scientific Reports*, 14 (2024), article 22589, doi:10.1038/s41598-024-69513-1.

4 See, e.g., Dana Shugrue, 'Outsmarting the Saffron Struggle', CEA Global World, 9 July 2024 <https://www.ceagworld.com/greenhouse-produce/outsmarting-the-saffron-struggle/> [accessed 25 April 2025].

5 Steve Edwards, interview with the author, 19 May 2025. Edwards works with Vertically Pure, a British vertical farm doing trials on vanilla production.

6 An expression I have borrowed from Michael Pollan (*Food Rules: An Eater's Manual* (Penguin, 2009)).

7 Idea expanded by David E. Cooper, *Farming according to 'the Laws of Beauty': Aesthetics and Agriculture.*

8 Roy Jenks, 'How to Grow Saffron', Garden.com, 21 January 2025 <https://www.garden.com/plant-management/how-to-grow-saffron> [accessed 23 April 2025].

9 Mario Giaccio, 'Crocetin from Saffron: An Active Component of an Ancient Spice', *Critical Reviews in Food Science and Nutrition*, 44.3 (2004), pp. 155–72. doi:10.1080/10408690490441433.

10 R. A. Dixon, 'Vanillin Biosynthesis – Not as Simple as It Seems?', in *Handbook of Vanilla Science and Technology*, ed. by D. Havkin-Frenkel and F. C. Belanger (Blackwell, 2010), doi:10.1002/9781444329353.ch18iot.

11 T. J. Zachariah and V. A. Parthasarathy, 'Black Pepper', *Chemistry of Spices*, ed. by V. A. Parthasarathy, B. Chempakam, and T. J. Zachariah (CAB International, 2008), pp. 21–40.

12 Maria Diez, Flavio Moreno, and Edison Gantiva, 'Effects of Light Intensity on the Morphology and CAM Photosynthesis of Vanilla planifolia Andrews', *Revista Facultad Nacional de Agronomía*, 70 (2017), pp. 8023–31, doi:10.15446/rfna.v70n1.61736.

13 See, e.g., Sowmya Mani, 'Senior Shubha Bhatnagar Builds Smart Farms for Indoor Saffron', Better India, 19 April 2024 <https://thebetterindia.com/347639/senior-shubha-bhatnagar-shubhavni-smartfarms-indoor-saffron-cultivation-iot-aeroponics-kashmiri-kesar/> [accessed 23 February 2025]; 'About Us', Veles Farming <https://www.velesfarming.com/about-us> [accessed 25 April 2025].

14 See, e.g., R. H. Kanth, B. A. Khanday, and S. Tabassum, 'Crop Weather Relationship for Saffron Production', in *Saffron Production in Jammu and Kashmir*, ed. by F. A. Nehvi and S. A. Wani (Directorate of Extension Education. SKUAST-K, 2008), pp. 170–88.

15 For one such effort, see 'Engineering and Technology', Australian Vanilla Plantation, 2024 <https://www.vanillaplantation.com.au/vanilla-dome> [accessed 12 May 2025]; Rebekka Boekhout, 'The Dome: A Potential Game-Changer for Vanilla Supply', Vertical Farm Daily, 19 August 2021 <https://www.verticalfarmdaily.com/article/9347817/the-dome-a-potential-game-changer-for-vanilla-supply/> [accessed 12 May 2025].

19
The Human Element
Automation, Robot Waiters, and the Disappointing Future(s) of Perfect Food

Rebecca Mazumdar

In 1995, actor Sam Waterston starred in a *Saturday Night Live* comedy skit as a spokesperson for a fictional insurance company. The two-minute commercial features a group of friends – all senior citizens – chatting in a living room. When one of them reveals that a friend's death wasn't covered by their life insurance policy, another one shares a hot tip: Old Glory Insurance covers you in case of a robot attack. The friends agree that the threat of a robot attack is terrifying; one proclaims, 'I don't even know why the scientists make them!', and another gasps, 'Oh! They're everywhere!' Waterston's deadpan monologue, which includes the matter-of-fact explanation that robots 'eat old people's medicine for fuel', overlays a montage of scenes featuring clumsy, bulky robots attacking the elderly. But for only four dollars a month, Old Glory policyholders can have the peace of mind that they're covered – in Waterston's words – 'For when the metal ones decide to come for you. And they will'.[1]

Thirty years later, reluctant to identify with the elderly woman who faints into her pile of leaves when a robot staggers into her yard, I find myself in an unthinkable position. On a day like any other, I'm at my local supermarket to get provisions for the week when a robot turns the corner and starts rolling down the aisle toward me. Marty, a 6′7″ 'mobile autonomous robot', rolls through Stop and Shop supermarkets to check stock on the shelves and look for floor hazards and spills. Shoppers have described his appearance, which features a flat heavy base supporting a singular tall pillar, as 'a cross between a hammerhead shark and a marital aid'. According to those who developed the robots, Marty isn't involved in surveillance or theft aversion, but that doesn't make him (or his vacant googly eyes) any less disturbing.[2]

I no longer shop at Stop and Shop.

I do, however, realize that the robots really are everywhere. Most recently, they've appeared as waitstaff (and even chefs) at restaurants, and at the risk of straining a muscle from clutching too tightly at my pearls, I'd like to use this space to examine the motivations and potential consequences of highly automated foodways. In other words, what are the actual costs and benefits of reducing the number of people involved in one's food preparation – and not by self-sufficiency, or locavorism, but by replacing so-called menial tasks with robots and automation? What are we losing when we remove the human element from our foodways? Reports of robot sightings aren't difficult to find. In March of 2024, *Forbes* published a prediction from a restaurant consultancy group 'that as much as 82% of restaurant roles could potentially be supplanted by robots, with up to 31% of those roles being devoted to food preparation'. Hygiene, efficiency, accuracy, and increased revenue are among the benefits noted in the article. The robots' novelty may also appeal to 'curious customers'. Author Daphne Ewing-Chow reviews five automated restaurants around the world, concluding, 'Embracing automation could enhance the quality and consistency of dishes while also opening doors to culinary exploration. But the question still remains – can a robot ever replicate the beauty of human artistry and creativity?'[3]

Meghan McCarron reported on the trend for *The New York Times* in August 2024. She notes, 'So far, and perhaps surprisingly, much of the automation currently being rolled out in the United States is on the human-facing, service side – robotic bussers, A.I.-powered drive-throughs and the ever-proliferating touch screens'. McCarron notes that restaurants often turn to robot servers for either cost-saving or labour-saving reasons. Robot servers can carry dishes to and from tables, sparing the human server of some of the most physically intensive parts of the job. McCarron also notes that robot chefs and other kitchen-based robotics have been harder to implement because those spaces are generally more chaotic and multilingual than the front of house. Reader comments on McCarron's article offer a glimpse of the debate, with some folks preferring humanless experiences out of concern for hygiene or because of introversion, with others pointing out that socializing with servers at their regular restaurants is part of their motivation to dine out in the first place. Another reader commented that because of their disability and their visibility as a minority, they'd prefer the unbiased attention of a robot 'any day'.[4]

In fall 2024, Florida International University (FIU) announced that it had installed the first fully robotic kitchen at a US college. Since robots only feed on the medicine of the elderly (that is, according to Old Glory Insurance), college students weren't at all concerned with the robot's presence on campus. The Beastro at FIU can create up to seventy different customizable meals, all within about four minutes, and it can prepare four at a time. It also cleans itself and compiles data to determine menu changes. And, rather than eliminating jobs, the Beastro has created four new positions for kitchen staff who will help operate the robot and give the completed meals to students.[5]

Before the Covid-19 pandemic, predictions of robotic foodways were cautiously optimistic. Writing for *The Atlantic* in 2018, Alana Semuels explains that rising labour costs and dropping . . . er, robot(?) costs, alongside the routine nature of many food-service tasks, made the restaurant and hospitality industries particularly prime for automation. Semuels also points out that, historically, new technology like automation creates more jobs than it eliminates, though she also notes that those new jobs often require more education and specialization than the eliminated jobs did, making them inaccessible for those displaced by the robots.[6] This reality is a significant contributor to the inequity that automation helps sustain, a point I'll return to later.

In the words of Carolyn Steel, 'Food and technology are twin pillars of our evolution'.[7] Robots may just be the next logical chapter in a long story about industrialized food, one which I'll attempt to gloss here. Initially, industrialized food offered consistency and safety. After all, it's harder for botulism to find its way into an aluminium can than into grandmother's basement stash of preserved carrots. By the early 1900s, foods like Jell-O (the sweet and wobbly gelatine dessert) began to appear on shelves, signalling a new era of progressivism and industrialization and widening the divide between eaters and any unpleasant details concerning the origins of their food.[8] After the World Wars, American consumers witnessed (and enabled) the rise of fast-food franchises, which offered side dishes of speediness, consistency, and hygiene.[9] (You might recall that hygiene, efficiency, and accuracy were some of the reported benefits of robots in restaurant settings.) The rise in availability of packaged food in the 1950s, coupled with misguided (and often misinterpreted) guidelines about nutrition and nutrients in the ensuing decades, has led to an average American diet comprised almost entirely of highly processed foods.[10]

During World War II, American corporations began to advertise for future purchases, promoting items that wouldn't be available until the war was over. Once wartime production could shift back to consumer products, the homemaker could return to her usual routine and – advertisers hoped – her usual brands. That anticipatory narrative informs the two ads for Fels-Naptha laundry soap which appeared in issues of *Good Housekeeping* in 1945. The text in one ad explains that Fels-Naptha soap will remain difficult to find at local shops as long as the plant is under orders to 'spend most of its time making soap for fighting men'.[11] The text in the second ad praises 'all these good-natured, patriotic women' who will resume their purchases of Fels-Naptha once the war is over, without holding any grudges for the current scarcity.[12] In such ads, consumers are encouraged to believe in an imagined future that will retain familiar details but also offer exciting new quantities and conveniences. This type of future resembles what Warren Belasco calls 'a classical future [which] is a continuation and elaboration of the progress of the past'.[13]

Within the same promissory structure of the ads mentioned previously were encouragements to homemakers to plan entire kitchen remodels without their husbands' help, even though companies admitted that they couldn't fulfil any such orders until after the war was over. A 1945 ad for a new, modernized kitchen from Crosley asks homemakers to conflate faith in a future defined by increased consumerism with both skilful homemaking and patriotic duty. A 'dream kitchen' is available with 'virtually no bother, upset, or troubles', but only when Crosley's '100% war production' is over.[14] But this kitchen, too, fits into Belasco's classical future: it's familiar, just better. Another possible future would soon make itself known, though.

After the war, when the production of consumer goods resumed and the possibilities of the dawning atomic age sprinkled everything with radioactive pixie dust, these promissory ads assumed a more technological, push-button style. Whirlpool's miracle kitchen of the future, the setting of Nixon and Khrushchev's 1957 kitchen debate in Moscow, boasted a push-button control centre and a fully automated dishwasher.[15] Vegetables could be chopped by an automated gizmo, and the floors were cleaned by something not at all unlike a Roomba. The incorporation of automated appliances and labour-saving gadgets makes the Whirlpool kitchen sound very much like science fiction. This type of future aligns with what Belasco calls a 'modernist future', which is defined by 'a distinct break with the past [and which] is a radically new vision based on the very latest technologies and scientific breakthroughs'.[16] In the printed images introducing these spaces to the American consumer, appliances have to be labelled, since it isn't immediately clear what they are. Also, something important is missing from this kitchen: people. The 1945 Crosley ad shows the homemaker's hands holding plans for her new kitchen. It's clear that the kitchen will be literally and figuratively in her hands. That assurance is missing from the 1957 push-button miracle.

The robotic trajectory of our food prep spaces began under the cloak of patriotism and scientific processes, and in this, our own post-pandemic world – one in which renewed and intensified fears of contagion saw some folks bleaching their Amazon delivery boxes – reducing the number of people handling their food became, for many people, a matter of life and death. Moreover, our fascination with food futures has almost always involved a conflation of technology and progress, even if plenty of sceptics have insisted on challenging the 'progress' part of that equation.

And here's where speculative fiction can give us risk-free glimpses into our possible futures, holding up mirrors that project both possibility and, often, caution. Ray Bradbury's 1950 short story 'There Will Come Soft Rains' imagines a future in which one of these mid-century push-button fantasies becomes the domestic reality. Set seventy-six years in the future, in the California of 2026, the McClellan house is fully automated. It anticipates the needs of its residents, announcing upcoming anniversaries and bill payments. It cooks breakfast without any human prompting: 'In the kitchen the breakfast stove gave a hissing sigh and ejected from its warm interior eight pieces of perfectly browned toast, eight eggs Sunnyside up, sixteen slices of bacon, two coffees, and two cool glasses of milk'.[17] When the food remains uneaten, the house also consumes the waste: 'An aluminum wedge scraped [the table scraps] into the sink, where hot water whirled them down a metal throat which digested and flushed them away to the distant sea'.[18]

As the tour of this miraculous house continues, the reader learns that the family has died in a recent atomic war, having had the misfortune of being in the yard when a blast occurred nearby. When the family dog returns, the door opens to his familiar whine, and he's admitted to a space empty of any human comfort. Readers learn about the family *in absentia* via the house's programmed activities, which it keeps performing in an otherwise silent and decimated suburb: there's an unattended bridge game, story hour in the desolate nursery upstairs, and a lonely cigar and a poetry reading in the evening. Meanwhile, the dog passes away in the living room, having been cruelly prohibited from reaching the delicious-smelling pancakes and syrup the kitchen prepares for nobody.

Eventually, the house succumbs to a fire. A branch falls through a window, spilling a chemical solvent in the kitchen, and the sprinkler system fails because no one was around to refill the house's depleted water tanks. The story warns us of the dangers of the over-automation of the home. The McClellan house performs actions usually attributed to a caring parent – such as cooking meals, hosting friends, and entertaining the children – with the routine nature of those programmed activities made evident in their post-nuclear persistence. But the house can't provide love or empathy. It can't comfort the dying dog; it can only sweep its carcass away.

The story also portrays *in extremis* the consequences of displacing the human element with a robotic one. The fiery death of the house re-creates that of the unfortunate McClellans, whose shadows are burned onto the side of their house in the atomic flash:

> The entire west face of the house was black, save for five places. Here the silhouette in paint of a man mowing a lawn. Here, as in a photograph, a woman bent to pick flowers. Still farther over, their images burned on wood in one titanic instant, a small boy, hands flung into the air; higher up, the image of a thrown ball, and opposite him a girl, hands raised to catch a ball which never came down.[19]

Both family and house are destroyed by fire – one from human aggression, the other from human negligence. Arguably, however, the *habitus* that would have made that house a home – the shared moments, the emotions, and the routine gestures of comfort that can create feelings of safety, belonging, and peace – were destroyed long before the war began, having been delegated to automated 'memory tapes' and 'electric eyes'.[20]

Bradbury's story reminds us that beneath the novelty and humourous curiosity around robots, there lies a more insidious problem. Whether entirely intentional or not, the long relationship between food and automation has been aided by both misogyny and class-based oppression. Promoting automated food as 'cleaner' and 'more consistent' than food made at home, which statistically speaking has been predominantly made by women, requires us to accept that consistency (as in, predictable sameness) is a more valuable part of a foodway than family, or cooperative kitchen work, or commensality, or eating intuitively or locally or sustainably. We're being asked to accept that sameness is more valuable than love.

Moreover, until the jobs created by restaurant robots are as accessible as were the entry-level service positions those robots eliminated, the introduction of these labour-saving measures will disproportionately displace vulnerable people: those who have lower levels of education, or those with family-care obligations who need flexible or nontraditional work schedules, or those in lower socioeconomic classes, or those in impoverished neighbourhoods, or those in rural areas where entry-level service jobs are some of the few remaining opportunities for employment. This automation trajectory insists that the transactional, financial component of food consumption is ultimately more important than the human experiences that food makes available to the consumer. In the United States, where access to health care is tied to employment, the rise of automation in entry-level service roles will ensure that, in the immortal words of Bono, 'the rich stay healthy [and] the sick stay poor'.[21]

I want to turn now to another story from Bradbury's 1950 collection *The Martian Chronicles* called 'The Long Years'. The story takes place on Mars, ostensibly just a few months before the events in 'There Will Come Soft Rains'. In it, a retired astronaut named Hathaway lives a quiet life on Mars with the robots he built and programmed to keep him company after his wife and three children died. Nineteen years later, when a rocket of Earthmen finally lands, Hathaway is so overcome with happiness that he has a heart attack and dies. His robot family lives on, a foursome of ageless, loving family members who never learned how to cry or to be lonely, because Hathaway never wanted them to feel those things. Just before he dies, Hathaway toasts his 'wife and children, without whom I couldn't have survived alone. It is only through their kindness in caring for me', he says, 'that I've lived on, waiting for your arrival'.[22]

Hathaway has built and programmed the perfect robot. This is fiction, after all. His robots provide the human element he craves and which has enabled his survival. He treats them with love and lives with them as family for almost twenty years. Those robot family members are so advanced, so realistic, that when the rescue crew leaves Mars, they find that they can't shut them down. They can't power them off, even though like the McClellan house in 'There Will Come Soft Rains', there aren't any more humans for them to serve. One crew member even admits that powering off the robot family would 'be murder!'[23]

And yet, Hathaway died because his heart couldn't take the excitement, the pure exhilaration, of seeing actual humans again.

Bradbury's stories, like much science fiction, revise and secularize the jeremiad. Originally a genre of sermons, the jeremiad reminded congregants of their promise to God, identified the ways they had failed to meet their obligations, and offered ways to return to the covenanted path. Science fiction, in contrast, offers what David Seed describes as an anti-jeremiad, or a 'negative prophecy where dreaded outcomes are envisaged and therefore hopefully deferred, in such a way that the reader is induced to ponder on present signs of disaster'.[24] Bradbury's texts ask us to look at our present circumstances for the foundations of any consequences we want to avoid. He asks us to look our robots in their creepy, vacant eyes and ask ourselves which aspects of our humanity are worth protecting.

About the Author

Rebecca Mazumdar is an Associate Professor of English at the New York City College of Technology. She teaches classes in gothic literature and gender studies, and her research focuses on foodways, food justice, and food and gender.

Notes

1 'Old Glory Insurance – SNL', *Saturday Night Live*, season 21, 1995 <https://youtu.be/g4Gh_IcK8UM?si=eEUTDutaJNiQvU9N> [accessed 24 May 2025].

2 Joanna Weiss, 'The Supermarket Superstardom of Marty the Robot', *Boston Magazine*, 12 September 2024 <https://www.bostonmagazine.com/news/2024/09/12/marty-the-robot-stop-and-shop/> [accessed 24 May 2025].

3 Daphne Ewing-Chow, 'Here Are Five Global Restaurants Staffed by Robot Chefs', *Forbes*, 31 March 2024 <https://www.forbes.com/sites/daphneewingchow/2024/03/31/here-are-five-global-restaurants-staffed-by-robot-chefs/> [accessed 25 May 2025].

4 Meghan McCarron, 'Do Robots Love Their Customers? Automated Restaurants Face Human Issues', *New York Times*, 13 August 2024 <https://www.nytimes.com/2024/08/13/dining/ai-robots-restaurants.html> [accessed 25 May 2025].

5 Gisela Maria Valencia, 'Welcome to the Future: Robot Chef Whips Up Tasty Meals at FIU', *FIU News*, 6 September 2024 <https://news.fiu.edu/2024/welcome-to-the-future-robot-chef-whips-up-tasty-meals-at-fiu> [accessed 25 May 2025].

6 Alana Semuels, 'Robots Will Transform Fast Food', *Atlantic*, January/February 2018 <https://www.theatlantic.com/magazine/archive/2018/01/iron-chefs/546581/> [accessed 25 May 2025].

7 Carolyn Steel, *Sitopia: How Food Can Save the World* (Chatto & Windus, 2020), p. 13.
8 Anna Zeide, *U.S. History in 15 Foods* (Bloomsbury, 2023), pp. 87–99.
9 For more on the importance of cleanliness in the advent of White Castle, America's first fast-food hamburger chain, see Adam Chandler, *Drive-Thru Dreams: A Journey through the Heart of America's Fast-Food Kingdom* (Flatiron Books, 2019). For a detailed history of the way fast food has come to define American cuisine and American life, see Eric Schlosser, *Fast Food Nation: The Dark Side of the All-American Meal* (Mariner Classics, 2001).
10 Kristen Lawless, *Formerly Known as Food: How the Industrial Food System Is Changing Our Minds, Body, and Culture* (St. Martin's Press, 2018), p. 3; Michael Pollan, 'Unhappy Meals', *New York Times Magazine*, 28 January 2007 <https://www.nytimes.com/2007/01/28/magazine/28nutritionism.t.html> [accessed 26 May 2025].
11 Fels-Naptha, 'Not yet, But–', *Good Housekeeping*, September 1945, p. 63.
12 Fels-Naptha, 'All Is Forgiven!', *Good Housekeeping*, November 1945, p. 63.
13 Warren Belasco, *Meals to Come: A History of the Future of Food* (University of California Press, 2006), p. xii.
14 Crosley, 'This String Is All I Need to Plan My NEW Kitchen!', *Good Housekeeping*, September 1945, p. 120.
15 'The 1957 RCA Whirlpool Miracle Kitchen of the Future: Living Tomorrow', Rare Historical Photos <https://rarehistoricalphotos.com/whirlpool-miracle-kitchen-1957/> [accessed 25 May 2025].
16 Belasco, *Meals to Come*, p. xii.
17 Ray Bradbury, 'There Will Come Soft Rains', in *The Martian Chronicles* (Bantam, 1950), pp. 166–72 (p. 166).
18 Bradbury, p. 167.
19 Bradbury, pp. 167–68.
20 Bradbury, p. 167.
21 Paul David Hewson and others, 'God Part II', in *Rattle and Hum* (Island Records UK, 1988).
22 Bradbury, 'The Long Years', in *Martian Chronicles*, pp. 156–66 (p. 163).
23 Bradbury, p. 165.
24 David Seed, *American Science Fiction and the Cold War* (Fitzroy Dearborn, 1999), p. 9.

20
Beyond the Lemon
A Comprehensive Exploration of Elements of Sourness in Persian Cookery

Nader Mehravari

Persian cuisine is renowned for its vibrant tapestry of flavours, with souring agents as a central and defining element.[1] Over the centuries, Persians have cultivated a unique and unparalleled appreciation for tangy tastes. This affinity has led to the discovery, development, and masterful use of a wider range of souring agents than found in any other global food cultures.

Souring agents are ubiquitous in Persian cookery.[2] Although often masterfully hidden in a range of savoury and sweet creations, at times, however, they are visible team players. The skilful use of sour flavours is a hallmark of Persian cuisine, contributing to its unique character and appeal. Whether it's the tangy marinade for grilled meat, the tartness of a yogurt-based dip, or the refreshing acidity of a summer drink, sourness plays a vital role in creating the memorable and flavourful experiences of Persian cuisine.

The range of souring agents employed in Persian cuisine is truly impressive. These range from the gentle tang of common and uncommon citrus fruit to such zesty berries as sumac and barberry, from sour fruit such as pomegranates and sour cherries to the sharpness of unripe grapes and unripe plums, from the fruit of leguminous trees such as tamarind to fermented dairy products such as yogurt and kashk – not to mention possibly the sourest culinary ingredient in the world, *ghareh-ghoroot* (a twice-cooked, strained, fermented dairy product with an extraordinarily sour flavour).[3]

This extensive assortment of souring elements is used in a variety of forms and shapes, including fresh, juiced, dried, whole, ground, pickled, fermented, jammed, syruped, distilled, and inspissated.

Persian cooks use this vast arsenal of ingredients across the entire Persian cookery landscape, including warm and cold cooked savoury food items (rice dishes, soups, stews, braises, porridges, grilled meats, dumplings, stuffed dishes, patties, egg-centric dishes, etc.), raw savoury dishes (yogurt dishes, salad-like dishes, etc.), confections, desserts, frozen sweets, hot and cold beverages, snacks, garnishing, and dry and liquid flavouring used by the diners at the table.

The goal of this paper is twofold: first, to explore an overarching, and as complete as possible, range of fundamental elements of sourness used in the Persian cookery landscape, their sources (plant-based, animal-based, or chemically synthesized), their preparation techniques, their uses across all categories of Persian savoury and sweet food items, whether used by cooks in the kitchen or by the diners at the table, and more; and second, to offer a generic approach and framework that can be employed by others to explore the other foundational elements of cookery in other food cultures.[4]

Nomenclature and Overall Approach

A comprehensive exploration of sourness in Persian cuisine requires distinguishing between 'individual souring agents', used as ingredients by cooks or diners, and the 'fundamental sources' from which they are derived. For example, fresh unripe green sour grapes are the fundamental source of at least six distinct souring agents:

- Whole fresh unripe green sour grapes (e.g. used by cooks to flavour stews)
- Verjuice (e.g. used by diners to flavour green salads)
- Whole dried unripe green sour grapes (e.g. used by cooks to flavour soups)
- Ground dried unripe green sour grapes (e.g. used by diners to sprinkle on grilled meat)
- Unripe green sour grape unsweetened molasses (e.g. used by cooks to flavour stews)
- Unripe green sour grape pickles (e.g. used by diners as an accompaniment)

My investigation began by compiling an initial list of individual souring agents used in Persian cooking. This list underwent a multifaceted analysis, including:

- Categorizing individual souring agents by their fundamental source
- Identifying typical culinary uses of souring agents across Persian cuisine
- Cross-referencing individual souring agents with their culinary applications
- Visualizing the relationship between fundamental sources and the culinary techniques used to create the final souring agents

Finally, several unique and less common Persian souring elements are highlighted with additional details.

Detail Listing and Categorization of Individual Souring Elements

Our investigation into souring elements in Persian cuisine began by compiling a comprehensive list of individual souring agents used by cooks based on a range of sources, including historical cookbooks, contemporary culinary texts, interviews with home and professional cooks, and drawing on my own over forty-year study of Persian culinary history, principles, and practices.[5]

Next, we developed a classification system to group these agents. After considering several options, such as categorization based on chemical properties and regional usage, we chose a scheme centred around the fundamental sources of the souring agents, shown in Figure 1, as the most useful for exploring these elements from a culinary perspective. These fundamental sources include a wide range of plant-based, animal-based, and chemically synthesized food items.

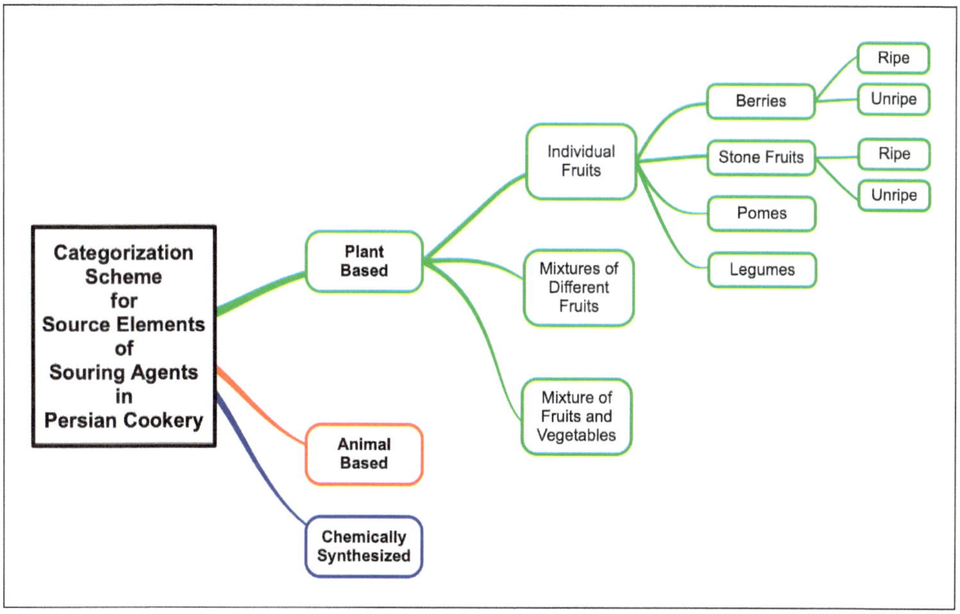

Figure 1. Categorization scheme for source elements of individual souring agents in Persian cookery.

Next, we applied the selected categorization scheme to the comprehensive list of individual souring agents that we had compiled. The results are shown in Figure 2.

Most Typical Uses of Souring Agents in Persian Cookery

The next key step in our exploration was to identify the most typical ways in which souring agents are used in Persian cookery. There are two facets to this question:

1. Whether the individual souring agents are used by the cook in the kitchen or by the diners at the dining table
2. What class (type, category) of dishes the individual souring agents are used in

The result of that analysis is shown in Figure 3.

Cross-Reference between Individual Souring Elements and Their Uses

Our next step was to develop an extensive cross-reference between individual souring agents and their uses in Persian cuisine. This was achieved by carefully relating almost all of the individual forty-eight souring agents identified earlier (Figure 2) with one or more of the uses as identified in the previous section (Figure 3). The results of that exercise are shown in Figure 4.

Higher-level categorization		Source elements	Individual souring elements
Plant based	Berry	Sumac	Whole dried sumac
			Ground dried sumac
		Barberry	Whole dried barberry
			Dried barberry jam
		Pomegranate	Pomegranate juice
			Fresh pomegranate aril
			Dried pomegranate aril
			Ground dried pomegranate aril
			Pomegranate unsweetened molasses
		Grape	Wine vinegar
		Tomato	Tomato paste
	Unripe	Unripe sour green grape	Whole fresh unripe green sour grape
			Verjuice – juice of fresh unripe green sour grape
			Whole dried unripe green sour grape
			Ground dried unripe green sour grape
			Unripe green sour grape unsweetened molasses
			Unripe green sour grape pickle
		Unripe sour green plum	Whole fresh green sour plum
			Whole dried green sour plum
			Pickled green sour plum
			Green sour plum unsweetened molasses
	Drupe	Cornelian cherry	Fresh Cornelian cherry
			Dried Cornelian cherry
			Cornelian cherry jam
			Cornelian cherry pickle
		Sour cherry	Fresh sour cherry
			Sour cherry juice
			Dried sour cherry
			Pickled sour cherry
			Sour cherry syrup
	Citrus	Lime	Lime juice
			Whole dried lime
			Ground dried lime
			Dried lime interior segment
			Dried lime interior segment jam
		Lemon	Lemon juice
			Pickled lemon
		Seville orange	Seville orange juice
			Seville orange unsweetened molasses
	Pom	Apple	Apple cider vinegar
	Leguminous	Tamarind	Tamarind paste
	Mixture	Various fruit	Lavashak – Persian-style fruit rollup
		Fruit & vegetable	Pickle
Animal based	Milk		Yogurt
			Strained yogurt
			Kashk
			Ghareh-ghoroot
Chemically synthesized			Citric acid powder

Figure 2. Categorization of individual souring agents.

	Uses of Persian souring agents	Used by the cook in the kitchen	Used by the diners at the table
a	As an ingredient in cooked warm savoury dishes	X	
b	As an ingredient in cooked cold savoury dishes	X	
c	As an ingredient in raw savoury dishes	X	
d	As an ingredient in confections	X	
e	As an ingredient in frozen sweets	X	
f	As an ingredient in compotes	X	
g	As an ingredient in jams	X	
h	As an ingredient in pickles	X	
i	As an ingredient in snacks	X	
j	As an ingredient in hot beverages	X	X
k	As an ingredient in cold beverages	X	X
l	As a garnish for cooked warm savoury dishes	X	
m	As a garnish for cooked cold savoury dishes	X	
n	As a garnish for raw savoury dishes	X	
o	As a liquid flavouring at the table		X
p	As a ground dry flavouring at the table		X
q	As an accompaniment at the table		X
r	As a breakfast component		X
s	As a snack		X

Figure 3. Uses of Persian souring agents.

Culinary Techniques Utilized to Develop the Individual Souring Agents

The final step was to study the relationship between the base sources for souring agents and culinary techniques utilized to develop the actual souring agents used by the cooks. The specific culinary techniques considered included:

- Used fresh and whole without any processing
- Juiced
- Dried whole
- Dried ground
- Pickled
- Fermented
- Jammed
- Syruped
- Distilled
- Inspissated
- Cooked and then dried

The results of this part of our exploration are shown in Figure 5.

				a	b	c	d	e	f	g	h	i	j	k	l	m	n	o	p	q	r	s
Plant based	Berry	Sumac	Whole dried sumac	X																		
			Ground dried sumac	X	X	X													X			
		Barberry	Whole dried barberry	X						X	X	X	X									
			Dried barberry jam									X		X							X	
		Pomegranate	Pomegranate juice	X	X	X									X	X	X					
			Fresh pomegranate aril			X																
			Dried pomegranate aril	X																		
			Ground dried pomegranate aril	X															X			
			Pomegranate unsweetened molasses	X	X	X									X			X				
		Grape	Wine vinegar	X	X	X												X				
		Tomato	Tomato paste	X																		
	Unripe	Unripe sour green grape	Whole fresh unripe green sour grape	X						X												
			Verjuice – juice of fresh unripe green sour grape	X	X	X					X							X				
			Whole dried unripe green sour grape	X																		
			Ground dried unripe green sour grape	X	X	X																
			Unripe green sour grape unsweetened molasses	X	X	X								X								
		Unripe sour green plum	Whole fresh green sour plum	X					X	X					X							X
			Whole dried green sour plum	X																		
			Pickled green sour plum																	X		
			Green sour plum unsweetened molasses	X	X	X								X								
	Drupe	Cornelian cherry	Fresh Cornelian cherry																			X
			Dried Cornelian cherry																			X
			Cornelian cherry pickle																		X	
		Sour cherry	Fresh sour cherry																			
			Sour cherry juice											X								
			Dried sour cherry	X																		X
			Pickled sour cherry																	X		
			Sour cherry syrup											X				X				

Plant based	Citrus	Lime	Lime juice	X	X	X	X	X	X	X	X	X	X	X				X				
			Whole dried lime	X																		
			Ground dried lime	X	X	X													X			
			Dried lime interior segments	X						X		X										
			Dried lime interior segments jam																		X	
		Lemon	Lemon juice	X	X	X	X	X	X	X	X	X	X	X				X				
			Pickled lemon																	X		
		Seville orange	Seville orange juice	X	X	X	X	X	X	X	X	X	X	X				X				
			Seville orange unsweetened molasses	X	X	X								X								
	Pom	Apple	Apple cider vinegar	X	X	X												X				
	Leguminous	Tamarind	Tamarind paste	X																		X
	Mixture	Various fruit	Lavashak – Persian-style fruit rollup									X										X
		Fruit & vegetable	Pickle																	X		
Animal based	Milk		Yogurt	X	X	X	X													X		
			Strained yogurt	X	X	X	X													X		
			Kashk	X														X				
			Ghareh-ghoroot	X											X							X
Chemically synthesized			Citric acid powder						X				X									

Figure 4. Cross-reference between individual souring agents and their uses.

a = Ingredient in cooked warm savoury dishes
b = Ingredient in cooked cold savoury dishes
c = Ingredient in raw savoury dishes
d = Ingredient in confections
e = Ingredient in frozen sweets
f = Ingredient in compotes
g = Ingredient in jams
h = Ingredient in pickles
i = Ingredient in snacks
j = Ingredient in hot beverages
k = Ingredient in cold beverages
l = Garnish for cooked warm savoury dishes
m = Garnish for cooked cold savoury dishes
n = Garnish for raw savoury dishes
o = Liquid flavouring at the table
p = Ground dry flavouring at the table
q = Accompaniment at the table
r = Breakfast component
s = Snacks

			Forms, shapes, and processing methods												
Higher-level categorization		Source elements	Fresh			Dried			Pickled	Fermented	Jammed	Syruped	Distilled	Inspissated	Cooked & dried
			Whole	Juice	Other	Whole	Ground	Other							
Plant based	Berry	Sumac				X	X								
		Barberry				X					X				
		Pomegranate		X	X		X	X						X	
		Grape								X			X	X	
		Tomato													
	Unripe	Unripe sour green grape	X	X		X	X		X					X	
		Unripe sour green plum	X			X			X		X			X	
	Drupe	Cornelian cherry	X			X			X		X				
		Sour cherry	X	X		X			X		X	X			
	Citrus	Lime		X		X	X	X							
		Lemon		X					X						
		Seville orange		X										X	
	Pom	Apple								X			X		
	Leguminous	Tamarind			X										
	Mixture	Various fruit													X
		Fruit & vegetable							X						
Animal based		Milk								X					X
Chemically synthesized		Citric acid powder						X							

Figure 5. Relationship between the base sources for souring elements and culinary techniques utilized to develop the actual souring agents used by the cooks.

A Sampling of Unique and Less Common Persian Souring Elements

In addition to the systematically exhaustive analysis of all the souring agents used in the Persian cookery landscape, I felt it was important to highlight several of the more unique and uncommon Persian souring agents.

Ghareh-Ghoroot

- Fundamental source: Milk

- Production process: An extraordinarily sour ingredient which is a fermented dairy product that has gone through two different lengthy stages of cooking and one stage of straining. It is one of the sourest culinary ingredients anywhere. Milk is first fermented to produce yogurt. A mixture of yogurt and water is heated, causing it to curdle. The curdled mixture is strained, separating the solids from the liquid whey. The liquid whey is then boiled for an extended period, resulting in a dark blackish-brown paste, which is *ghareh-ghoroot*. See Figure 6 for the colour changes of whey as it boils down to form *ghareh-ghoroot*.
- Primary usage: Beyond its role as a flavouring agent for meat braises and soups, it's a cherished snack for many Iranians.

Limoo-Parak

- Fundamental source: Lime fruit
- Production process: *Limoo-paraks* are the dried interior segments (endocarp) of lime with the thin membranes removed (Figure 7). It is the most concentrated form of lime sourness. The process begins by first fermenting and drying whole fresh Persian limes, resulting in what is known as *limoo-omani*, which has a unique umami earthy flavour and comes in a range of colours from pale brown to dark brownish black. These dried whole limes are then cracked open, and the interior dried segments of the dried lime are separated.
- Primary usage: Souring agents for meat braises and soups. They are also used as an ingredient in sour tisanes.

Figure 6. Colour changes of ghareh-ghoroot *in its second cooking stage; it gets sourer as it gets darker.*

Figure 7. Limoo-parak*: the dried interior segments (endocarp) of lime with the thin membranes removed.*

Rob-é-Ghooreh

- Also known as: Unripe green sour grape unsweetened molasses
- Fundamental source: Unripe green sour grapes
- Production process: Inspissation is the process of increasing the viscosity of a fluid, typically by boiling, resulting in anywhere from a highly viscous but still pourable liquid to a paste. It is an ancient and popular practice in the Persian culinary landscape. In Persian, the resulting thickened syrup or paste is called *robe*. The process is applied to a wide range of fruits and vegetables, including pomegranates, sour plums, and tomatoes. There is no single-word English name for *robe*. While English-language texts often categorize it as a 'molasses', traditional *robe* contains no added sugar.
- Primary usage: Souring agent for meat braises and soups.

A Generic Approach and Framework

The specific steps used to explore the elements of sourness in the Persian culinary landscape, as described previously, can easily be generalized to form a generic approach and framework that can be employed by others to explore the other foundational elements of cookery in different food cultures. Figure 8 depicts such a generic framework and its encompassing steps. The italicized texts in Figure 8 represent examples of how the components of this generic framework correspond to steps used to explore the elements of sourness in the Persian culinary landscape.

Conclusions and Closing Thoughts

Persian cuisine boasts a rich and diverse tapestry of flavours, with souring agents playing a pivotal role. This exploration delved into the unique significance of sourness in Persian culinary traditions, showcasing the vast array of souring agents used, their diverse applications in both sweet and savoury dishes, and their preparation techniques. Through

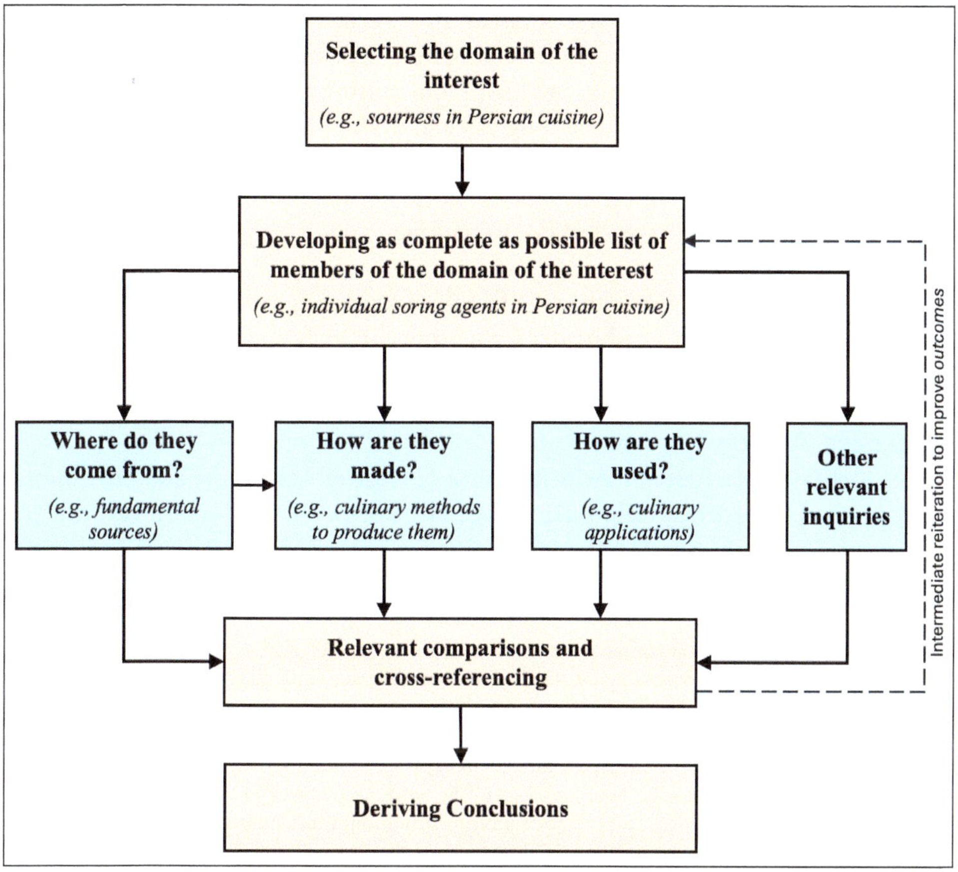

Figure 8. A generic approach and framework.

meticulous research, approximately fifty distinct souring agents were identified while highlighting their fundamental sources, their culinary uses, and the intricate relationship between raw ingredients and finished products.

Some of the key findings include:

- Persian cuisine employs approximately fifty distinct souring agents, with even more regional variations, showcasing the breadth of their culinary tradition.
- These souring agents are derived from a variety of sources, including plants, animals, and synthetic processes, with the majority being plant-based.
- Unripe fruits play a pivotal role in the creation of souring agents, highlighting the ingenuity of Persian culinary practices.
- The versatility of certain fruits is evident in their ability to yield four to six individual souring agents, demonstrating the depth of flavour manipulation in Persian cuisine.
- Pomegranates, unripe grapes, unripe green plums, sour cherries, and limes stand out as foundational ingredients, collectively contributing to over half of the most popular souring agents.

- Souring agents are integral to a vast array of Persian dishes, spanning from hearty stews and delicate pastries to refreshing beverages and flavourful condiments.
- The transformation of base ingredients into souring agents involves a diverse range of culinary techniques, including juicing, drying, jamming, pickling, grinding, fermenting, distilling, and inspissating.
- Citrus-based souring agents emerge as the most versatile, finding applications across nearly all categories of Persian cuisine.

This paper not only illuminates the specific role of souring agents in the Persian culinary landscape but also proposes a broader framework for analyzing foundational elements in any food culture.

About the Author

Nader Mehravari is a Research Associate at the College of Agriculture and Environmental Sciences, University of California, Davis. His work explores the history, principles, and practices of ancient and contemporary Persian cookery and associated foodways.

Notes

1 Margaret Shadia, 'The Sometimes Sweet but Mostly Sour Flavour of Many Persian Dishes Today', in *Spicing Up the Palate: Proceedings of the Oxford Food Symposium on Food and Cookery 1992*, ed. by Harlan Walker (Prospect Books, 1993), pp. 243–46.

2 Bita Arabian and Beata Nazem Kelley, 'The Sours in Persian Cuisine', *Modern Persian Food Podcast*, 9 September 2021 <https://modernpersianfood.com/episode-52-the-sours-in-persian-cuisine> [accessed 21 October 2024]; Gilgamesh, 'Sour and Sweet: The Secret Ingredients of Persian Cuisine', Iran Tour <https://web.archive.org/web/20250218030350/https://irantour.tours/iran-blog/sour-and-sweet.html> [accessed 12 May 202].

3 Vanniya Sriangura, 'The Sour Side of Thai Cuisine', *Bangkok Post*, 8 December 2023 <https://www.bangkokpost.com/life/social-and-lifestyle/2701211/the-sour-side-of-thai-cuisine> [accessed 21 October 2024].

4 Arva Ahmed, 'Persian or Indian Recipes That Use Unusual Sour Ingredients', *Frying Pan Adventures*, 31 October 2020 <https://www.fryingpanadventures.com/the-deep-fried-podcast/persian-or-indian-recipes-sour-ingredients> [accessed 21 October 2024]; Mishan, Ligaya, 'How Sourness Has Come to Dominate Our Dining Habits, and Our Discourse', *New York Times Style Magazine*, 14 May 2019 <https://www.nytimes.com/2019/05/14/t-magazine/sour-food.html> [accessed 21 October 2024]; Sriangura, Vanniya, 'The Sour Side of Thai Cuisine', *Bangkok Post*, 8 December 2023 <https://www.bangkokpost.com/life/social-and-lifestyle/2701211/the-sour-side-of-thai-cuisine> [accessed 21 October 2024].

5 Nader Mehravari, 'How to Stock a Persian Pantry', Serious Eats, 1 December 2022 <https://www.seriouseats.com/how-to-stock-a-persian-pantry-6834907> [accessed 21 October 2024]; Mehravari, Nader, 'Persian Comfort Food: A Foodways Exploration', *Digest Journal of Foodways and Culture*, 9.2 (Fall 2022) <https://scholarworks.iu.edu/journals/index.php/digest/article/view/35795/38855> [accessed 21 October 2024]; Margaret Shaida, *The Legendary Cuisine of Persia* (Lieuse, 1992).

21
Food as Transmission

Jennie Moran

Last November, I had an opportunity to travel up the last 150 centimetres of my large intestine. Because of my family medical tapestry, I am a regular visitor to the colonoscopy clinic.

Being generally of resilient physical condition, I approach these investigations with low-key stoicism. I gave birth when I was nineteen. I have worked for many years in kitchens. I broke my foot on the way to host a dinner in an art gallery and proceeded unflinchingly with the site-specific conceptual banquet. I have more than once witnessed the little surprise of seeing the top of my finger appear on a chopping board next to some diced shallots. I am painting a picture for you of a competent operator with little time for histrionics. My physicality conveys this, but you can't see me here. When I was in my early twenties, I hated looking strong. I begrudged my large breasts and sturdy shoulders. I believed the overall parcel attracted needy men. Here is a body built for performing acts of hospitality. I will hold you up and feed you things. Picture me kneading dough and telling you everything is alright – folding you into me while you leak luxurious, existential tears onto my floury bosom.

There is a soft spot, though, a weakness, something that will cause me to crumple like a ball of tinfoil, and that is the prospect of being penetrated by a needle of any kind. So, to relate this back to the colonoscopy – because I know you are eager to get back on topic – the ingesting of several litres of osmotic laxative I face with quiet deep breaths. The idea of a camera travelling up my bum is strangely acceptable. Whatever, have at it. The prospect of a well-meaning medical practitioner in a faded teal outfit telling me to take deep breaths while they insert a canula into the back of my hand is, however, unthinkable. It invokes terror. I panic and cry, and sometimes I pass out altogether. Early in my career as a colon muse, I gleaned that the only reason for interacting with a needle in the endoscopy suite is for the purposes of sedation. Henceforth, I resolved to face the ordeal lucid. I don't disclose my rationale, so declining sedation gives the nurses the mistaken impression that I am extremely hard ass (apologies). I am ready with my wits about me for this intimate inward adventure.

The Mouth as Threshold

There is more to this, of course. I have some extra curiosity about the gut. I have spent the past twelve years feeding people on a large scale. Looking through our till records, I estimate six hundred thousand times I exchanged food for money. Over these years, I thought a lot about food as a tool for unconditional welcome as I wrestled and wrangled with the laws of hospitality – ancient, strict, and thoroughly unreasonable: 'Say YES to who or what turns up'.[1] I interacted with a good proportion of these six hundred thousand hungry wayfarers myself and, over time, developed the skill of prescribing food like medicine, of gauging by looking at someone what it was they needed from me: 'The more mouths there are, the more mouths that need to be fed. The more mouths there are however, the more that also want to speak and be understood'.[2] There is a complex intimacy to feeding people. The 'biological individual' standing before me is trusting me to send a parcel of nourishment across the threshold of their mouth and down into their gut, where it will be 'incorporated' into their inner selves.[3]

There is important context around this act. Feeding people is an act of hospitality. Hospitality is the practice of allowing strangers into our lives. It is responding to the rupture that is the arrival of the stranger with a willingness to suspend suspicion. It is a high-risk contract that was authored not by us but by our primordial ancestors. There is no physical record of this contract, no ancient manuscript to visit, no pilgrimage to be done. This codex we carry internally in our human DNA. It manifests as a willingness to engage with the interrupting stranger. We preserve this bond by activating it, by opening the door. Through these acts of hospitality, we become diligent scholars of our coexistence, learning by doing, the habit of hands that stretch out instinctively to the stranger. Tiny gestures to our fellow human and more-than-human beings. Overtures. Yesses. Maybes. These extending tendrils are proof that we still carry the knowledge of this beautiful practice, even unconsciously, even without fully understanding why or where it comes from.

I can see the mesmerizing beauty of these countless unconscious yesses. It is like a whispering murmuration. The effect is immense. The accumulation of these *gut* instincts transmitting and dancing produces a force. Here is proof that this code we carry inside us is resilient and has not been forgotten. Here is hope. Philosopher Emmanuel Levinas articulates this beautifully as a permanent readiness to receive the stranger: 'I am always automatically a host bearing responsibility for the strange Other, who is my guest [. . .] as if I had to do with the other before knowing him, in a past that has never taken place'.[4] I am moved by this line. It has a fizzy truth about it that I want to cling on to. Here Levinas is suggesting that we already have a connection to this stranger, but we don't know what it is or when or why it was formed. We will probably never understand it. And yet, sometimes we manage to accept it. Here is the unseen code. Here, in a hidden, inaccessible place within us exists a propensity for active coexistence. This is unconditional welcome, and it is beyond reason or language. Imagine us – born ready and waiting. Preprogrammed and predisposed towards hospitality. Look for it now in your own cells. Find it way down there in our collective humanity.

For Levinas, this responsibility to the stranger supersedes all other ethical concerns. In classical society, hospitality (*xenia* in Greek, *hospitium* in Latin) takes precedence over kinship or marriage in matters of affection and obligation. The bonds formed between host and guest do not even dissolve with death. The connection is passed down along to the descendants of the original partnership. The rites of initiation into this bond involve a series of elements – a solemn declaration between both parties. 'I make you my *xenos*', an exchange of gifts, a handshake, and finally, to render the bond indissoluble – a feast.[5]

Dinner Lady

This is where I come in. I am in the business of feasting and of yesses. I am here to close the hospitality deal, to bond people together for generations to come. I believe that the mesmerizing tangle of connections we form while we are alive is our most precious legacy: the lives we touch, the people who interrupt us, the sparks we ignite. This beautiful, tangled mesh is spun gold. Hospitality is how these connections are formed. I am attempting to administer unconditional welcome in edible form to aid this process. This could never work if I saw the six hundred thousand stomachs in front of me as mere receptacles. I must believe that there is an exchange. When the stranger's mouth opens faithfully to accept food from the host, it creates a channel from the inner self to the world.[6] In exchange for aliment, I am offered glimpses of the inner self of the mouth's owner. These stories are wordless and untranslatable, from a place and time that I have never been. The resonance is unfamiliar yet recognizable. They are intellectually indigestible. Yet the desire for these stories is like some sublime craving, never fulfilled, only deepened: 'It nourishes itself one might say, with its hunger'.[7]

When the stranger's lips part to allow the food offering in, they are engaging actively with the rules of hospitality. These directions are not issued with words. Like the stories that emerge, they are separate from language. This simple physical gesture of the open mouth is connected to the profound and ancient predisposition towards coexistence. Something that comes from 'a past that has never taken place'.[8] To open one's mouth in good faith to the host is to reactivate the primordial code of hospitality. I witnessed six hundred thousand of these openhearted overtures. I want to describe for you the impact of all these yesses. I am reaching for large language, terms like 'sublime', 'transcendence', but they are too shouty. I refresh myself on the definition of a 'hyperobject' – entities massively distributed in time and space, of such vast temporal and spatial dimensions that they defeat traditional ideas about what a thing is.[9] This gets me close. Timothy Morton, who coined the phrase, uses it in relation to climate change, plutonium, and polystyrene – things that invoke dread and can never be seen in their entirety. I invite you into the thought exercise of imagining hospitality as a benevolent hyperobject. I feel its immensity but have no means of describing its form. It is more powerful than we who authored it, if, in fact, we did author it. Perhaps we, as a species, just tuned into the practice of hospitality as a means of 'sharing the curved surface of the earth'.[10] It is potentially always everywhere. I stood surrounded by these hundreds of thousands of open mouths and felt an activation of something unknowable.

Carnal Hospitality

This is a roundabout way of contextualizing for you the desire to be lucid for my colonoscopy. I wanted a visual reference for this inner world. I wanted to see where these open mouths lead to. I wanted to be better placed to contemplate the landscape of incorporation, the place where my edible hospitality endeavours end up. I wanted to look into my own entrails like a haruspex, seeking explanations for things I don't understand. In *The Birth of the Clinic*, Foucault helps with the incongruous intersection of perception and observation. I *perceive* the effect of food as a tool for unconditional welcome. I *observe* intestines. There is a danger that the observation, the cold fact of the image, will override my own embodied knowledge, much like watching the film version of a beloved book. I proceed nonetheless, with faith in the resilience of the sculptural inner world I have preconceived. I am eager to penetrate this interior with my gaze. Modern medicine has brought the private universe of the body from darkness into light, into a space 'of constant visibility'.[11] I wasn't so naive as to look for truth or proof of anything, merely to receive impressions.[12] I know that the subject of my investigation, the ancient manuscript of unconditional welcome, is invisible, silent, and limitless.

What I did see on my internal archaeological tour was a lone chia seed which had stubbornly superglued itself on to the wall of my gut. I felt somehow quite exposed to be caught eating superfoods, outed as a millennial. Then I was struck by the invasiveness of the procedure. I had initially refused to entertain any histrionics regarding the camera's point of entry. We are all adults here. These people are professionals. And indeed, they were. However, as the experience unfolded, I was doing the maths on the distance the camera had travelled up, and soon I didn't need any calculator because I could feel it trying to get around a tight corner in my intestine, next to my ribs. It occurred to me that this 'technical armature of [the] medical gaze' was in the midmost part of me, my very most internal territory.[13] It had breached the house of my emotions. This is where I feel things. This, it turns out, is where I keep my heart. This tender pouch is where life hits me, where I process fear and anxiety. Excitement too. This is where my instincts are formed. In the gut. Vital warnings are issued here. There is a full kaleidoscope of butterflies living in this visceral biome, probably also feeling disgraced by the chia seed.

Are you still with me, or have you checked out of this colonoscopy? It's almost finished.

Lying there, I had the most intriguing double sensation, on the one hand, of reeling from the invasion in the depth of my being and, on the other, of a tender protectiveness as I watched the drama unfold on the screen next to me, as though it was something entirely separate. I will say a little more about this second sensation, it being the more interesting and lasting one. Through travelling inside my own gut, I glimpsed the mechanism which converts food to continuance. I saw inside this strange, ridged tube which awaits offerings from the mouth, ready to welcome unconditionally the new arrival, ready to 'say yes to who or what turns up'.[14] I saw something beautiful and vulnerable about this action. It reminds me of the anecdote in Aruna D'Souza's essay 'Imperfect Solidarities' about her child's declaration of love for their pet chameleon, not a particularly cute animal or an obvious candidate for anthropomorphizing. 'How can you love him? He doesn't seem to be very lovable?' she asks. 'But Mom, I've been keeping him alive for all this time. I've been taking care of him. How can you not end up loving something that you have to take care of?'[15] D'Souza's essay is making a case for care before empathy. She is proposing that it is not necessary to like, love, or even understand something to behave with kindness towards it. Solidarity can come from a commitment to care. This feels connected to Levinas: 'I am always automatically a host bearing responsibility for the strange Other, who is my guest'.[16]

My pulsating intestine was not lovable or cute. The tenderness I felt towards it was unexpected. Peristalsis is the complex coordination of muscles and nerves, a movement often described as wavelike because it contracts and relaxes in a continuous pattern to move food forward.[17] This perpetual dance is not subject to the whims or fancies of conscious thought. I cannot will it to stop or start. I had the feeling of discovering something more-than-human in my own insides: an unchoreographed operation of species survival, indifferent to the human container that surrounded it.

The examination draws to a close. It has been quite something. I have had an epiphany of sorts and witnessed the removal of three polyps with miniature hedge trimmers. I make a note to rethink my approach to sedation for future procedures. I dress myself and am led to a recovery area. I am told they have run out of tea – and I lose all composure, immediately shattering my mysterious toughwoman status. Hospital recovery rooms are up there with funerals in the hierarchy of nonnegotiable hospitality. To withhold tea here is bad. It was eventually administered, you will be relieved to know.

Holistic HACCP

The food business I operated also functioned as an art project. This required a conceptually laden approach to running the business. During the setup phase, I was acutely aware that, unlike most art projects, this one had the capacity to inflict real harm on those who engaged with it. My work was not to be exhibited on walls or pedestals but placed inside the viewer's mouth and sent into their gut, to be incorporated into their inner selves. I prepared for this by undertaking in-depth HACCP training to protect the inner selves of my community from microbial harm.[18] During this study, I learned about the various types of bacteria potentially present in food and how to be inhospitable to them. I encountered intriguing images from microscopes and learned about the unspeakable intestinal woes they can inflict. I was particularly mesmerized by *Bacillus cereus*, which has the unsettling capacity to grow back from the dead (reheated rice syndrome). Learnable facts assuaged some of my anxiety.

Around this time, I encountered an essay by Anna Meigs exploring the role of food in the Hua people in the Eastern Highlands of Papua New Guinea. In this community, food is formally employed as a means of transferring 'vital essence' between entities and organisms. This transmission permeates human and more-than-human life-forms. It travels through the body but also through soil, plants, water, and fire. Food is used as a conductor of this vital essence. It features prevalently in folktales and is one of the most common topics in remembered dreams. There is a strict set of guiding principles that dictates what, when, how, and by whom food is prepared, wholly separate from microbial/HACCP-related concerns. Central is the belief that all foods contain the vital essence of their producer and can be both positively and negatively charged. If the relationship is one of warmth and friendship, the food promotes strength and growth. If there is hostility, or even ambivalence, the food will deplete the eater: 'Food [. . .] does not nourish by virtue of its innate properties'.[19] Instead, food nurtures through 'the connections it makes with the contagious properties of the surrounding world'.[20] We act in agreement with this principle whenever we buy free-range chicken. It is our naive attempt to safeguard ourselves against the embodied misery of the creatures we are ingesting.

I am in full accord with the Hua tribe on this subject and was both relieved and emboldened by this perspective. The anxiety I had in converting an art project to a food business was wholly unrelated to HACCP. It was in fact connected to the immense responsibility I felt for charging food with safe, non-miserable, empowering things. I was trying to rise to the ludicrous challenge of administering unconditional welcome through food, on a massive scale. I was parcelling up bits of the 'surrounding world' in edible form and placing them into the mouths that stood open in front of me for incorporation into their inner selves, into their most interior place. I was the stranger at the threshold that is the mouth, asking to be accepted in. This hospitality is a complex force. The root of the word is *ghostis* and refers neither to the host nor the guest but to the fizzy relationship of trust that forms between the two parties.[21] It is a dynamic, reversible movement. Now my guests have become hosts to the stranger at their mouths.

Here, the Hua tribe had articulated beautifully the gravity of the act of feeding people by situating related rules and systems at the core of their society. These rules are not governed by things visible through a microscope. They are governed by ancient Indigenous wisdom and a belief in things we know to be of vital importance but cannot see. I fantasized about an alternative health inspector whose job it was to visit commercial kitchens to ensure that food is prepared with good intentions for the eater. After all, if the food is made with warmth and friendship, it will, by nature, require responsible provenance and good hygiene practices. I was blessed to work with chefs who used food as a vehicle for joy, pleasure, and kindness. I witnessed this transmission through their culinary endeavours and marvelled at this powerful tool for active coexistence.

Elemental

Through this action of offering food to the stranger, we trespass into the realm of wordless stories, told in the 'dark underside of the body lined with endless unseeing dreams'.[22] We become 'earthlings',[23] a species cobbling together a means of sharing the curved surface of the earth.[24] We rely not on language and facts but on ancient primordial contracts, unwritten but internalized, guiding this predisposition towards coexistence. Yes, we eat for continuance, strength, and restoration, but what and how we eat is the poetry of our survival. We eat for identity. We eat to remember that we are a species – alive, thrumming, pulsing. We eat food grown in soil, knowing on some level that we are that same thing as that soil. Food is a crucial element of hospitality. It is a tool for unconditional welcome and for the profoundly beautiful suspension of suspicion that we engage in when we open our doors and mouths to strangers.

Thank you for the hospitality of your attention, dear reader. This is one of the most unreasonable requests that we can make of our fellow humans in this era of oversaturation. Thank you, also, for coming to the endoscopy unit with me.

About the Author

Jennie Moran is an Irish visual artist who uses the philosophy, culture, and infrastructure of hospitality to create opportunities for shared connection. She runs Luncheonette, a nomadic emergency response unit which uses food to make places better.

Notes

1 Jacques Derrida and Anne Dufourmantelle, *Of Hospitality* (Stanford University Press, 2000), p. 7.
2 Michael Cronin, *Eco Translation: Translation and Ecology in the Age of the Anthropocene* (Routledge, 2017), p. 4.
3 Claude Fischler, 'Food, Self and Identity', *Social Science Information*, 27.2 (June 1988), pp. 275–92 (pp. 276, 278).
4 Emmanuel Levinas, *Is It Righteous to Be? Interviews with Emmanuel Levinas*, ed. by Jill Robbins (Stanford University Press, 2001), p. 52.
5 *Oxford Classical Dictionary*, ed. by Simon Hornblower and Antony Spawforth, 3rd edn, rev. (Oxford University Press, 2003), p. 612.
6 Fischler, 'Food, Self and Identity', p. 278.
7 Emmanuel Levinas, *Totality and Infinity: An Exteriority*, trans. by Alphonso Lingis (Duquesne University Press, 2007 (1961)), p. 34.
8 Levinas, *Is It Righteous to Be?*, p. 52.
9 Timothy Morton, *Hyperobjects: Philosophy and Ecology after the End of the World* (University of Minnesota Press, 2013), p. 1.
10 Immanuel Kant, *To Perpetual Peace* (Hackett Publishing, 2003 (1795)), p. 15.
11 Michel Foucault, *The Birth of the Clinic: An Archaeology of Medical Perception*, trans. by A. M. Sheridan Smith (Pantheon Books, 2003 (1973)), p. x.
12 Jean Senebier, qtd in Foucault, *Birth of the Clinic*, p. 108.
13 Foucault, p. 121.
14 Derrida and Dufourmantelle, *Of Hospitality*, p. 77.
15 Aruna D'Souza, *Imperfect Solidarities* (Floating Opera Press, 2024), p. 83.
16 Levinas, *Is It Righteous to Be?*, p. 52.
17 'Peristalsis', Cleavland Clinic, 28 April 2022 <https://my.clevelandclinic.org/health/body/22892-peristalsis> [accessed 15 May 2025].
18 Hazard Analysis Critical Control Point is a systematic preventive approach to food safety from biological, chemical, and physical hazards in production processes.
19 Anna Meigs, 'Food as Cultural Construction', in *Food and Culture: A Reader*, ed. by Carole Counihan and Penny Esterik (Routledge, 1997), pp. 95–106 (pp. 97–98).
20 James Frazer, qtd in Meigs, p. 97.
21 Émile Beneviste, *Indo European Language and Society* (Faber and Faber, 1973), p. 71.
22 Foucault, *Birth of the Clinic*, p. xi.
23 Cronin, *Eco Translation*, p. 5.
24 Kant, *To Perpetual Peace*, p. 15.

22

Sheltering from and with the Elements

Island Vignettes Past and Present

Hanika Nakagawa

Introduction

In the present, I am conducting my PhD research, consulting with Elders and community groups on Tokunoshima, with plans to move among the neighbouring islands. Everyday past meets present: my past as a child on the islands, my father's past growing up there, my grandparents' pasts, the lands' past, the past of my master's research. These are all blended into multigenerational stories or vignettes, snapshots of lives lived, lives being lived, and lives to come. In keeping with Indigenous Amami worldviews, time on the island can feel simultaneous, with past, present, and future all existing in every space, in every conversation. Time on the island can also seem still in comparison with time outside the island that is relentlessly moving, forcing its way through housing, schooling, family lines, work rhythms, and supply chains.[1] This time slippage also creates friction among elements.[2] Owing to friction, infrastructures wear, habits shift, by-products disappear; yet, the elements remain, if in different forms.

Tokunoshima is one of the main islands of the Amami Archipelago, located roughly six hundred kilometres south of mainland Japan in the East China Sea between Kyushu and Okinawa. The Amami islands are home to the Indigenous Shimanchu, or 'island people', whose cosmologies and foodways are distinct from those of mainland Japan. Historically, island communities maintained diverse subsistence systems combining wet rice cultivation, farming, household livestock, fishing, and forest foraging. Although the islands were repeatedly colonized, first through tributary relations with the Ryukyu Kingdom, later through conquest by the Satsuma domain in 1609, and eventually through incorporation into the Japanese nation-state, Elders recall that local communities remained largely food self-sufficient until the disruptions of World War II and the postwar restructuring of agriculture and trade.[3] These historical transformations form the background against which the elemental relations described here unfold.

Based on interviews with Indigenous Elders aged 60 years or more on Tokunoshima, I examine how relations with fire, freshwater, seawater, and salt structure food and memory on Tokunoshima. Through multigenerational vignettes, I argue that shifts in elemental infrastructures, from *irori* hearths to gas ranges, from rice paddies to export sugarcane, from communal wells to bottled water as a commodity, reorganize not only material sustenance but also cosmologies.[4] My method is simple: I stay close to how Elders narrate food, water, heat, and weather as everyday facts that carry moral and practical weight. These are not 'just' stories but practical histories of what it has taken to live within elements and to shelter from elements that do not stay in their lanes. This is also one of the foundational insistences of food studies; foodways are never merely nutritional but are simultaneously social, historical, and political.[5] As Annemarie Mol suggests, 'Eating and feeding are entangled'; I acknowledge this while also casting light on how identity is also entangled.[6]

In one of my research meetings, the Elders of the village of Tokuwase came together with me as audience in explaining how there are many household gods. Tokuwase is home to the Nanseitogyo (southwest sugarcane factory), atop a hill and only one kilometre from the beach.[7] The noteworthy god dwelling in each home is Higamisama (fire god), to whom offerings of all the food and tea were given, much in the same way as the ancestors who live in the decorative Shinto or Buddhist altars at home receive food and tea every day. Higamisama are important and respected, since they protect each house. Worshipping Higamisama in past homes involved placing three stones at the kamado; some Elders now place three rocks beside their gas burners. As Elder Susumu Machida explained, all beings are gods. He identified fire, water, each of the pests that eat the fields, the rice that grows there, and so on. Elder Takefumi Tsukawa explained that rivers are beings and gods themselves. Mizugami sama (water god) was celebrated at the household level; islanders would make small white mochi *dango* (sweet rice balls), like those made for tsukimi (moon gazing), and they would give *dango* to mizugami sama on a moonlit night while saying a prayer. Elder Chiemi Nakagawa does not remember what her mother said to mizugami sama, but she remembers that they ate the *dango* after offering it, and that was the end of the worship. At the times of *tsukimi*, islanders would also place a branch of *susuki* (Japanese pampas grass) in a vase and offer *dango*. This too was not a gathering like a festival; it was a family gathering.[8] These practices endure, if worn a bit, because they are portable. A household can change its stove from traditional to modern and still remember its gods.

Colonial taxation regimes, wartime scarcity, and postwar agricultural policy reshaped island food systems, yet everyday relations with elemental infrastructures, specifically here fire, freshwater, seawater, and salt, continued to structure how food was produced, preserved, and remembered. In keeping with this understanding, in the sections that follow, I examine island lifeways element by element, speaking about fire (preservation and risk), freshwater (fields, wells, and daily life), seawater and salt (typhoons, supply, and contamination), and finally, communal energy transitions that stitch these together into a social ecology of food.

Fire

Other parts of the island – for example, the communities of Metegu or Todoroki – are rice-growing, having easy access to water. Each piece of land dictated what could be grown on her; families with more land and irrigation grew rice; wheat, sesame, and soy grew in drier places; all people grew *imo*. These were not simply 'crops'; they were an arrangement of elements, soil, water, sun, labour, and storage. Modernization and globalization have changed the physical arrangement of elements. New home construction has exchanged floor plans centring a sunken hearth (*irori*) that separated rooms for rooms with gas ranges located by an external wall. Food preparation thus moved from the centre to the periphery. In that architectural shift, smoke also moved. When fire relocated to a sealed burner, smoke ceased to be an everyday by-product; preservation by smoke shifted from routine to specialized technique. Traditionally, cooking was done in the *irori* over a wood fire that also heated homes. Household pots had handles allowing them to be hung from a hook over the hearth. The hook was attached to a shelf where fresh firewood would be dried, and slices of tofu were laid to smoke for about a week; the tofu was preserved with a miso coating, then layered into a clay vessel called a *kame*. Now that cheese has been introduced to the island and is sold in grocery stores, Elder Chiemi Nakagawa recalls it tasted like a kind of cheese and was one of her favourite foods:

> We did grow wheat at our homes – not everyone, but those who could, did. Of course, some people had more land, while others had less. Those who had more land would grow rice, *imo*, wheat, sesame, and soy. We grew soy at home. Only twice per year, during Obon [a festival for welcoming back ancestors who have died] and New Year's, we took the soy and processed it to make tofu. Making tofu took the whole day. The grandmas would start in the morning setting out the soybeans to absorb water to make them soft, then cooking them . . . because it is of no use making them soft and not cooking them. After making them soft, we used things called *hikiyusu* [a stone mill] to turn the soybeans into a powder. It is a grinder made of stone. The lower stone was about this tall [about fifteen centimetres], and it was this round stone. On top of that, there was another stone, and in between where the two stones fit together are these little notches that are fit together like this [she demonstrates with her hands]. You take that . . . place it on top of a table, and on the upper circular stone was this handle to turn it, and you would take the handle and spin it round and round and feed in the soybeans from the hole. And where the teeth of the two meet is where the beans grind into a powder – the soybeans, that is. That is then taken and steamed and placed in a big pot and boiled; that is how tofu was made. Yeah, doing that, then it was boiled; it took the entire day to make tofu. That tofu goes bad quickly, right, so we took that and cut it into pieces about the size of the tofu found in stores today. Then we thinly sliced them and placed them on a large woven basket to dry, shaking off all the water. Then you took that, and I mentioned before that there was the *irori* where we kept the fire going and did the cooking.[9]

Today, this tofu dish is no longer available for infrastructural reasons; households do not routinely grow and mill soy and wheat for subsistence; the *irori* has been replaced by gas stoves; smoke is no longer produced as a daily surplus; tending a central fire plus sourcing many days of wood no longer fits life rhythms. In a folklore museum beside an *irori* that once worked, Chiemi Nakagawa expressed wistful regret: 'I wish I had learned how to do that'. I find myself thinking that the loss is not only personal but architectural. It is difficult to learn a preservation technique that depends on a hearth when the hearth is already an exhibit.

Fire is still respected for many legitimate reasons, but fire is also feared. Fire is a concern on Tokunoshima, particularly in some small inlet regions on the east side of the island where houses are placed closely together. Fire can burn down a house, a community, and their food stores and even spread to the fields. The same element that preserves can also erase. Protection is never pure; it always carries risk. Every meal is a risk, the risk of using fire, the risk in the gathering process, and a taste risk; not everything prepared is guaranteed to be enjoyable.[10]

Freshwater: Everyday Politics of Access

Diversified subsistence systems once characterized many communities across the Amami Archipelago. Before World War II, island households typically combined farming, fishing, livestock raising, and forest foraging in ways that provided a high degree of local food autonomy.[11] Over the lifetimes of the Elders on Tokunoshima, there have been many changes to the spaces and purposes of buildings and landscapes which have, in turn, changed human relationships with the elements. New buildings also bring new relations with wind and water. Elder Yoko Nao grew up in the community of Todoroki, a land known for its water and rice fields. Todoroki in Tokunoshima-cho is a lush, green place, filled with rice fields. She remembers having rice fields across the whole land that had a golden shine when ready for harvest. The breezy homes of the past, shut only when a storm approached, were ideal for feeling the wind that had picked up water from the surrounding rice fields in the summer. The heat with a tinge of cool mist was something very different from the heat that exists now, retained in concrete buildings, paved roads, and dry fields of sugarcane. Elder Nao recalls how at peace she felt in her open, airy wooden home, napping in the breeze. Now, she explains, she is trapped in a concrete box that must be artificially cooled to prevent her from cooking in her own skin.

Freshwater organizes daily life, including how one bathes, how one drinks, what can be grown, and how schoolchildren move through the day. For several years, I have taught swimming lessons to elementary and middle school children in Inutabu on Tokunoshima. My volunteer work highlights agricultural change in an unusual way. The agricultural region surrounding Inutabu does not have the water resources to allow every school to fill a swimming pool to fulfil the mandatory national swimming curriculum. Therefore, Agon Elementary School rents a bus so that the students can attend swimming lessons at Inutabu Middle School; Inutabu Elementary School children walk across the street for their turn in the pool. Elementary school students have their swimming lessons in June, while the middle school students partake in July. The pool in Inutabu overlooks a field of sugarcane. Some mornings, I was lucky enough to see the sprinklers turn on while the buses arrived. One morning, the kids who drove there from Agon smiled and said, 'That's why we get to come here, to water the food'. Then, they quickly turned away, having noticed a friend waving to them from the second-floor window. A swimming curriculum becomes irrigation. A bus route overcomes scarcity. Freshwater infrastructure is not a background condition; it is part of what children learn, holistic knowledge of stewardship of lands, waters, air, and all our kin.[12] Wells, too, are food infrastructure; they are the foundation beneath tea, rice, and the feeling of a 'good life'.

Freshwater: How Household Waste Becomes Food Again

A feeling of sturdiness and permanence comes with current concrete homes. But Tokunoshima is hard on structures. Termites and snails attack wooden frames; sea breezes mount a salt assault even on modern steel and concrete. In these conditions, 'the home' is always a negotiation with elements. Moreover, there was always an element of fear in traditional homes. The *habu* (vipers) like places where there are prey, particularly mice. My Obaachan tells me two of her three encounters with the aggressive *habu* were in the washroom built with stones with gaps and holes for snakes to hide. The human waste attracted mice, she explained, which is why it also attracted the *habu* more than the house did, though she also has heard stories of *habu* entering into the home.

Sulphur was burned to smoke the *habu* out of the crevasses so families could whack it or stab it in the head. The *habu* would then be cooked by my great-grandma Uto Motoi (*Uba gua* = adorable aunt) or Masu Moti (who my Obaachan called Amma, meaning mother; biologically, Uto and Moti were both her aunts). After Uto Motoi cooked the *habu*, it would be fed to the pigs, who were later eaten by the family. Human waste, eaten by mice, eaten by *habu*, eaten by pigs, eaten by humans, with the chemical element sulphur playing an important role. An elemental economy, in which food sovereignty depends on a circulating chain of conversions, tying sanitation, predators, livestock, and family survival together.

Another use of sulphur was as a treatment for asthma, a problem that has plagued our maternal family. Elder Michio Hisaeda, my uncle/cousin/big brother, recalled hearing about a family visit to Torishima, a volcanic island about ninety kilometres off the coast of Tokunoshima. They were seeking a remedy for asthma for the youngest sister, Uto; bathing and breathing in the volcanic rocks was known to be an effective remedy. The family brought home basins and rocks from Torishima to Tokunoshima that they had purchased from the craftspeople who lived there. Toku, as the eldest sister, was charged with carrying the basin made of volcanic rock up the mountain's path, all the way back home. Many years later, Michio told me that he helped move the volcanic rock basin (by car) when Toku moved in with Toshiko (Michio's mother, also my great-aunt). Michio smiled, informing me that the basin currently sits proudly in front of Toku's grave; she grew attached to it, using it in her kitchen. The volcanic stones were cherished by Uto also, who still brought them with her into the shared bath decades later. Elements travel. Volcanic stone became basin, then kitchen object, now memorial. Infrastructure can vanish, leaving behind only the objects and the stories that once made them ordinary. Torishima is no longer inhabited.[13]

Seawater

During typhoons, elements come to life. Typhoons dictated the landscape and are intertwined with eighteenth-century sugar extraction.[14] When sugar production intensified, mountains and land dedicated to subsistence needs were carved up to meet the demands of the Satsuma domain, taxes.[15] During typhoons, fire threat from the dry and daily sun fades slightly, giving way to water, wind, and salt. When a typhoon hits, fruits sail off trees, shingles blow from roofs, and seawater spray reaches kilometres inland. Salt, a compound made up of the elements sodium and chlorine, causes more harm to fruit and other crops than high winds do. Salt must be washed from fruit and plants as soon as possible; therefore, freshwater rain is welcomed during and after a typhoon. Despite being dangerous, typhoons have a purpose. Elders told me the typhoons are necessary, cleansing the land. I understood their words to reference a spiritual cleansing, something I can sense today venturing outdoors after a typhoon. The island climate encourages all life to grow, including mosses and green fungi that colonize the concrete. However, after the wet salty air settles, I see how the elements cleansed the pathways and the steps.

In the present time, while I helped my Obaachan prepare for a coming typhoon, I began to think about what had changed. Some things were obvious: homes are no longer built in gullies for protection, we must close storm shutters that are now made of metal, and we move the ornamental plants, carefully displayed in beautiful pots lined on the concrete path and tended daily, to a safe place on the steps by the door. But the most consequential change is logistical. Knowing a storm is coming, my Obaachan and I must stock up on groceries, as ships cannot move at all if ocean waves exceed five metres. Ships might not arrive for five days or even a week or more. At the end of the typhoon, it takes time for ships to resume. Hardy plants left outdoors are watered to wash away salt. Farmers tend crops in the fields. Bulls are calmed and fed. Shelves in grocery stores are restocked with items not grown on Tokunoshima. Typhoons reveal a layered infrastructure, local crops and livestock replaced by imported foods and shipping schedules. Typhoons make dependency visible. In 2021, Tokunoshima and neighbouring Amami-Oshima were designated UNESCO World Natural Heritage Sites, a status that protects biodiversity but also places new regulatory limits on how island residents access forests, rivers, and coastal resources.[16] Resulting vulnerabilities are made visible during typhoons, reflecting long histories of agricultural restructuring in the Amami islands, where colonial taxation regimes and later national policies redirected production toward export commodities such as sugarcane, reducing local food autonomy.[17]

Salt

I explained my PhD project about food sovereignty and food memories to the ladies who come to learn ikebana from my Obaachan. My explanation stirred a conversation about research; one of their grandchildren came to Tokunoshima for the summer, and they made salt together as the grandchild's independent school project. It was a well-done project, receiving praise from the school. Then, surprisingly, she said there was a more important element to the project than making salt. She insisted I conduct this experiment so that I could bear witness to the amount of plastic that comes to the surface when boiling down and creating sea salt for human consumption. She told me it was enough to make them think about whether the salt they produced was safe for consumption. The world's oceans are all connected, she explained, an element that was not discussed deeply in the elementary school classroom. Maybe, the ladies pondered aloud, industrial salt producers have more tools available to mitigate the degree of plastic in their product. However, other ladies chimed in saying eliminating plastic must be impossible, as it breaks down at the micro level. Salt is not only seasoning but an index of planetary contamination and scale. A household can boil seawater; a household cannot filter a global ocean. The fire that once condensed seawater into taste now also condenses seawater into evidence. What the salt experiment revealed was not only contamination but also what Rob Nixon calls slow violence, environmental harm that accumulates gradually and often invisibly, appearing first as a faint film on the surface of boiling seawater.[18]

I am a highly educated sociologist in Western educational terms, but a blank slate, knowing nothing on the land in Tokunoshima. What I bring is curiosity. Making sea salt, I learned, is energy intensive but not terribly difficult. Boiling seawater to make salt used to be done on the beach; some places had a person who specialized in making salt over an open fire. As I understand it, salt making had to be completed in one day because it took a lot of firewood. This is elemental infrastructure again; salt depends on fire. When fire changes form (from wood to gas), salt making changes in cost, timing, and meaning. A practice that once fit within local rhythms becomes recalculated against household budgets and fuel access.

Communal Life and Energy Transitions

Elder Sakiko Yasuda lived in a small community group with three other families; her mother was charged with making rice for them all. There was not enough to have a rice-based meal every meal. So, she explained, 'My mother would make around five bentos. The grandfather and people would go to work, so for those people a bento lunch was necessary'. The rice in the bentos was paired with eggs or something pickled in miso. Back home, the children would mostly eat *imo* with rice in rotation. This vignette is a small economy of scarcity and planning; rice rationed into portable meals, miso and pickles as durable companions, *imo* as dependable backbone. In many places, as Sidney W. Mintz reminds us, staple foods are never 'just' nutrition; they are structured by labour, trade, and power, by what counts as everyday and what counts as enough.[19]

Elder Nakagawa mentioned that in her village, three families took turns filling the bath, and I asked Elders Yasuda and Machida what it was like for them. Chiemi Nakagawa explained how families who tended wet rice fields would often fill their baths with water from the stream that irrigated the field, then dump used bathwater, the grey water, into the rice field after it had cooled.[20] Yasuda said her family had a *goemonburo* and that they were a big family. They also filled their tub not from the river but from a well. Yasuda explained to me that a *goemonburo* was a proper bath, and some who did not have a proper bath would heat water in a drum can. Elder Machida responded that the need for firewood did not change; a basin for bathing is a basin. Her thoughts led Elder Yasuda to think deeply about well water. The Yasuda family's well water was known for being delicious, and many people came to get water for their tea there. Yasuda remembers that on the twelfth day of O-shougatsu, when midnight struck, everyone would get ready for *ichiban cha* (the first tea). It was competitive, a race, and the kids would race to the Yasuda well. Nowadays, people eat soba first thing, but back then, it was tea everyone drank as the first thing to cross their lips. Yasuda reflected that around middle school, life started to get easier with gas coming into the homes. At first, there was no electricity, so oil was a necessity for the lamps. Small shops that sold different oils for the lamps also appeared. The shops sold the oil by weight, meaning that customers would bring a container to be weighed from home. All cooking oils and lamp oils were sold in this way. Until she was in grade six, there was absolutely no electricity, and she explained the oil lamps were similar to those now used for camping. Once electricity came, Yasuda explained it was only on at designated times. These are not simply nostalgic details. They are the infrastructure of attention, time, and possibility. Cooking fuel, lighting oil, electricity schedules, and water sources braid together into what households can reliably do, such as how long food can be preserved, how often rice can be cooked, when people can gather, and how children remember 'normal'.

Conclusion: What Elemental Infrastructures Teach Us

The Amami islands have experienced successive regimes of colonial extraction, from the sugar taxes of the Satsuma domain to postwar agricultural restructuring, yet elemental infrastructures continue to shape everyday food practices in ways that cannot be fully captured by political histories alone. I began this paper with time slippage, the sensation that island time holds layers at once, while outside time insists on linear progress. Elemental infrastructures are where those temporalities collide. Hearths become burners, but the memory of smoke persists. Rice fields shrink or shift, but the bodily memory of mist remains on the skin. Seawater arrives as spray and as a shipment route, then returns as salt, which is a seasoning, preservative, and now troubling evidence of plastic. If there is a through-line in these vignettes, it is not that Tokunoshima is either fragile or resilient. It is that island life repeatedly requires people to shelter from the elements while also sheltering with them, treating fire, water, wind, salt, sulphur, and stone as companions whose gifts and dangers must be negotiated.

Typhoons sharpen this lesson. They make the island's dependencies legible: what can be grown and fed locally (crops, bulls, pigs, *imo*), what must be imported and therefore stockpiled, and how quickly 'abundance' can become a question of docking schedules. Yet Elders' insistence that typhoons cleanse the land is not a romantic denial of destruction. It is a way of locating meaning in recurring disruption. The storm is a force that rearranges, strips, reveals, and demands repair. Therefore, community knowledge becomes infrastructure too, almost a physical being that engages residents in a theoretical task to consider whether changes to the landscape made to accommodate human food production prevent other beings from being born there.[21] Food sovereignty, in this consideration, is less an endpoint than a practice of calibration, adjusting meals, fuel, water use, and preservation to changing conditions while retaining a sense of what counts as good food, safe water, and enough. The now-rare smoked miso tofu shows what is lost when by-products disappear. The swimming pool bus route shows how scarcity is taught without being named. The salt experiment shows how global contamination collapses into a pot on the stove. The bentos, wells, shared baths, oil lamps, and designated times show that elemental infrastructures are also moral infrastructures; they produce habits of care, restraint, pride, and sometimes grief.

Famines, the Elders told me, occurred from time to time, but famines exist everywhere. Support from other places is welcome; but just because the peoples living in a place are recipients of aid at one point in time, this fact alone does not mean there is not normally abundance. The island's story is not a simple arc from self-sufficiency to dependency. It is a continual reweaving of what the elements allow. And because humans are also elements, made up of what we have eaten, becoming elements in the soil when we are dead and buried, the chapter returns to a final, quiet point; infrastructures change, but the elemental fact of eating, preserving, sharing, and remembering remains the most durable shelter we build.

About the Author

Hanika Nakagawa is a PhD student at Dalhousie University in Halifax, Canada. Her research focuses on Indigenous food sovereignty, specifically in Tokunoshima, Japan.

Notes

1 Anna Lowenhaupt Tsing, *Friction: An Ethnography of Global Connection* (Princeton University Press, 2005).

2 Tsing, *Friction*; Anna Lowenhaupt Tsing, *The Mushroom at the End of the World: On the Possibility of Life in Capitalist Ruins* (Princeton University Press, 2015).

3 Tom Monaghan, 'Amami and Amami: Confecting Sugar Islands in the East China Sea', in *Entire of Itself? Towards an Environmental History of Islands*, ed. by Milica Prokić and Pavla Šimková (White Horse Press, 2024), pp. 171–90; Hiroto Takamiya and others, 'Subsistence Strategies in the Ryukyu Islands during the Gusuku Period', *Asian Perspectives*, 55.2 (2016), pp. 247–78.

4 Hanika Nakagawa, 'Drinking Sugar's Waste: Pollution and Dispossession in Tokuwase, a Community in Tokunoshima, Japan', *South Pacific Studies*, 46.2 (forthcoming 2026), pp. 61–73.

5 See, for example, *Food and Culture: A Reader*, ed. by Carole Counihan and Penny Van Esterik, 3rd edn (Routledge, 2013).

6 Annemarie Mol, *Eating in Theory* (Duke University Press, 2021), p. 110, doi:10.1515/9781478012924.

7 徳之島町, *世界自然遺産登録記念徳之島町史 民俗編シマの記憶* (*History of Our Daily Lives*) (徳之島町, 2022).

8 Susumu Machida, interview with the author, Tokunoshima, 2022; Takefumi Tsukawa, interview with the author, Tokunoshima, 2022; Chiemi Nakagawa, interview with the author, Tokunoshima, 2022.

9 Chiemi Nakagawa (translation by the author).

10 David Sutton, *Bigger Fish to Fry: A Theory of Cooking as Risk with Greek Examples* (Berghahn Books, 2021).

11 Monaghan, 'Amami and Amami'; Takamiya and others, 'Subsistence Strategies', pp. 247–78.

12 Enrique Salmon, 'Kincentric Ecology: Indigenous Perceptions of the Human–Nature Relationship', *Ecological Applications*, 10.5 (2000), pp. 1327–32, doi:10.2307/2641288.

13 Michio Hisaeda, interview with the author, Tokunoshima, 2022.

14 佐野静代, *外来植物が変えた江戸時代* (*Foreign Plants That Changed the Edo Period*) (吉川弘文館, 2021).

15 佐野, *外来植物が変えた江戸時代*.

16 UNESCO World Heritage Convention, 'Amami-Oshima Island, Tokunoshima Island, Northern Part of Okinawa Island and Iriomote Island', World Heritage List entry no. 1574, 2021 <https://whc.unesco.org/en/list/1574/> [accessed 5 March 2026].

17 Monaghan, 'Amami and Amami'.
18 Rob Nixon, *Slow Violence and the Environmentalism of the Poor* (Harvard University Press, 2011), p. 2.
19 Sidney W. Mintz, *Sweetness and Power: The Place of Sugar in Modern History* (Viking, 1985).
20 Chiemi Nakagawa.
21 Mol, *Eating in Theory*, p. 104.

23
The Five Tastes, Five Colours, and Five Cooking Methods in *Shōjin Ryōri*
Historical Development and Modern Practice in Japanese Zen Temple Cuisine

Saori Nishida

Introduction

Shōjin ryōri is the traditional vegetable cuisine of Buddhist temples in Japan and literally means 'food for spiritual training'. Rooted in Buddhist teachings, it avoids animal products, alcohol, and pungent vegetables like onions and garlic, which are believed to disturb the practitioner's mind. Yet *shōjin ryōri* is more than just a plant-based diet; it is a spiritual and philosophical discipline that governs not only what to eat but also how to cook and how to eat.[1] The right mindset and attitude are considered integral to its practice.

This culinary philosophy shaped the tea ceremony, which later influenced *kaiseki ryōri* (multicourse cuisine), and is now often explained through the principle of 'five tastes, five colours, five cooking methods' (*gomi goshiki gohō*), a framework guiding temple cooks to prepare nutritionally balanced and aesthetically harmonious meals.[2] This paper examines the historical development of this concept, tracing its origins, evolution, and integration into *shōjin ryōri* and broader Japanese cuisine. The analysis focuses on the distinct historical trajectories of the five tastes, five colours, and five cooking methods and their relationship with the Chinese Five Element Theory. By adhering to historical accounts, this study analyzes how this principle was formed over centuries.

Historical and Philosophical Foundations of *Shōjin Ryōri*

Understanding *shōjin ryōri* requires examining its etymology and its place within Buddhist philosophy. The term *shōjin* signifies 'devotion to Buddhist practice', and *ryōri* refers to cuisine. *Shōjin ryōri* avoids ingredients (animal products as well as pungent vegetables such as garlic and onion) believed to stimulate worldly desires and hinder spiritual training.

The principles of *shōjin ryōri* were formalized in the mid-thirteenth century by the Zen master Dōgen, especially in his works *Instructions for the Zen Cook* (*Tenzo Kyōkun*) and *The Dharma of Taking Food* (*Fushuku Hampō*).[3] While studying in China, Dōgen encountered local temple cooks and realized that Japanese monks of his time had little respect for the role of the cook.[4] Before the arrival of Zen Buddhism, daily chores such as cleaning, cooking, and eating were not considered spiritual training; therefore, practitioners often tried to minimize the amount of time spent on these tasks. Although only experienced monks could work in the temple kitchen due to their impact on the community's well-being, temple cooks once delegated their tasks to others, often spending their time studying, meditating, or resting. Dōgen elevated cooking and eating to spiritual practices, emphasizing the importance of food as a medium for self-discipline and gratitude.[5] Since then, the role of the temple cook has been regarded as one of the most sacred in the temple.[6]

The philosophy of *shōjin ryōri* is underpinned by guiding principles:

- Respect for nature: using seasonal and local ingredients to harmonize with the rhythms of nature. The body and the land are considered inseparable; eating from one's environment supports both.
- Minimizing waste: using every part of the ingredient, in line with the Buddhist view that all living beings are complete and balanced.
- Gratitude and interconnectedness: recognizing the countless lives and efforts involved in producing food – from nature to farmer to cook.
- Mindful preparation: cooking is a spiritual act. Full attention and care should be given to every step while appreciating the opportunity to serve others.
- Mindful consumption: eating with restraint and awareness to avoid greed and waste. Food should be finished entirely to honour its life-giving role.

These principles influenced broader Japanese gastronomy. Sen no Rikyū, a pivotal figure in formalizing the tea ceremony in Japan, integrated *shōjin* ideals into his philosophy, which later inspired *kaiseki ryōri*, the multicourse cuisine emblematic of Japanese haute cuisine.[7] UNESCO's 2013 recognition of *Washoku*, traditional Japanese dietary culture, as an Intangible Cultural Heritage highlights the lasting impact of these *shōjin* values, particularly sustainability, waste reduction, seasonality, and balance.[8] In contemporary times, temples and culinary literature increasingly describe *shōjin ryōri* and Japanese cuisine (especially *kaiseki ryōri*) through the framework of the five tastes, five colours, and five cooking methods, solidifying its recognition as a fundamental concept within Japanese cuisine.[9]

The Five Tastes, Five Colours, Five Cooking Methods: Core Principles of Japanese Cuisine

The five tastes, five colours, and five cooking methods constitute a fundamental concept in Japanese cuisine, comprehensively addressing taste, visual presentation, and cooking methods to achieve harmony and richness in food.[10]

Origins Rooted in Yin-Yang Five Element Theory

The concept of the 'five tastes, five colours, five cooking methods' is deeply influenced by the ancient Chinese Yin-Yang Five Element Theory (*Wu Xing*). This philosophy posits that all phenomena are composed of five elements – wood, fire, earth, metal, and water – which interact to explain the workings of the universe, changes in nature, and even human health and physiological functions. The Five Element Theory established correspondence relationships with various aspects, including tastes, colours, and seasons. This philosophy was later introduced to Japan, where it evolved into culturally distinct interpretations, forming the bedrock of its food culture.

The Five Tastes

The traditional five tastes – bitter, sour, sweet, pungent/spicy, and salty – have ancient roots. The earliest mention of these five tastes, paired with the five elements, appears in the *Book of Documents* (*Shujing*, compiled by the third century BCE), one of the oldest Chinese texts on the Five Element Theory. Based on each element's characteristics, water was paired with salty taste, fire with bitter, wood with sour, metal with pungent/spicy, and earth with sweet.[11]

Within Buddhist philosophy, particularly in the context of temple cuisine, the 'six tastes', which include the traditional five tastes plus *tanmi*, subtle taste, were mentioned in ancient Buddhist scriptures such as the *Nirvana Sutra* (*Mahāyāna Mahāparinirvāṇa Sūtra*, established around the fourth century AD). The subtle taste refers to the inherent, natural flavour of ingredients in their purest form, and preparing a meal with subtle taste has been considered the essence of temple cuisine.[12] Therefore, the subtle taste can be characterized as an overarching flavour guideline, encouraging temple chefs to prepare a meal by respecting the natural flavours of ingredients and bringing out the original taste of each ingredient. The reference to six tastes demonstrates that the Five Element framework had already informed Buddhist conceptions of flavour.

The taste associated with the metal element is 辛 (pungent/spicy), a sharp taste that stimulates the tongue. This category includes pungent vegetables (such as green onions, garlic, and leeks), spices (such as ginger, chili, pepper, *wasabi*, and mustard), and herbs (such as *shiso* and coriander).[13] While Buddhist dietary restrictions prohibit the use of pungent vegetables due to their strong aromas and stimulating qualities, temple cooks often incorporate spices and herbs to evoke the pungent taste without violating these principles.

While the Five Element framework traditionally links pungent flavour to the metal element, in contemporary Japanese explanations of the five tastes, 'pungent' is often replaced by or supplemented with *umami*, the scientifically recognized fifth basic taste, reflecting both modern nutritional science and the characteristics of Japanese cuisine, which rarely emphasizes strong spiciness.[14]

The Five Colours

The five colours – green (blue), yellow, red, white, and black – are integral to the aesthetic balance of Japanese cuisine today. Their original application in the Five Element Theory, however, was not related to food. *Yellow Emperor's Inner Canon* (*Huang Di Nei Jing*), a foundational text of traditional Chinese medicine from the Warring States period (475–221 BCE), expanded the system by associating the five elements with five organs, five colours, and therapeutic dietary recommendations. The five colours referred to the

Table 1. Facial complexion, associated organs, and flavours in traditional Chinese medicine based on an analysis of the principles in Yellow Emperor's Inner Canon.

Colour of the face	**Associated organ**	**Associated flavour**
White	Lungs	Pungent/spicy
Red	Heart	Bitter
Green	Liver	Sour
Yellow	Spleen	Sweet
Black	Kidneys	Salty

colours of the facial complexion, which were used in conjunction with the pulse to diagnose disease. For instance, a white (pale) facial complexion was considered a sign of lung disease, and the lungs desire pungent flavours.[15]

This diagnostic approach informed dietary recommendations, where specific tastes corresponding to the diagnosed condition were prescribed as medicinal food. Chinese medicine holds that illness is caused by an excess or deficiency of the five tastes, and it is fundamental to maintain a balance of the five tastes in the daily diet in order to maintain and improve health.[16] This theory laid the foundation for medicinal cuisine. During the Han dynasty (206 BCE–220 CE), Chinese medicine reached a certain degree of perfection, and although there have been some changes over the past two thousand years, the tradition has basically been carried on in one form or another. The idea that a balanced diet leads to a healthy body, and that food can prevent and cure sickness, is still strong today not only in China but also in Japan and Korea.[17]

While the Five Element Theory was introduced to Japan during the early centuries of Chinese cultural transmission and influenced various aspects of culture, including court ceremonies, the application of five colours to food was primarily limited to special ritualistic and formal meals.[18] These ceremonial dishes often used colours for their auspicious or decorative qualities rather than for their inherent appeal as food. Historical records indicate that interest in the colours of food was less emphasized compared to the vibrant dyes used in costumes and visual aesthetics, even during the Heian period (792–1185), when aristocratic culture flourished and formality and beauty were emphasized.[19] Achieving a full five-colour scheme with natural ingredients was challenging in earlier times, and therefore, preparing five-coloured ingredients was assumed to be limited to ceremonial, formal, and special dining occasions. Evidence of adherence to the five colours can be found in the cookery books by court cooks in the Muromachi period (1336–1573), which is widely regarded as marking the origin of classical Japanese cooking.[20] By the eighteenth century (late Edo period), food colouring techniques had developed, with evidence describing efforts to create a vibrant array of colours using natural colourants such as purple shiso leaves for red, Gardenia fruits for yellow, mustard greens for blue, and ash for black.[21]

In the temples, the meals consumed by practitioners, both historically and in contemporary times, were and remain typically humble and earthy in appearance.[22] Five colours had never been a guiding principle in preparing daily practitioners' meals at temples. In the late fifteenth century, restaurants started being subcontracted by temples to cater for special occasions and large numbers of guests or for writers and artists unrelated to Buddhist activities. *Shōjin ryōri* at Daitokuji Temple in Kyoto is a typical case of the former, while that of Hida Takayama is a typical case of the latter: both introduced more refined cooking techniques and produced elegant and colourful dishes that differ from those of temples.[23]

The five colours became particularly important in formal Japanese dining, manifesting in the aesthetic principle of using ingredients of five different colours to create visual harmony that reflected cosmic balance. This aesthetic approach continues to influence modern Japanese plating techniques and food presentation.

The Five Cooking Methods

The five cooking methods – serving raw, simmering, grilling, frying, and steaming – form the foundational cooking methods in Japanese cuisine.[24] Although Chinese culinary traditions include a vast range of cooking methods, the specific concept of five cooking methods is not articulated in Chinese texts describing their own cuisine or the Yin-Yang Five Element Theory. When the five cooking methods are mentioned in Chinese literature, it is exclusively in the context of explaining Japanese cuisine, alongside the five tastes and five colours.[25] Furthermore, Chinese culinary practices generally avoid raw food, with most dishes being cooked.[26] This stark contrast suggests that the concept of five cooking methods, particularly the emphasis on raw preparation, is a unique development within Japanese culinary philosophy. A clear articulation of five cooking methods in ancient Japanese culinary texts is scarce. Based on available sources, a formalized articulation of this concept in culinary guides for professional chefs appears to date from the late twentieth century, as seen in works such as Abe Kokyu's 1984 *Traditional Culinary Protocol*.[27] This suggests that the integration of five cooking methods into the broader 'five tastes, five colours, five cooking methods' framework likely emerged as a recent systematization, perhaps intended to provide a comprehensive and appealing structure for Japanese cuisine, perhaps even leveraging the established appeal of the Five Element concept for broader understanding. The connection between the five cooking methods and the five elements often lacks clear historical or philosophical basis, suggesting a later, perhaps artificial, alignment. Similar to the five colours, the formalized concept of five cooking methods, as a codified principle, does not appear to have been consistently applied or explicitly articulated in historical temple culinary practices, even though the individual cooking methods themselves were integral.

The development of simmering as a cooking technique in Japan was closely tied to the introduction of *shōjin ryōri*. While bronze and iron implements arrived from China around 300 BCE, metal pots remained extremely scarce due to iron's priority use for weapons and farming tools, with iron pots only becoming more widespread around the fourteenth century. Prior to this period, Japanese cooking consisted primarily of grilled foods, dried provisions, or raw dishes. Through *shōjin ryōri*, however, the technique of simmering and stewing in *dashi* using *kombu* and vegetables (such as daikon and shiitake) emerged as a revolutionary cooking method, introducing hot, liquid-based dishes that could be served immediately at the table – a significant departure from traditional Japanese culinary practices.[28]

Vegetable oils appeared in Japanese history as early as prehistoric times, initially used for lighting and fuel. By the end of the eighth century, sesame seeds and oil extraction techniques had been imported from China, and oils like sesame, hemp, and perilla were paid as taxes.[29] Sweet rice flour snacks fried with sesame oil were also introduced from China, marking the origin of frying techniques. Sesame production gradually expanded nationwide, and as a result, these fried sweets were often served to the aristocracy and also offered to temples and shrines.[30]

The arrival of Zen *shōjin ryōri* in the thirteenth century brought recipes that used various vegetable oils like sesame, soybean, rapeseed, and camellia for frying tofu and wheat gluten. When oil is used, the aromatic compounds produced during cooking are retained in the oil, making the flavour more complex and enhancing the richness of the flavour, adding variety to otherwise light-tasting vegetable foods.[31] By the end of the sixteenth century, the principles and techniques of *shōjin ryōri* had influenced the development of *kaiseki* and other forms of formal Japanese cuisine, which incorporated fish and poultry while retaining the aesthetic and philosophical foundations of temple cooking. It was not until the Edo period, with the increased production and wider availability of more affordable oils like rapeseed oil, that using oil in cooking became common among the general populace.[32]

Nowadays, temples offering *shōjin ryōri* to their visitors actively promote their cuisine, saying their meals follow the 'five tastes, five colours, five cooking methods' concept, and many academic journals and media also describe this concept as if it is a part of Japanese culinary philosophy. The concept is believed to create meals that are nutritionally balanced, aesthetically pleasing, and health-promoting.

Discussion: Reconsidering Culinary Frameworks

While the 'five tastes, five colours, five cooking methods' framework is widely recognized as a fundamental principle of Japanese cuisine, academic discussions exist regarding its historical scope and the period of its conceptual formation.[33] Food culture historian Tokura Tsunenobu argues that the 'five tastes, five colours, five cooking methods' of *shōjin ryōri*, and similar theories espoused by chefs, are often 'over-interpreted' to apply to Japanese cuisine as a whole.[34] He contends that such narratives are not always based on historical evidence. This observation suggests that the clear framework of 'five tastes, five colours, five cooking methods' was not consistently applied to all Japanese cuisine from ancient times. Instead, it is more likely that this concept was systematically organized and popularized within the context of specific cuisines (such as *shōjin ryōri* and *kaiseki ryōri*) or through modern culinary education and promotional activities.

Conclusion: The Evolving Framework of Japanese Culinary Principles

This paper has examined the historical formation of the 'five tastes, five colours, five cooking methods' concept, a principle now widely associated with *shōjin ryōri* and traditional Japanese cuisine. While often presented as an ancient, unified framework, this analysis reveals a more complex and distinct evolutionary path for each component.

The five tastes, deeply rooted in the ancient Chinese Five Element Theory and further elaborated by the Buddhist 'six tastes', represent the most historically consistent element.

Their association with the elements and their application in medicinal contexts have a long and continuous lineage. In contrast, the five colours originated as a diagnostic tool in Chinese medicine, and their application to food in Japan was initially symbolic and ritualistic. The conscious pursuit of a vibrant, five-colour palette in cuisine, often through artificial colouring techniques, appears to be a later development. The five cooking methods, particularly the emphasis on raw preparation, emerge as a uniquely Japanese culinary development. Historical evidence suggests that its formal articulation as a cohesive 'five methods' concept, and its explicit linkage to the five elements, is a relatively recent systematization, primarily appearing in culinary guides from the late twentieth century.

These findings suggest that the comprehensive 'five tastes, five colours, five cooking methods' framework, as understood and promoted today, is largely a modern construct. While individual elements and their underlying philosophical influences have deep historical roots, their integration into a singular, overarching principle for all Japanese cuisine, including *shōjin ryōri*, is a more recent phenomenon. This study contributes to a nuanced understanding of Japanese culinary history by disentangling the historical trajectories of these elements and challenging the often-simplified narratives surrounding their origins.

About the Author

Saori Nishida is a researcher exploring the philosophical foundations of Japanese food culture. Drawing on nearly a decade of professional experience promoting Japanese cuisine in Europe, her research examines how culinary philosophy influences relationships between food, culture, and sustainability.

Notes

1 Shoyo Yoshimura, interview with the author, 10 August 2021.

2 Naoyuki Yanagihara, '日本料理における精進料理について' ('Shojin Ryori in Japanese Cuisine'), *Journal of Cookery Science Japan*, 54.1 (2021), pp. 66–69; Lily Crossley-Baxter, 'Japan's Ancient Vegetarian Meal', *BBC*, 9 September 2020 <https://bbc.com/travel/article/20200907-shojin-ryori-japans-ancient-vegetarian-meal> [accessed 1 May 2025].

3 Takayuki Tokuno, '曹洞宗における「食」と修行' ('Food and Practice in Soto Zen Sect'), *Journal of Religious Studies*, 90.2 (2016), pp. 81–105.

4 Dōgen, *典座教訓・赴粥飯法* (*Instructions for the Zen Cook: The Dharma of Taking Food*), trans. by Shohachi Nakamura, Rikizan Ishikawa, and Toshiyuki Nakamura (Kodansha, 1991).

5 Isao Kumakura, *日本料理の歴史* (*History of Japanese Cuisine*) (Yoshikawa Kobunkan, 2007), p. 60.

6 Rryokyu Miyoshi, *身近な食材を使って* (*Cooking with Readily Available Ingredients*) (Eiheiji, 2014), p. 209.

7 TEN, *おいしい精進料理* (*Tasty Shojin Ryori*) (Oizumi, 2005), p. 131.

8 Ayako Ehara, 'What Is "Washoku" on UNESCO's Intangible Cultural Heritage List? Its Characteristics and Promotion', *Journal for the Integrated Study of Dietary Habits*, 28.1 (2017), pp. 3–5.

9 Naoe Imura, '戦略的地域資源としての和食・日本料理・京料理' ('Managing Washoku, Japanese Cuisine and Kyoto Cuisine as Strategic Regional Resources'), *Kyoto Management Review*, 37 (2020), pp. 119–37.

10 Japanese Culinary Academy, *Introduction to Japanese Cuisine: Nature, History and Culture* (Shuhari Initiative, 2015), p. 95.

11 Brian May and Takako Tomoda, 'The Origins of the *Wu Xing*', *Journal of the Australian Chinese Medicine Education and Research Council*, 4.2 (1999), pp. 14–23.

12 Shoyo Yoshimura, *精進料理考* (*Thoughts on Shojin Ryori*) (Shunjusha, 2019), p. 27.

13 Toshiko Tejima, '薬食同源：薬膳を楽しむには' ('Medicine and Food Have the Same Origin: How to Enjoy Medicinal Foods'), in *健康づくりとくすり* (*Health Promotion and Medicine*) (Nagasaki University, 1995), pp. 225–35.

14 *Umami: The Fifth Taste* (brochure), Umami Information Center, 2014, p. 16.

15 Shohachi Nakamura, 'Food Culture and Thought of Five Elements in China', *Journal for the Integrated Study of Dietary Habits*, 11.3 (2000), pp. 209–15.

16 Tamiko Yamazaki, '中医栄養学における五味について' ('The Five Tastes in Traditional Chinese Medicine Nutrition'), *Bulletin of Obihiro Ohtani Junior College*, 38 (2000), pp. 71–81.

17 Nakamura, 'Food Culture'.

18 Kazuhiko Takamiya, '色から見た食文化' ('Food Culture Seen through Colours'), *Japan Society of Cookery Science*, 36.2 (2003), pp. 177–83.

19 Kazuhiko Takamiya, '食物の配色' ('Food Colour Scheme'), *Journal for the Integrated Study of Dietary Habits*, 1.1 (1991), pp. 52–56; Takeo Inoue, 'A Historical Study of the Development of Food Colorings and Its Regulation in Japan', *Japanese Journal of History of Pharmacy*, 51.2 (2016), pp. 75–85.

20 Nao Watanabe and others, 'Recreation and Use of the Menu for the Banquet Held for Motonari Mouri during the Warring States Period Recorded in the Motonari-kō Yamaguchi go-Gekō no Setsu Kyōō Shidai', *Journal of Food Culture of Japan*, 18 (2022), pp. 27–38.

21 Takamiya, '色から見た食文化'.

22 Yoshimura, interview; Daisuke Nomura, interview qtd in Crossley-Baxter, 'Japan's Ancient Vegetarian Meal'.

23 On Daitokuji Temple in Kyoto, see 'Interview with Hiroko Tsuda, Daitokuji Ikkyu', Murasaki Style Project, n.d. <https://murasaki-style.rental-space-kyoto.com/syoukai/ikkyu/> [accessed 1 May 2025]; 'Daitokuji Ikkyu', Hyakumikai, n.d. <https://www.kyomeibutuhyakumikai.jp/shop/daitokuji-ikkyu.html> [accessed 1 May 2025]. On Hida Takayama, see Kosenkaku, '究極の"お・も・て・な・し"料理日本の四季を楽しめる『精進料理 慶月』リニューアルオープン' ('The Ultimate "Omotenashi" Cuisine: Enjoy the Four Seasons of Japan at "Shojin Ryori Keigetsu" Reopening'), PR Times, 26 September 2018 <https://prtimes.jp/main/html/rd/p/000000001.000037825.html> [accessed 1 May 2025]; 'Kakusho', Takayama City, 18 December 2021 <https://www.city.takayama.lg.jp/kurashi/1000021/1000119/1000847/1000848/1000897.html> [accessed 1 May 2025].

24 'The Washoku Way: Japan's Nuanced Approach to Food', Ministry of Agriculture, Forestry and Fisheries, n.d., p. 21 <https://www.maff.go.jp/j/shokusan/gaisyoku/pamphlet/> [accessed 1 May 2025].

25 Based on an examination of Chinese-language sources using the search term '五味五色五法' ('five tastes, five colours, five cooking methods').

26 Xiang Dong, 'A Comparative Study between Food Cultures in China and Japan', *Studies of Regional Policy*, 9.4 (2007), pp. 69–74.

27 Kokyu Abe, *庖丁軌範：りょうりのしきたり* (*Traditional Culinary Protocol*) (Japan Art, 1984).

28 Japanese Culinary Academy, *Introduction to Japanese Cuisine: Nature, History and Culture* (Shuhari Initiative, 2015), p. 68.

29 Kyoko Ohashi, 'Cooking Oil & Fat and Their Use Seen in Publications during the Meiji Period – Centering on a Novel Shokudoraku', *Bulletin of the Institute of Modern Culture, Showa Women's University*, 803 (2007), pp. 84–93.

30 Miki Matsunaga, 'Research of Gastronomic Culture: About the History of Development of the Cake', *Bulletin of Nakamura Gakuen College and Nakamura Gakuen Junior College*, 42 (2010), pp. 203–11.

31 Toshihide Nishimura, 'Koku Perception Involved in Food Palatability – Its Formation Mechanisms and Application in Cooking Process', *Journal of Cookery Science Japan*, 57.3 (2024), pp. 159–65.

32 'History of Tempura', Showa Sangyo <https://www.showa-sangyo.co.jp/knowlege/tempura/learn/> [accessed 1 May 2025].

33 Japanese Culinary Academy, *Introduction to Japanese Cuisine*, p. 95.

34 Tsunenobu Tokura, '日本飲食文化史中兩類「出鱈目」' ('Two Types of "Detarame" in the History of Japanese Food Culture'), *Liberty Times Net*, 23 November 2020 <https://ec.ltn.com.tw/article/breakingnews/3359586> [accessed 1 May 2025].

24
An Essential Element of Harmony and Hegemony
Royal Navy Foodways in the Age of Fighting Sail

Blake Perkins

Perceptions of Royal Navy foodways during the Revolutionary and Napoleonic Wars have been skewed by fiction and myth that describe official indifference to the quality of nutrition. As a result, the conventional wisdom holds, crews of warships endured hardship and disease. 'The experience of naval life in the 18th century', Andrew Lambert explains, 'has often been portrayed as one of suffering in something little more than a floating concentration camp, where an unwilling crew, raised by the press-gang, was systematically beaten, starved and terrorized into doing their duty'.[1]

It takes scant insight to question those depictions of shipboard life in the Royal Navy. Its polyglot personnel formed the most formidable force of the era.[2] British warships remained on station for unprecedented durations to enforce the blockade of continental Europe and operate in the Baltic, Caribbean, and Mediterranean Seas as well as the Atlantic and Indian Oceans. Provisions 'had to be moved over long distances, at a time when it took six weeks to reach the Mediterranean and up to six months to get to India, and when unarmed transport vessels had to be convoyed under naval escort or risk capture by hostile ships'.[3]

The Victualling Board Spins Magic

Those ambitious operations were possible because the Royal Navy was confident in the superior competence and morale of its people. Jonathan Dull maintains without exaggeration that those attributes

> were not so much the product of battle as they were the result of superb organization and administration, the culmination of developments going back to Pepys almost a century and a half earlier. British admirals, officers, and sailors deserve much of the credit, but also important were, for example, the thousands of employees of the Victualling Board who baked the navy's biscuit, brewed its beer, prepared its salted meat, and made the casks to store the food and drink.[4]

The board was an independent subsidiary of the Admiralty, sufficiently autonomous that its offices stood at Tower Hill, far to the east of the Admiralty itself in Whitehall.

Individual Initiative

Officials down the chain of command – from the Lords of the Admiralty and Victualling Board to captains and their pursers – spent an inordinate amount of effort to ensure that their people received a substantial and varied diet combining fresh with preserved foods. Throughout the era, the most frequent purchase by ships in foreign waters was beef, and the first among the enumerated duties of a ship's purser approaching port was the preparation of contracts 'for fresh beef and vegetables – lime juice', followed by 'beer and water [. . .] pudding bags'.[5] In addition to items subject to spoilage, captains shipped vegetables that could keep, for officers and ratings alike: cabbage and onions, along with roots, especially turnips.[6]

Some 40 percent of the food consumed by the fleet that blockaded Brest during 1804 and 1805 was fresh.[7] Naval leaders knew that nutrition was essential to the efficient functioning of sailors engaged in arduous physical labour. The navy, therefore, developed a complex global chain of seaborne transport.

At the outset of the era, the Victualling Board had encountered persistent deficiencies in quality from private contractors. Its solution was the production of those vital commodities – the biscuit, beer, and salt meat – itself. The board operated its vast principal yard, or outport, at Deptford. Its output was prodigious. The ovens alone baked over twelve tons of biscuit a day.[8]

At its height during the French Wars, the Royal Navy deployed some 140,000 men at sea and became the largest buyer of agricultural products on the British market.[9] The effort expended by the board was a spectacular success. In February 1804 and again in March 1805, for example, Nelson informed St Vincent, 'We have not a sick man in the fleet'.[10]

No other seaborne force supplied its fleets so well. French and Spanish operations foundered for lack of food or from putrid stores, and scurvy remained rife aboard their warships long after the British had very nearly eradicated the disease. Before Trafalgar in 1805, the Mediterranean fleet had blockaded Toulon for twenty-two uninterrupted months before pursuing the combined French and Spanish fleet back and forth across the Atlantic. On the eve of battle, the Physician of the Fleet reported 110 deaths and 141 hospitalizations on account of scurvy among the complement of 7000. In contrast, several of the Spanish ships, at sea for only three months, suffered over 200 cases and one more than 300.[11]

With the eventual exception of the East Indies station, the board shipped all preserved foods in bulk across the globe from its yards in the British Isles to ensure quality control and accurate accountings. Ships typically carried preserved provisions to last from four to six months, but fleets needed to buy drink and fresh food locally after a relatively short time at sea.[12]

Equality before the Mast

Officers did, invariably, call their crews the people, an index of egalitarianism insofar as it went, and on most warships, it went pretty far, at least when it came to foodways and leisure time. The chief purser of the Royal Navy declared unequivocally in 1788: 'Each

Person on Board a King's Ship, from the Captain down to the Cook's Boy, has the same Allowance' of food and drink, although in practice, captains might compensate boys for their allowance rather than allow them to drink it.[13]

The doctrine of equality also applied in situations of scarcity, which the sources indicate were rare.[14] The 1745 Admiralty *Regulations and Instructions Relating to His Majesty's Service at Sea* provide:

> It is left to the discretion of commanders of squadrons to shorten [. . .] the allowance of provisions according to the exigence of the service, taking care that the men be punctually paid for the same. The like power is given to captains of single ships in cases of absolute necessity, but the purser is strictly charged not to supply any officer at whole allowance while the rest of the company are at short; but all are to be equal in point of victualling.[15]

So nutrition was a right, not an indulgence, and seamen were to be compensated for short rations, which were tolerable only *in extremis*, while officers had to share in the hardship.

Comparative Comestibles

Life was contingent in a brutal age, but less so in the British senior service, where a sailor was entitled to his 'three square meals' each day, a term that may be derived from the square trenchers that the ratings used to eat their food. 'Three square meals' may have gained currency in connection with the Royal Navy because a lot of landsmen could not get them.[16]

Even within the service itself, food might be better afloat than ashore. Captain Marmaduke Wybourn of the Royal Marines reported from '<u>this</u> vile town' of Alexandria in July 1801 that he 'had salt beef, a little stinking rum, and water a little saltish [. . .] but on board [. . .] the Arabs supply the Markets very amply so that we shall soon get fat again'.[17]

William Todd, a corporal who served at sea with the Royal Marines on the frigate *Thetis* during 1757 and 1758, confirmed the favourable comparison in his journal; on shipboard, he 'could Et, Drink & sleep full as well if not better than ashore'. On first embarking, he wrote:

> I like very well being here & so doth all our Soldiers for the Beer we have [. . .], so they are quite Merry on Board the Thetis Man of War [. . .]. They always has plenty of the Best. [. . .] I live very well, having Commonly Beef, Pork etc, platter fulls standing besides me & always a Large Rum Bottle.[18]

Equal Rights

Admiralty regulations remain 'the best source of information about the food served to the crew'.[19] The regulation establishing the amounts 'Of the Provisions' includes a table specifying what each man was entitled to receive on a weekly basis (see Table 1).[20]

It is one thing to enact a regulation, another to enforce it. Evidence indicates that the navy did. During an interval of short rations in 1796, Admiral John Jervis, eventually Earl

Table 1.

	Biscuit (lb)	Beer (gallons)	Beef (lb)	Pork (lb)	Pease (pint)	Oatmeal (pint)
Sun	1	1		1	½	
Mon	1	1				1
Tues	1	1	2			
Wed	1	1			½	1
Thurs	1	1		1	½	
Fri	1	1		1	½	1
Sat	1	1	2			
Total	7	7	4	3	2	3

St Vincent, who demonstrated a 'domineering and ruthless personality' combined with a 'great concern for the welfare of his men', issued a heartfelt order to his ships:

> The commander in chief derives great satisfaction from the prospect that he has, by the late arrival of victuallers from England, of soon restoring the respective officers and ships' companies of the fleet to whole allowance of provisions and he will cause the short allowance money to be paid as soon after the fleet anchors at Gibraltar as the lists can be cast by the Agent Victualler.[21]

As early as 1729, César de Saussure, the unlikely Swiss passenger aboard the frigate *Torrington* bound from Portsmouth to Lisbon, displayed apparent surprise at the equality of victualling. 'Rations are all alike', he remarks. 'Officers receive the same food, no difference being made between them and the men'. He marvelled at the amount, 'as much as, or more than, can be eaten or drunk in a day', and found the biscuits 'very much to my taste, and they reminded me of nuts'. De Saussure concludes that 'sailors [. . .] are well fed'.[22]

Decades later, Corporal Todd corroborates the consistency of the regulatory ration:

> Our provisions served out every day is: Monday – Bargue, what the men will take, 4 pound of Bread, 1 ½ pound of Suffolk Chees & 1 pound of Butter to each Mess for 4 Men; then Tuesday – Bread the same, 1 ½ pound of flower, ½ pound of either Sewit or Raisins to mixt with their puddings & 4 pound of Beef per Mess; Wednesday – Pease Boyl'd into soop, Each Mess may fetch what they please, Chees & Butter as Monday; Thursday – 4 pounds of Pork & Pease Soop, what they please Each Mess; friday – Bargue, one pound of Butter & no Chees; Saturday – flower & Beef the same as Tuesday; Sunday – Pork & Pease as Thursday. This is the prvisons allow'd on board.[23]

De Saussure describes a similar substantial allowance:

> A pound of cheese is allowed to every four sailors, half a pound of butter for breakfast, the same for supper. Each sailor eats one pound of boiled beef three days in the week for dinner, together with a pudding made of flour and suet. On two other days he eats boiled salted pork with a pudding of dried peas and the remaining [two] days pea soup and salt fish or bargow [*sic*].[24]

Samuel Leech, who began his career in 1810 at the age of 12 as a powder monkey aboard HMS *Macedonian*, describes shipboard fare

> composed of hard sea biscuit, fresh beef while in port, but salt pork and salt beef at sea, pea soup and Burgoo, or, as it was sportively called, skillagallee, was oatmeal boiled in water to the consistency of hearty pudding. Sometimes we had cocoa instead of burgoo. Once a week we made 'duff', or pudding.[25]

Modern sources calculate that those quantities provided seamen on Royal Navy ships some four thousand to five thousand calories a day, a robust amount then as now.[26]

A Thousand Words

The pictorial record reinforces contemporaneous chronicles. The great caricaturist James Gillray alone executed more than 120 realistic naval studies, drawings, and watercolours from life. They are rendered in meticulous detail. The images, Jim Sherry notes, 'were created in 1793 and 1794 as part of two projects [. . .] neither of which had anything to do with caricature'. Those images remain 'the greatest number of surviving Gillray drawings'.[27] Among many of those drawings, Figures 1 and 2 depict muscular, hardy sailors from the Royal Navy.

Images by Isaac and George Cruikshank as well as lesser lights follow suit (Figures 3 and 4). The most famous depictions of Royal Navy sailors may be Rowlandson's realistic series from 1799, and the most famous of those may be the ship's cook. As Figure 5 indicates, cooks might be sailors who had been disfigured by battle, but even so, the cook appears as robust as the other sailors.

Parodic prints indicate the same thing. Effective satire employs distortion to amplify reality, but without a measure of truth, the effect becomes insubstantial, and the great satirists of the age are famously effective. Satirical depictions, therefore, can be useful evidence, and

Figure 1.

Figure 2.

Figure 3.

Figure 4.

Figure 5.

satirical depictions of sailors are consistent with the life drawings. Figure 3, for example, portrays a healthy and muscular crew engaged in the ritual of 'crossing the line'. They are at the equator, so they have been at sea awhile and appear none the worse for that.

The Flexibility of Food

The regulations related to food were not rigid; deployment of fleets on distant stations required flexibility. The Admiralty stipulated general substitutions for various items, and substitutions were not uncommon:

> In case it may be thought for the service to alter any of the foregoing particulars of provisions in ships employed on foreign voyages, it is to be observed that a pint of wine, or a half a pint of brandy, rum or arrack, hold proportion to a gallon of beer. Four pounds of flour, or three pounds of the same with raisins, half a pound of currants or half a pound of suet beef pickled, are equal to a four pound piece of beef, or two round piece of pork with peas. Half a pound of rice is equal to a pint of oatmeal.[28]

On particular stations, commanders could consider other alternatives: 'For example, chick peas or lentils were issued in place of pease in the eastern Mediterranean or in India; sugar could be used instead of oatmeal, butter or cheese'. And so on and on, and squadron commanders and their captains also could improvise through the expenditure of 'necessary money' for nonstipulated substitutes purchased from local suppliers in distant waters.[29]

Quality Control

The quantity apparently was there; what about the quality? N. A. M. Rodger explains that an 'astonishing' – that is, minuscule – rate of spoilage for food shipped in cask, some 1 percent from 1750 to 1757, was 'achieved by great care in the manufacturing process (backed by continual experiment and development) and by using only the most expensive ingredients'. The Admiralty declared 'their intention that the seamen should be supplied with the best of everything in its kind', and in general, they were.[30]

Commanders improvised to supply their ships with fresh provisions. Shore parties hunted game, including wildfowl, and paid as well as foraged for all varieties of fruit eaten fresh or folded into the ubiquitous puddings served at sea.[31] Sailors and marines embarked in blockading squadrons went so far as to plant gardens on secure islands to obtain 'the luxury of fresh vegetables', sometimes, as a marine lieutenant found, 'to the astonishment and surprise of the natives'.[32]

'The most frequent of [. . .] purchases', according to Captain A. J. Griffiths, who served as a lieutenant onboard HMS *Culloden* at the Battle of St Vincent in 1797, was 'fresh beef', and warships joined merchantmen to ferry livestock to hungry fleets.[33] Returning from Yarmouth to blockade Dutch warships trapped in the Tegel during 1805, the frigate *Flora* shipped 'some thirty or forty live bullocks and an abundance of vegetables in bags, so that it was with good reason that the North Sea station was called by sailors a full belly station'.[34] 'Captains often took it on their own initiative to buy fresh fruit and vegetables', while Nelson 'exploited the British protection of Sicily to buy thirty thousand gallons of lemons. [. . .] Eventually the Sicilian lemon gardens supplied the whole navy'.[35]

Culinary Custom

By the eighteenth century, it had become an unwritten practice to serve fresh vegetables with fresh meat. In 1757, the custom was codified, although it would disappear from print but not practice by 1806:

> Some of the elder Pursers [. . .] representing, that it had been their constant Practice, as often as their respective Ships were victualled with Fresh Meat, to boil such a Quantity of Greens and Roots with it, as to give sufficient Satisfaction to the Men.

The language chosen by the Admiralty recognizes that although they might be a treat for them, the ratings, conservative in their foodways, could become leery of fresh vegetables if the bonus cost them their dried pease:

> And that, to give them no Room to murmur on Account of the saving of Pease by boiling fresh Meat, they boil Pease for their *Monday* Dinners, besides their allowed Oatmeal for Breakfast; and proposing [though not literally requiring], if the Greens and Roots furnished them should not be judged sufficient for the Men, to increase the same to any Quantity that should be prescribed; all Commanders are to take Care, that their respective Pursers do comply with what is said [by the elder pursers cited], by furnishing a sufficient Quantity of Roots and Greens to the Seamen, with their fresh Meat; or to represent to us, if they fail therein.[36]

So fresh provisions might amount to a customary supplement to, rather than substitute for, the statutory allowance that sailors held dear.

Other Egalitarian Elements

Officers issued orders to ensure that food was served out equitably to their crews, sometimes to the disadvantage of the officers themselves. St Vincent's standing orders as commander of the Mediterranean Fleet during 1796–98 include detailed instructions. When captains ordered the killing of cattle, he ordered officers to take the head and tongue as their share in order to reserve the choicer cuts for the men, 'he taking a head every slaughtering by way of example'.[37]

The captain's orders issued by John Fyffe for HMS *Indefatigable* on 8 August 1812 do not quite require as much abnegation by the officers but do take pains to ensure fairness:

> A master's mate, boatswain's mate, quartermaster and captain of the forecastle are always to be present at opening and cutting up salt provisions or fresh beef [. . .] are always to be served out to the messes without partiality [. . .] and no choice pieces of salt provisions allowed to be taken by or for officers.[38]

Warships carried hooks, lines, and skeins (Figure 6). Admiralty regulations stipulated that 'the fish which shall be caught shall be distributed daily to such persons as are sick or upon recovery', an apparent (and effective) treatment for scurvy:

Figure 6.

> And if there be any surplus the same shall be distributed by turns amongst the messes of the officers and seamen without favour or partiality and gratis, without any deduction in their allowance of provisions on that account.[39]

As with the vegetables, fresh fish was a supplement rather than substitute for the stipulated foodstuffs. According to Francis Liardet, who served in frigates from 1809 to 1816, some captains went further to ensure fairness:

> Even in very strict ships, we have known the word passed for the men to get their fishing lines up. [. . .] In close harbours we have known the same officers [. . .] give a boat for line-fishing, sending a man from each mess for that purpose.[40]

For Liardet, fishing also served a recreational purpose. 'Line- and seine-fishing', he thought, 'makes a little variety in the sameness of a sailor's life, besides giving him a good fresh meal'.[41] Wybourn enjoyed fishing, even under the eyes of enemy warships 'ready for sea'. During blockade duty with HMS *Blake* in 1809 and *Marlborough* in 1812, he fished off the deck while 'watching the Dutch fleet', at one point quipping that 'we [. . .] are now catching the Enemy's fish'.[42]

Similarly, in distributing antiscorbutics, St Vincent ordered that 'no consideration is to be had to the officers in preference to the men but they are to be equally apportioned to the number of persons victualled, with a reserve for the sick'.[43]

Custom further contributed to the equitable distribution of food. Each sailor took his turn as 'mess cook' for a week. The cook would whistle while mixing flour with raisins and suet for puddings, 'as it is impossible to clandestinely eat raisins while whistling'. He would lower their table stowed on lines above the deck, arrange benches, collect mugs and trenchers, prepare some of the food (including pastry for duff and puddings), collect it from the galley when cooked, and then serve his mates. The cook carved the meat with his back to the table, calling 'Who shall have this?' to unseen messmates. Those unwritten customs held sway throughout the navy. After dinner, he 'washed the mess kids, &c, stowed the implements and hoist the table back to the deckhead'.[44]

Offshore at the Zoo and in the Market

Captains shipped livestock and the fodder they required into the reach of long journeys, despite the considerable risk that the people would adopt them as pets. Poultry was sufficiently ubiquitous to merit a standard Admiralty coop pattern, introduced in 1815 for permanent installation amidships. They were brought on deck during daylight for the benefit of the birds.

Local populations eager to trade, often undeterred by nominally belligerent status, rowed bumboats selling all manner of food alongside warships to not only British officers and their pursers but also crewmen eager to buy fresh produce, dried fruit, and, in the Mediterranean, charcuterie.[45] Wybourn describes the spectacle he witnessed from HMS *Ambuscade* off Naples during 1805: 'The ships were soon surrounded by boats, bands of Music welcoming us', and while 'other boats brought Fruits & Vegetables in abundance', others were filled with prostitutes, invariably, it seems, of all ages in Italian waters.[46] William Mark, a purser who would rise to the post of diplomatic consul, had recorded the same kind of thing as HMS *Hydra* approached Palermo in 1802. 'Boats came alongside', he observes, 'before we were well at anchor, with the fruits of the season, to be sold at cheap rates; this was a great treat'.[47]

Twenty-five years earlier in French waters, Todd observed a similar scene from *Thetis*: 'Plenty of Bom Boats came a Board of us from Portsmouth, al sorts of necessaries they sold us very Dear. I bought an Apple tart or two that I Eat very pleasantly'.[48] Ratings on the Moluccas station traded their rice ration, never much liked, for sweet potatoes, pumpkins, and yams.[49] The practice of trading with bumboats was so popular and prevalent that Fyffe embedded it in his standing orders: 'Bum boats will at all times be permitted alongside during mealtimes'.[50]

Much Depends upon Dinner

The navy understood that a good dinner eaten together is good for the morale of the diners and also understood that the chance to socialize was very near equal in importance to the food itself. 'Dinner', as Erin Maglaque recently wrote, 'is never just dinner'.[51] Captains could consider mealtime an occasion for ceremony. A landsman aboard HMS *Gibraltar* wrote during 1811 that dinnertime was 'noticed by the drums and fifes which play "Oh the Roast Beef of Old England"'.[52]

As early as 1786, the Duke of Clarence, commanding HMS *Pegasus*, issued a standard order: 'The people are never to be interrupted at their meals but on the most pressing occasion'.[53] Captain Edward Riou's standing orders from 1799 for HMS *Amazon* cover the subjects he considered paramount. Along with detailed instructions for sail handling, 'preparation for battle', 'the exercise of great guns', and other critical operations, they stipulate:

> The ship's company are never to be interrupted at their meals but on the most pressing occasions [. . .] and if the duty will admit they should never have less time allowed for their breakfast than three quarters of an hour and an hour and a half for dinner.[54]

Riou's orders would become a template, and throughout the navy, dinners became inviolate. In 1805, Sir Edward Pellew issued standing orders for all ships of the East Indies squadron. Among other things, his orders stipulate that 'one hour and a half will usually be allowed' for dinner at noon, adding that 'the Commander in Chief is extremely particular, that the Meal times of the people are not to be broken in upon'.[55]

More Depends upon a Mess

The fact that captains might maintain morale in the confined conditions of the wooden world remains a minor marvel. HMS *Victory*, for example, sheltered nearly nine hundred people within a structure 186 feet long and about 53 abeam. In his *Observations on Some Points of Seamanship*, Griffiths asks and answers the question:

> Can any reason be assigned why the crew should not be as happy as the necessity of subordination etc will admit? Ten thousand may, why they should. Leave them as much as you can to themselves with propriety and the more you do so, the more content they will be.[56]

Those freedoms extended to the determination of foodways. Along with the food itself, the experience of dining contributed more than any other factor to the happiness of the crew. The navy divided its ratings into messes of four to twelve men who shared a table, depending in part on the size of the ship (Figure 7).[57]

Samuel Leech offers a sound assessment of the mess: 'The crew of a man of war is divided into little communities of about eight, called "*messes*". These eat and drink together, and are, as it were, so many families'.[58] So much so that Leech describes his mess as something of a quotidian philosophical society: 'We eat [*sic*] together, drank together, joked, sang, laughed, told yarns; in short, a perfect union of ideas, feelings, and purposes, seemed to exist among all hands'.[59]

The union itself was very nearly communal, an incongruous building block for the disciplined, autocratic hierarchy that was the shipboard population. It was a refuge from those rigours – as Matthew Brenckle observes, 'the only democratic institution on board a warship, a group in which there was true equality regardless of age or station'.[60]

The concept of community was in effect codified by W. H. Smyth in his *Sailor's Word-Book* and by the popular slogan he cites: 'MESSMATE. A companion of the same mess, hence comrades in many ways; whence the *saw*: "Messmate before a shipmate, shipmate before a stranger, stranger before a dog"'.[61]

The mess or 'little community' was the basic element of operational organization, the organic unit that facilitated the functioning of the battleship. Messes in turn cooperated to form 'the disciplined collective workforce of what was [. . .] the most complex artefact in existence; in this sense, there was nothing else in the contemporary world quite like a ship-of-the-line'.[62]

Sailors cherished a stable mess, an indication that dinner companions formed the foundation for the cohesion and morale required to man and fight a warship. Griffiths understood and left his people to choose and change their messes themselves. Under his

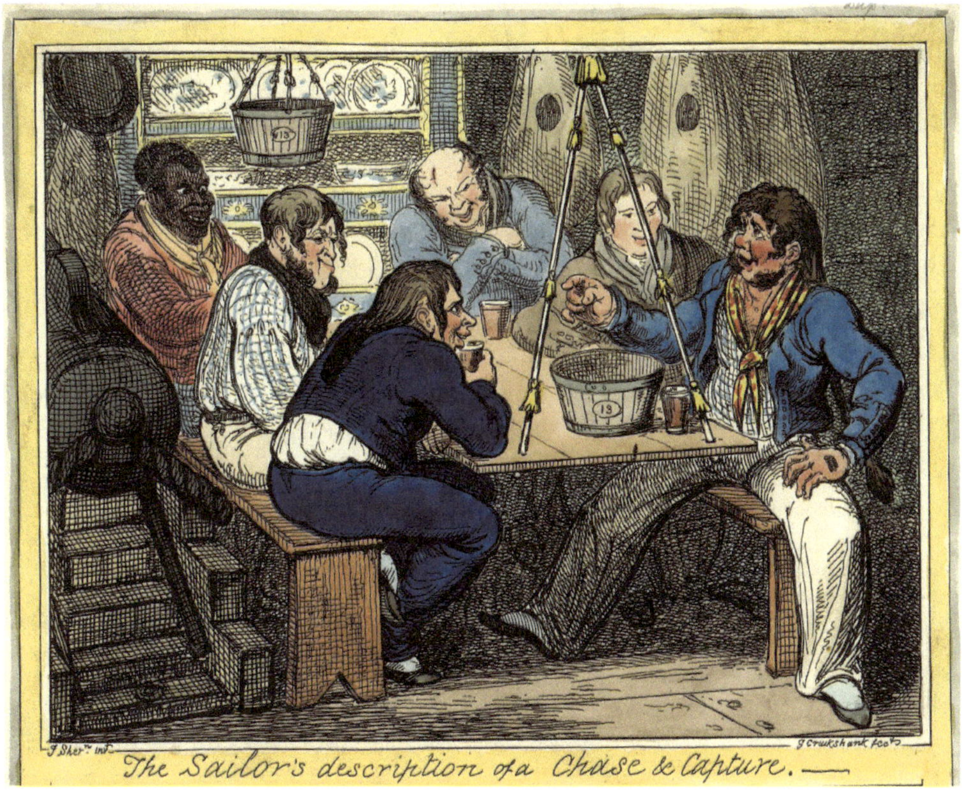

Figure 7.

commands, the 'crew ever messed as they liked', a predilection he also found useful to deter malfeasance:

> So satisfied was I of the misery of obliging men to mess together that among my punishments was a drunken, a dirty, a blackguard and a thieves mess. Instead of being numbered, they were called by these names at the grog tub, the cook's coppers etc.[63]

He found the disciplinary device more effective than the lash.

Monitoring the messes enabled an incoming commander to assess morale onboard his ship. Griffiths 'always had a mess book kept for me and regularly corrected every month. Experience proved it one of the best possible means of acquiring knowledge of the real character of the crew. [. . .] You cannot know too much of those you command'.[64]

The mess enabled a captain or fleet commander to gauge and maintain the morale as well as efficiency of his people. The messing system itself, a system that gave an astonishing amount of agency to the sailors themselves, was, along with the food they consumed, the essential element of organization that propelled the Royal Navy to global dominance.

About the Author

Blake Perkins conducts research and writes about culinary history. His primary but not only interest in that respect is the evolution of British foodways and their impact on historical culinary practices.

Acknowledgements

Grateful thanks to Lee Anzick, Hailey Morales, and Jasmine Sykes-Kunk at the John Hay Library of Brown University for their friendly and knowledgeable research assistance.

Notes

1 Andrew Lambert, 'Life at Sea in the Royal Navy of the 18th Century', BBC History in Depth, 17 February 2011 <https://www.bbc.co.uk/history/british/empire_seapower/life_at_sea_01.shtml> [accessed 14 January 2025]. The press accounted for at most some 15 percent of Royal Navy crews at various times during the eighteenth century, most of them seamen (see, for example, Joachim Voth and Guo Xu, 'Patronage for Productivity: Selection and Performance in the Age of Sail', Guo Xu, 17 January 2020, p. 10, n. 14 <https://guoxu.org/docs/patronageproductivity_vothxu_jun19.pdf> [accessed 11 February 2025]). Neil Faulkner notes, 'Ideally, the press gangs went for skilled sailors; at a pinch, pretty well any unfortunate would do' ('Nelson, Navy, Nation: The Secret of Victory', The Past: Military History, 7 January 2014 <https://the-past.com/feature/nelson-navy-nation-the-secret-of-victory/> [accessed 22 February 2025]).

2 'Only 51% of the men in Nelson's Navy were English: 19% were Irish, 10% Scottish, 3% Welsh, and the remainder were foreigners' (Faulkner, 'Nelson, Navy, Nation'). See also *Shipboard Life and Organisation, 1731–1815*, ed. by Brian Lavery (Ashgate, 1998), pp. 449, 451; and John Keegan, *The Price of Admiralty* (Viking, 1989), p. 47.

3 Roger Knight, Martin Wilcox, and James Davey, 'Sustaining the Empire: War, the Navy and the Contractor State, 1793–1815', Royal Museums Greenwich <https://www.rmg.co.uk/about-Wilcoxus/sustaining-empire-1793-1815> [accessed 16 September 2021].

4 Jonathan Dull, *The Age of the Ship of the Line: The British & French Navies, 1650–1815* (Bison Books, 2009), pp. 174–75.

5 A. T. Griffiths, *Observations on Some Points of Seamanship* (J. J. Hadley, 1824), pp. 13, 221, 222.

6 See, for example, David Fictum, 'Salt Pork, Ship's Biscuit, and Burgoo: Sea Provisions for Common Sailors and Pirates', *Colonies, Ships, and Pirates*, 24 January 2016 <https://csphistorical.com/2016/01/24/salt-pork-ships-biscuit-and-burgoo-sea-provisions-for-common-sailors-and-pirates-part-1/> [accessed 15 January 2025].

7 Janet Macdonald, 'Two Years Off Provence', *Mariner's Mirror*, 2.4 (November 2006), pp. 443–54 (p. 450).

8 Janet Macdonald, *Feeding Nelson's Navy: The True Story of Food at Sea in the Georgian Era* (Chatham Publishing 2004), pp. 56, 184.

9 Ferdinand Mount, 'Salt Spray', *London Review of Books*, 5 December 2024, p. 14; N. A. M. Rodger, *The Wooden World: An Anatomy of the Georgian Navy* (Naval Institute Press, 1986), p. 82; N. A. M. Rodger, *The Command of the Ocean: A Naval History of Britain, 1649–1815* (Penguin, 2004), p. 307; Roger Morriss, *The Foundations of British Maritime Ascendancy: Resources, Logistics and the State, 1755–1815* (Cambridge University Press, 2011), p. 277; Knight, Wilcox, and Davey, 'Sustaining the Empire'.

10 Macdonald estimates the rate of illness a still remarkable 5 percent ('Two Years Off Provence', p. 450).

11 Martin Wilcox, 'This Great Complex Concern: Victualling the Royal Navy on the East Indies Station', *Mariner's Mirror*, 97.2 (4 November 2006), pp. 32–48 (p. 33); Morriss, *Foundations*, p. 294; Macdonald, 'Two Years Off Provence', p. 444. T. Marmaduke Wybourn cites evidence from Alexandria in 1801: 'The Market is large and well supplied. A Sheep only costs a Dollar, Fowls 36 for a Dollar, Geese four, eggs 750 a Dollar, fruit in great abundance, and almost for nothing' (*Sea Soldier* (Parapress, 2000), p. 43).

12 'A Treatise on the Office of a Purser on Board His Majesty's Ships' (unpublished MS 1788), cited in Ben Kinmont, *Gastronomy*, Catalogue 16 (Sebastopol, CA, 2018); *Shipboard Life and Organisation*, p. 113.

13 Kinmot, *Gastronomy*.

14 Captain Marmaduke Wybourn of the Royal Marines complains about only three brief stretches of short rations from 1797 to 1813, one of them ashore and another during quarantine (*Sea Soldier*, pp. 13, 47, 66). According to Roger Morriss, 'After the American War of Independence, shortages on foreign stations were unusual. On the contrary', because the victualling system had become so efficient, 'abundance was more of a concern than deficiency' (*Foundations*, pp. 301–02).

15 *Shipboard Life and Organisation*, pp. 17–18.

16 Blake Perkins, 'A Note about the Origin of the Term "Square Meal"', *Petits Propos Culinaires*, 132 (August 2025), pp. 45–60.

17 Wybourn, *Sea Soldier*, pp. 44, 47 (emphasis in original).

18 *The Journal of Corporal William Todd, 1745–1762*, ed. by Andrew Cormack and Alan Jones (Sutton, 2001), pp. 27, 24, 25.

19 *Shipboard Life and Organisation*, p. 5. Not so the officers: in their case, private diaries and cooks' accounts are crucial to understanding what they ate. See, for example, César de Saussure: 'Officers [. . .] club seven or eight together to buy wine, punch and fresh provisions' (*A Foreign View of England* (J. Murray, 1902), p. 364).

20 The 'banyan' or meatless days appear anomalous, but the practice was pragmatic, unrelated to frugality. It allowed use of the boilers for washing the linen and boys (H. T. A. Bosanquet, 'Banyan Day', *Mariner's Mirror*, 3.1 (January 1913), p. 89).

21 *Shipboard Life and Organisation*, pp. 205, 206, 212.

22 de Saussure, *Foreign View of England*, pp. 356, 363–64. 'Bargue' is an archaic term for porridge.

23 *Journal of Corporal William Todd*, p. 26.

24 de Saussure, *Foreign View of England*, p. 364.

25 Samuel Leech, *Thirty Years from Home* (Tappan & Dennet, 1843), p. 46.

26 Lambert, 'Life at Sea', p. 47; Macdonald, *Nelson's Navy*, pp. 177–78.

27 Jim Sherry, 'Gillray's Works', James Gillray: Caricaturist, 2026 <https://www.james-gillray.org/drawings.html> [accessed 14 January 2025].

28 *Shipboard Life and Organisation*, p. 18.

29 Knight, Wilcox, and Davey, 'Sustaining the Empire'.

30 Rodger, *Wooden World*, p. 85.

31 William Mark, *At Sea with Nelson* (Sampson Low, Marston & Co., 1846), pp. 141, 143; *Journal of Corporal William Todd*, p. 31.

32 Wybourn, *Sea Soldier*, p. 14.

33 Griffiths, *Observations*, p. 13.

34 *Shipboard Life and Organisation*, p. 67.

35 Mount, 'Salt Spray', p. 14.

36 Admiralty, *Regulations and Instructions Relating to His Majesty's Service at Sea*, 10th edn (1766), pp. 202–03. See also *Regulations and Instructions Relating to His Majesty's Service at Sea*, 9th edn (1757); and *Regulations and Instructions Relating to His Majesty's Service at Sea* (Wm. Wilkinson & Sons, 1806). A measure of the Admiralty's concern about morale and its connection to foodways, the understated usage of 'murmur' strikes a nearly adorable note here.

37 *Shipboard Life and Organisation*, p. 214.

38 *Shipboard Life and Organisation*, pp. 193–94.

39 *Shipboard Life and Organisation*, p. 17.

40 Francis Liardet, *Professional Recollections on Points of Seamanship* (Longmans, 1849), pp. 97–99.

41 Liardet, p. 98.

42 Wybourn, *Sea Soldier*, pp. 153, 138, 34.

43 *Shipboard Life and Organisation*, p. 210.

44 Leech, *Thirty Years from Home*, 46; *Landsman Hay: The Memoirs of Robert Hay 1789–1847*, ed. by M. D. Hay (Robert Hart-Davis 1953), p. 73 ('I [. . .] made the duff, attended to the sea-pie or lob-scouse, brought down the pease-soup and grog, washed up the traps, and made all snug'.); Macdonald, *Nelson's Navy*, pp. 111, 112.

45 Mark, *At Sea with Nelson*, 105; Macdonald, *Nelson's Navy*, pp. 118–19.

46 Wybourn, *Sea Soldier*, p. 86; Mark, *At Sea with Nelson*, pp. 105–06, 120.
47 Mark, *At Sea with Nelson*, p. 105.
48 *Journal of Corporal William Todd*, p. 35.
49 Peter Henry, 'Surgeon's Journal, HMS *Daedalus* (1802)', in *Shipboard Life and Organisation*, p. 528.
50 *Shipboard Life and Organisation*, p. 195.
51 Erin Maglaque, 'The Unpoetic Calorie', *London Review of Books*, 21 November 2024, p. 43.
52 Anon., 'Diary Kept on Board HMS *Gibraltar*, 1811', in *Shipboard Life and Organisation*, p. 462.
53 *Shipboard Life and Organisation*, p. 95.
54 *Shipboard Life and Organisation*, p. 151.
55 Sir Edward Pellew, 'General Orders, His Majesty's Ship *Culloden* (Bombay Harbour 19 March 1905)', in Northcote Parkinson, *War in the Eastern Seas* (George Allen & Unwin, 1954), p. 439.
56 Griffiths, *Observations*, p. 98.
57 *Shipboard Life and Organisation*, pp. 244–45.
58 Leech, *Thirty Years from Home*, p. 36. See also W. H. Smyth, *The Sailor's Word-Book* (Blackie and Sons, 1867), p. 477.
59 Leech, *Thirty Years from Home*, p. 147. Leech described his experience on USS *United States* following its capture of *Macedonian*; the organization of the fledgling US fleet was modelled meticulously on the Royal Navy.
60 Matthew Brenckle, 'Food and Drink in the U.S. Navy, 1794 to 1820', USS Constitution Museum, 2019, p. 26 <https://www.usscm.org/publications/food-and-drink-in-the-us-navy-1794-to-1820.pdf> [accessed 25 July 2023].
61 Smyth, *Sailor's Word-Book*, p. 478.
62 Faulkner, 'Nelson, Navy, Nation', p. 14.
63 Griffiths, *Observations*, p. 99.
64 Griffiths, p. 98.

25

The Paradox of Abundance

Water and Food Insecurity in Uganda's Rwenzori Region

Karen Salazar Ledezma and Paula Ibarra Flores

In Uganda, a country rich in rivers and lakes, one might expect water to be abundant and food to be secure. Yet the reality is quite the opposite. Many communities face daily struggles to access clean water and, by consequence, secure nutritious food. This study sets out to understand why such a paradox of abundance exists. It follows two key directions: first, to explore Uganda's broader challenges related to water and food insecurity, and second, to take a closer look at how these challenges affect the daily lives of the Bakonzo tribe in the Rwenzori region.

To understand this local experience, we draw on the data collected by Karen Salazar Ledezma for the GlobalRise survey in 2021: 'Nutrition Situation Analysis in the Community of Kirembe' (unpublished). This study provides insight into how people access, store, and treat water, as well as how traditional food is prepared, stored, and consumed in the community.

The first part of the article introduces key concepts such as water insecurity and its connection with food insecurity and food systems. It also presents the idea of the paradox of abundance in Uganda, with the existence of many water bodies but inaccessible resources at the community level. In the second part, the analysis takes place with the study case on the Rwenzori region to illustrate how this national issue plays out at the community level. Through the case of the Bakonzo tribe, the article examines how water scarcity directly affects nutrition and food culture, offering a grounded, real-life example of how infrastructure and policy gaps impact people's everyday lives.

The data featured in this article were collected by GlobalRise, a US-based nonprofit organization, between 2019 and 2020 in Kirembe Village, Cell 1, in Kasese District, western Uganda. The study focused on health, nutrition, and sanitation and involved 172 households and two focus groups with thirty-one participants.

By connecting national-level challenges with community-level realities, this paper aims to highlight the urgent need for integrated solutions that address both water and food insecurity. The story of the Bakonzo offers a powerful example of how water scarcity affects not just health and nutrition but also cultural traditions and everyday life.

Water Insecurity and the Paradox of Abundance

Understanding Water and Food Insecurity

Access to safe water, sanitation, and hygiene is recognized by the United Nations as a fundamental human right and an essential foundation for human health, dignity, and sustainable development. In 2010, the UN General Assembly passed a resolution affirming the human right to water and sanitation, underscoring the essential role that water plays in realizing all other human rights. This principle was further institutionalized in 2015 with the adoption of the 2030 Agenda for Sustainable Development, where Sustainable Development Goal 6 (SDG6) specifically aims to ensure universal and equitable access to safe and affordable drinking water for all and to reduce the number of people suffering from water scarcity.[1]

Despite this global commitment, water insecurity remains a critical and escalating issue. Global water demand has been increasing faster than population growth, exacerbating existing pressures on water systems. As of 2021, about 2 million people still lacked access to safe drinking water; approximately 720 million, around 10 percent of the global population, lived in countries experiencing high or critical levels of water stress; and roughly 3.2 billion people reside in agricultural regions characterized by high to very high water shortages. Of those, 1.2 billion live in severely water-constrained areas.[2] In response, the United Nations has called for significant investment, improved infrastructure, and institutional capacity building to address water insecurity and promote water sustainability.

UN-Water offers a comprehensive definition of water security:

> the capacity of a population to safeguard sustainable access to adequate quantities of and acceptable quality water for sustaining livelihoods, human well-being and socio-economic development, for ensuring protection against water-borne pollution and water related disasters, and for preserving ecosystems in a climate of peace and political stability.[3]

This definition highlights the centrality of water not only in environmental stability but also in political, economic, and social well-being.

Water scarcity, a key dimension of water insecurity, is shaped by both physical and systemic factors. The Food and Agriculture Organization (FAO) emphasizes that scarcity is a dynamic and context-specific phenomenon, determined by the balance between supply and demand and influenced by both environmental limitations and human consumption patterns.[4] Importantly, water scarcity is not only about the absolute availability of water: it also involves its quality and accessibility and the capacity of institutions to manage and distribute it effectively.

There are three primary dimensions of water scarcity. First, physical and quality scarcity occurs when natural water sources are insufficient due to drought, climate change, overuse or water quality issues that can make water unusable for drinking, agriculture, or sanitation. Second, water scarcity arises when governance and institutional systems fail to provide equitable access due to mismanagement, lack of regulation, or poor infrastructure. Third, infrastructural scarcity limits water access despite resource availability, because of inadequate investment in delivery systems, storage facilities, or treatment technologies. All these types of scarcity are relevant to our study case, as it reflects how systemic and logistical challenges can perpetuate water insecurity even in regions with sufficient natural water resources.[5]

Beyond basic consumption and hygiene, water plays a pivotal role in food security and the functioning of food systems. As the backbone of agricultural production, water is indispensable to every stage of the food supply chain, from farm to table. An adequate and reliable supply of water is essential to ensure all four pillars of food security defined at the 1996 World Food Summit: food availability, access, utilization, and stability.[6]

First, water ensures food availability by enabling agricultural productivity. Crops require water for growth, livestock require it for health and feed production, and fisheries depend on aquatic ecosystems sustained by adequate water flows. Without sufficient water, agricultural yields decline, increasing the risk of food shortages and economic instability, particularly in rural and agrarian economies.[7]

Second, water influences food access, both physically and economically. Reliable water sources enable households to engage in subsistence farming or small-scale food production while also supporting livelihoods and incomes, especially for rural communities. When water is scarce or inaccessible, communities may experience crop failures, income loss, or rising food prices, all of which reduce their ability to purchase or produce sufficient food. Furthermore, poor water access can force households to spend disproportionate time and resources securing water, diverting labour and income away from food acquisition and preparation.[8]

Third, water contributes to food utilization, which refers to the body's ability to absorb and metabolize nutrients. This element is essential for food preparation, hygiene, and digestion, and it is used in washing, cooking, and rehydrating food, as well as for maintaining utensils and keeping kitchen surfaces clean, practices that help prevent contamination and disease. Safe water is also necessary to maintain adequate sanitation, reducing the risk of waterborne illnesses that can compromise nutrient absorption and overall health. As highlighted by the MAL-ED study, poor environmental conditions, including unreliable water access, are closely linked to stunted growth and cognitive delays in children, primarily due to chronic intestinal infections and malnutrition.[9]

Finally, having a reliable supply of water guarantees the stability of not only crops but the whole food system. It makes it more resilient to any difficulty, whether environmental events, economic stresses, or social conflicts.

In addition to its role in food security, water is foundational to food systems, which encompass the complex processes and infrastructure involved in food production, transformation, distribution, and consumption. According to the FAO, enhancing the performance and resilience of food systems, particularly through sustainable water management, is essential to achieving SDG2: Zero Hunger.[10] In summary, the intersection of water insecurity and food systems is profound, because water is not only a human right but also a strategic resource for ensuring global food security, public health, and development. Addressing water insecurity, therefore, is not merely an environmental or technical challenge; it is a prerequisite for building sustainable and resilient societies.

Uganda's Paradox of Abundance

Despite its natural wealth and strong agricultural tradition, a significant proportion of Uganda's population remains food and water insecure. This contradiction exposes systemic weaknesses in infrastructure, governance, and resource management that prevent the country from converting natural abundance into sustainable and equitable access to essential services.

Agriculture remains the backbone of Uganda's economy, employing over 70 percent of the population and contributing significantly to livelihoods. However, the sector is burdened by persistent structural challenges; a major issue is the lack of adequate infrastructure, which limits farmers' access to markets and leads to uneven food distribution and considerable postharvest losses.[11] Changing dietary preferences and stringent export standards also require greater attention to food safety, traceability, and quality control, challenges that Uganda's current food system is not fully prepared to meet.

The 2021 UN Food Systems Summit emphasized the urgent need to transform global food systems; in response, Uganda established the National Food System Coordination Committee (NFSCC) and developed a Food Systems Transformation Action Plan to strengthen governance, improve access to nutritious food, and enhance environmental sustainability across the agricultural value chain.[12]

Water, central to agricultural productivity and human well-being, is both a resource and a challenge in Uganda. Despite vast freshwater reserves, lakes, rivers, and groundwater covering 17 percent of the country's surface area, many Ugandans lack reliable access to clean water. Only 5.83 percent of the country's renewable water resources are withdrawn, well below the SDG water stress threshold. Yet water scarcity is emerging in multiple regions due to poor infrastructure, urbanization, population growth, and weak regulatory enforcement. Water distribution is uneven, with entire communities relying on unimproved sources like ponds, streams, and unprotected wells. According to Lifewater International, 51 percent of the population lacks access to safe drinking water, and 32 percent face limited access, spending over thirty minutes collecting water daily due to distance or overcrowding.[13]

The link between water and food security is undeniable, and Uganda faces significant food security challenges, as evidenced by its 2022 Global Food Security Index score of just 47.7 out of 100, ranking ninety-third among 113 countries.[14] The country experiences persistently high levels of malnutrition: 10.4 percent of children are underweight, and 27.9 percent are stunted. Poor diet and lifestyle-related noncommunicable diseases contribute to roughly one-third of annual deaths. Food safety remains a major concern, with an estimated 1.3 million food-related illness cases annually and 14 percent of all disease linked to food contamination. Finally, food also represents an economic burden where households devote nearly 50 percent of their household income to food, with low-income groups spending as much as 57 percent compared to 41 percent among the wealthiest households.[15]

Uganda's paradox of abundance thus lies in its failure to convert natural and human capital into sustainable systems that ensure food and water security. Addressing these challenges requires integrated policy action focused on infrastructure development, water governance, agricultural modernization, and institutional reform. Only through coordinated investment and inclusive planning can Uganda unlock its full potential and ensure the well-being of its people.

To better understand how this paradox plays out on a local scale, the following section examines the case of the Bakonzo tribe in the Rwenzori region, where, despite being surrounded by abundant water resources, there is a persistent struggle in accessing clean and reliable water for their daily routines. Their experience vividly illustrates the disconnect between natural resources availability and actual access to the population, underscoring the very heart of Uganda's food and water security dilemma.

The Bakonzo Tribe: A Case Study

The Rwenzori Region and Its Dwellers

The Bakonzo (pl. Bakonzo, sing. Mukonzo), also called the Konjo people, are a Bantu-speaking people in Uganda who have lived in the hillsides of the Rwenzori Mountains for centuries. Indeed, to write about the Bakonzo, one must write of the Rwenzori Mountains, and vice versa.

It is in the local language (Lukonzo or Lukonjo) that the name of Rwenzori is born. As Eggermont, Van Damme, and Russell note, the word *Rwenzururu* means 'place of snow'. The mountain range is also locally known as *Gambaragara*, which means 'the great leaf in which clouds are boiled'.[16] Both names clearly describe the striking characteristics of Rwenzori's climate – wet and misty – and reflect the mountains' importance to the water supplies of the Bakonzo.

According to the latest National Population and Housing Census report in 2024, the Bakonzo tribe population was 1,104,462.[17] Living in mountainous terrain, the Bakonzo subsist by farming the hills. Their access to basic services such as water, electricity, schools, and health-care facilities as well as roads are severely limited.

Western Uganda's Rwenzori Mountains, which were declared a UNESCO World Heritage Site in 1994, give place to the Rwenzori region, encompassing the districts of Kasese, Kabarole, and Bundibugyo. Often called the Mountains of the Moon, the Rwenzori Mountains rise majestically along the Uganda–DR Congo border and are the third-highest mountain range in Africa.[18]

The importance of the Rwenzori Mountains can be highlighted from many points of view; however, in this article, we focus on the water resource capacity that these mountains provide not only to their own region but to the whole country. Three main water bodies surround the Rwenzori Mountains; on the north is Lake Albert, while in the south Lake Edward connects to Lake George.[19] There are many other smaller lakes in the lowlands which are very important water and food sources to the Bakonzo people.

Lifestyle, Cultural Practices, and Food Heritage of the Bakonzo

This section presents the results of a survey conducted in Kirembe Cell 1, a Bakonzo village located in the district of Kasese in the Rwenzori region. The data shown belong to a sample of 172 participating households.

For general context, Figure 1 shows the foods and preparations regularly consumed by the Bakonzo. The most popular dish among the Bakonzo is *bundu*, a dish prepared with a mix of cassava and millet flour usually accompanied by a stew – ideally, a meat stew made from goat and/or beef or a fish stew. However, since meat is usually expensive, bean stew is more likely to be not only a culturally accepted combination for their *bundu* preparation but also one of the main protein sources accessible in the Bakonzo diet.

Bundu is a preparation that deserves particular attention. Community members reported during focus group meetings that millet is becoming scarce, mainly due to recurrent droughts, reduced availability and access to seeds, lower crop yields, and limited capacity of farmers to implement mitigation strategies such as irrigation. As a result, millet cultivation is being displaced from local food systems. These changes impact the cultural heritage embedded in *bundu*, because the dish is in many cases prepared with cassava flour only, altering its traditional composition and diminishing not only its cultural but also its nutritional value.

Figure 1 also highlights that starchy foods such as cassava, green bananas (locally called matooke), and sweet potato are the base of the Bakonzo diet, while dairy, fruits, and vegetables are consumed with less frequency.

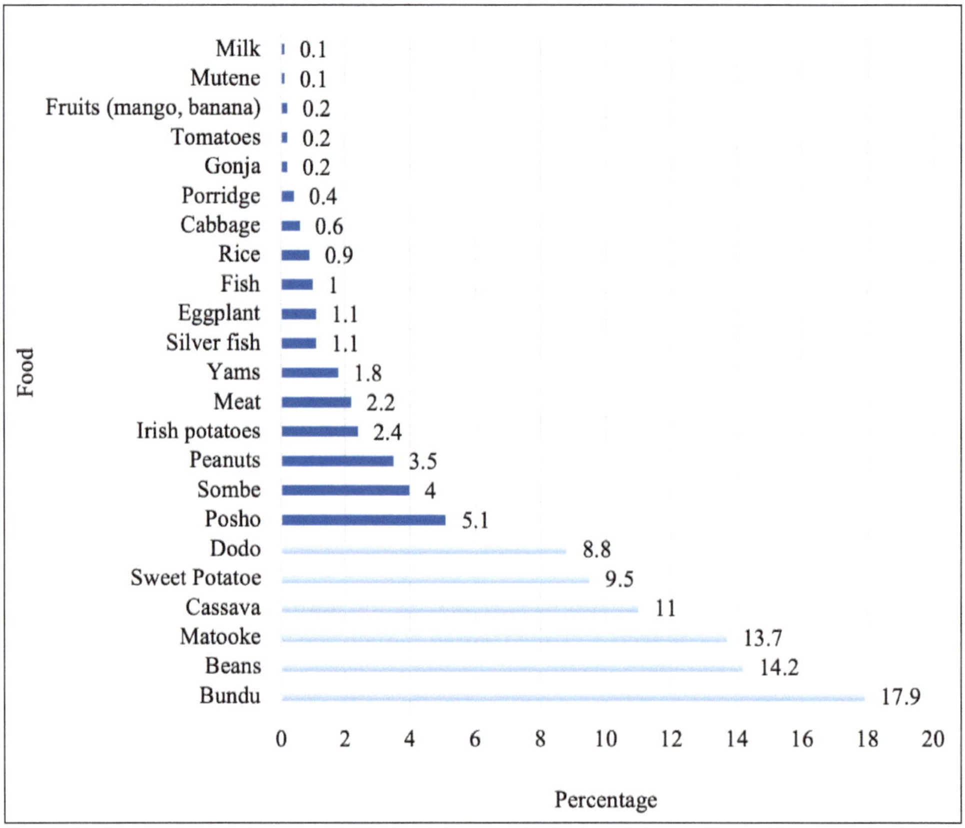

Figure 1. Foods and preparations regularly consumed in the interviewed households.
Note: 'Regularly consumed' was defined as eaten at least three days a week. All figures represent our elaboration of data from GlobalRise.

In addition to reporting regular consumption, households were asked about the foods they would like to include in their diet. The responses obtained represented not only a much higher diversity of foods and preparations but also the typology of foods (Figure 2). For example, meat, rice, and fish were the top three most desired additions. Meanwhile, foods like eggs, milk, and even spaghetti, rarely or never consumed, were mentioned as desirable.

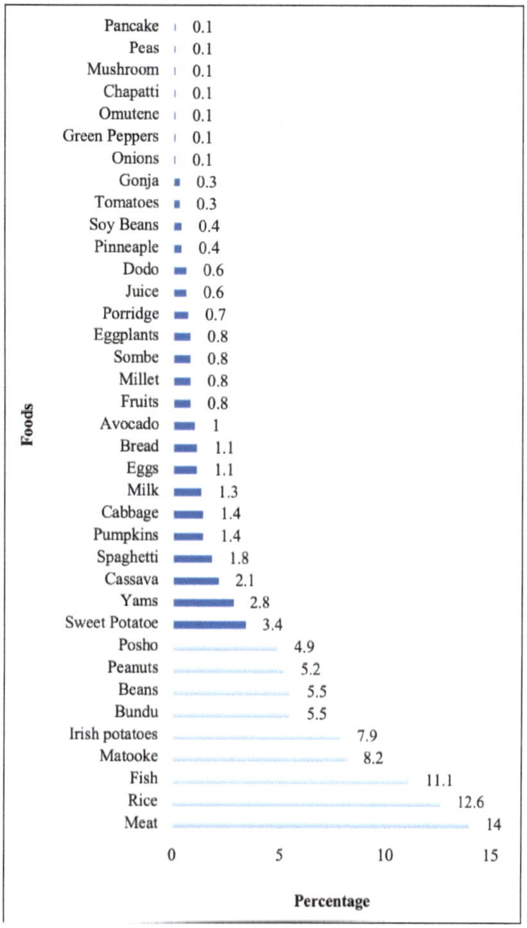

Figure 2. Foods and preparations households wish to include in their diet.

The total of foods and dishes reported to be regularly consumed in the households was twenty-three, but the total of foods and dishes wished to be added or increased in the household's diet was forty. These results show that there are differences in and limitations on the foods that household members would like to eat and their frequency versus those foods that they can actually access and their frequency. This access in most cases is limited for economic and/or physical reasons. In other words, if the money to purchase food is limited, the family gives priority to foods that are cheaper, or if the wished-for foods are not locally grown or are traded in the market farther away, then the family lets go of buying a determined food because it is not accessible at that moment and place.

Water Access, Availability, and Use

Moving into the cooking techniques the Bakonzo use for food preparation, the interviewed households indicated that the methods they utilize are frying (48 percent), boiling (49 percent), and steaming (3 percent). Also, the findings showed that 95 percent of the households rely on wood or charcoal for cooking purposes, including boiling water as treatment for safe drinking water.

As mentioned earlier, access to water is a significant challenge in rural communities of Uganda, and the Kirembe villages are no exception. In the surveyed community, households reported obtaining water for cooking, cleaning, and hygiene in two main ways: 34.7 percent of the households reported obtaining drinking water gathered from rain, while 65.3 percent collected it from a communal pipe or well (Figure 3). Communal pipes and wells are scarce and often nonfunctional or found in poor conditions.

Figure 4 shows a map of the households of Kirembe Cell 1 and the identified sources of water (taps). Twenty-one taps were identified in the community; at the time of the data collection, nineteen were partly working and two were completely nonfunctional.

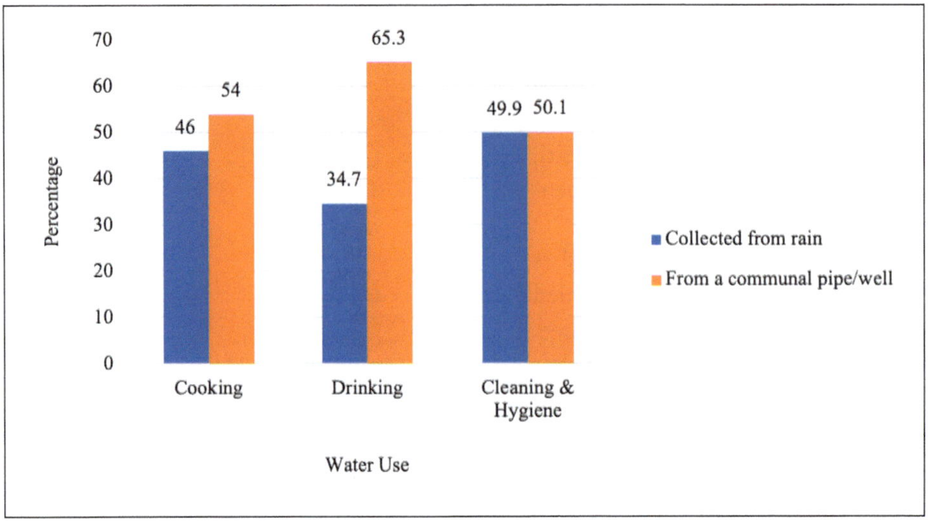

Figure 3. Sources of water for basic household needs.

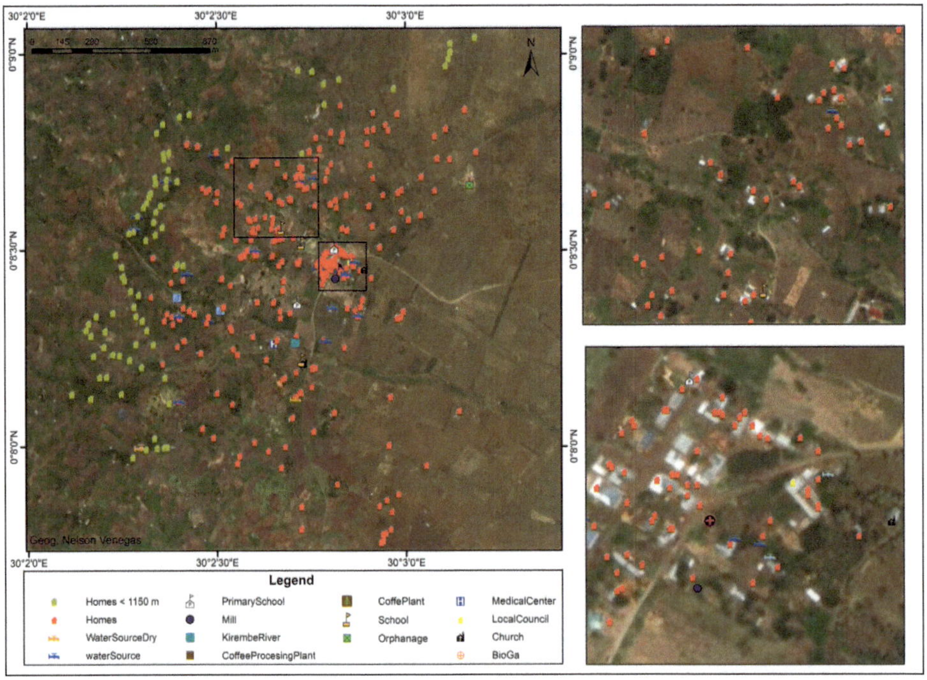

Figure 4. Map of Kirembe Cell 1 with identified taps (blue and yellow tap icons).

According to the community members, those nineteen partly working taps had water sometimes during the day or some days during the week; during dry season, the taps may go dry for days and weeks. In most cases, these identified taps have very low pressure, and filling up a twenty-litre jerrycan could take up to thirty minutes.

National data show that, in addition to limitations of water access, getting water implies walking long distances (three kilometres on average), spending an average of twenty-five minutes walking to and from the water source, waiting in lines for another twenty-five minutes, and carrying heavy containers to bring it home.[20]

The situation around water access forces many people to use water collected from rain as drinking water. Finally, most of the water, independently of the purpose, is stored in jerrycans (83–92 percent), as shown in Figure 5.

The Bakonzo and Waterborne Diseases

The surveyed households were also asked about episodes of diarrhoea experienced by members of the household. The data collected showed a strong correlation between water treatment practices and the prevalence of diarrhoea. Households which reported cases of diarrhoea in the past also reported that they were not treating water.

Figure 6 shows that in households where water was not treated, diarrhoea prevalence was 94 percent, compared to just 5.8 percent in households that did treat their water.

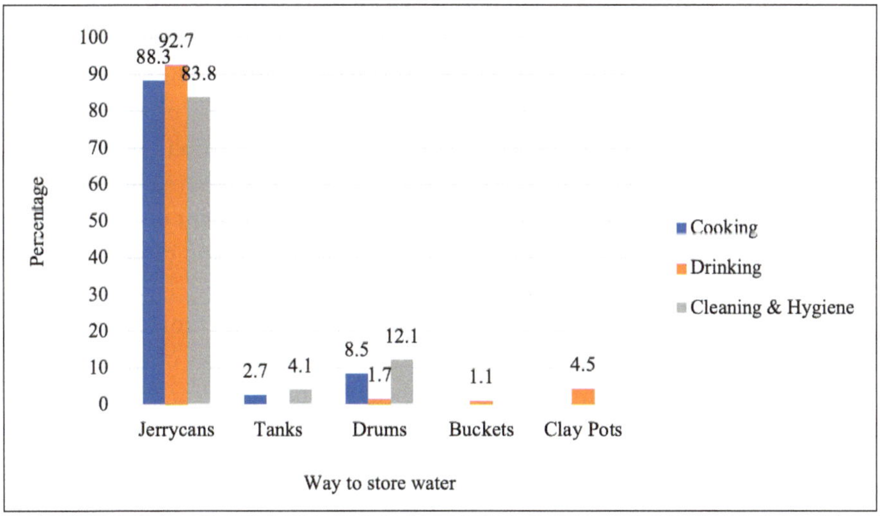

Figure 5. Means of storage for water in the households.

		Currently	Last week	A month ago	A long time ago	Other	Total	%
Treating water	Yes	0	1	0	9	0	10	5.8
	No	11	14	4	132	2	163	94.2
Total							173	100

Figure 6. Diarrhoea prevalence and water treatment in the surveyed households.

Unfortunately, these results show that the majority of households do not treat water. While diarrhoea has multiple causes, these findings in Kirembe suggest a significant link to untreated water. Finally, the surveyed households reported to treat water do so by mainly two methods: boiling (82 percent) and using chlorine tablets (18 percent), as shown in Figure 7.

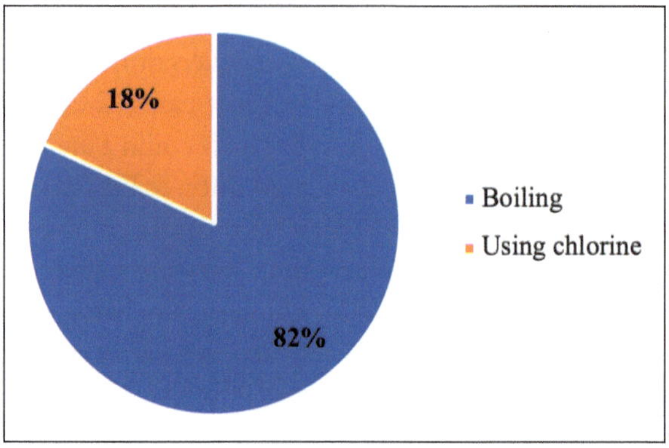

Figure 7. Type of treatment applied to drinking water.

Discussion

The results from this Bakonzo community analysis confirm the presence of a paradox. Despite the richness and abundance of water resources in the Rwenzori region, members of Kirembe Cell 1 remain water insecure. The lack of hydric infrastructure forces hundreds of households in the hilly Kirembe villages to rely on an inefficient water supply system. A community that relies on rainwater collection or partly functional water taps challenges the very notion of a system at all. It is the infrastructural scarcity, despite the environmental abundance, that exemplifies Uganda's paradox of abundance.

The implications of the water insecurity identified in Kirembe Cell 1 are both serious and multifaceted. Households spend a considerable amount of time and energy collecting water. Furthermore, water insecurity affects the safe use of water for sanitation, cooking, and drinking, all of which are fundamental to public health and food security. A striking outcome of the survey is the extremely high prevalence of diarrhoea among households that do not treat their drinking water: 94 percent of untreated-water households reported such cases. This finding underscores a clear and direct link between waterborne illness and inadequate water treatment.

Diarrhoeal disease affects nutrient absorption and contributes to chronic undernutrition, especially in children. These local results shed light on the root causes of the high stunting rates in the Bakonzo population. It can be confirmed that there is a clear link between water quality and sanitation with nutritional outcomes, reinforcing that access to safe water is not only a public health concern but a precondition for food utilization and improved nutrition.

Perhaps one of the most important aspects of this study is how water and food insecurity are impacting the Bakonzo's food cultural heritage. The results revealed that while households regularly consume only twenty-three food items, they express a desire for forty different items. However, monetary or physical limitations force a gap between aspiration and reality, suggesting an ongoing erosion of dietary diversity and culinary identity. At the same time, starchy staples dominate the Bakonzo diet, while fruits, legumes, and vegetables have a lower consumption frequency.

Notably, these findings illustrate the reaction chain from climate change–induced problems to diet switch. But it also allows a look at the adaptation capacity that communities have. In an environment where droughts are affecting farmers more and more, when the Bakonzo do not have a water infrastructure that allows them to implement irrigation systems, key crops like millet are being displaced, *bundu* is modified, and the result is not just changes in texture or flavour but also a change in the nutritional content of the staple dish for the Bakonzo. The decline of millet has altered not only the dish's cultural integrity but also its nutritional value.

Similarly, the cooking techniques identified in the community suggest challenges in food preparation and nutritional input deriving from water insecurity. For example, due to the lack of safe water for washing produce such as fruits and vegetables, cooking becomes the only way to ensure food safety. However, both boiling and frying require either water or oil, resources that may be limited or costly. Boiling vegetables, while effective for safety, involves the laborious task of fetching water, which consumes time and physical energy.

At the same time, frying increases the cost of food preparation, a serious concern in communities where economic limitations already shape dietary choices. Furthermore, both methods alter the nutritional profile of foods. Cooking can degrade certain heat-sensitive nutrients, particularly in vegetables, reducing their vitamin and mineral content. In contrast, eating raw vegetables contributes positively to dietary diversity and provides essential fibre, vitamins, and minerals in their most bioavailable forms.

During focus group discussions, participants frequently expressed frustration with the monotony of their current diets, emphasizing their desire to introduce not only diversity but quantity and traditional dishes into their everyday meals.

A final observation from the results concerns the gendered burden of water treatment in the community. Boiling water is closely associated with kitchen tasks and is largely shouldered by women. It is women who are responsible for fetching water, collecting firewood, and ensuring proper storage, adding significantly to their already demanding domestic workload. While chlorination is used by 18 percent of the participating households, this method introduces an economic consideration that some households may be unable or unwilling to take on. Additionally, focus group discussions revealed that many community members find the taste of chlorinated water unpleasant.

This gendered labour dynamic underscores how water insecurity not only affects health and nutrition but also reinforces existing social inequalities, particularly the unpaid workload borne by women in managing household water safety.

Conclusion

The community of Kirembe Cell 1 embodies the paradox of abundance in Uganda. The water insecurity experienced by households is due not to lack of resources but to a systemic failure in infrastructure, governance, and planning. This study reaffirms that water insecurity is directly related to food insecurity. The impact ranges from diarrhoeal disease to the erosion of traditional culinary practices. In addition, the consequences of water and food insecurity fall disproportionately on women, worsening gender inequalities. Local and regional governance should be directed to communities in the Rwenzori region to systemically address related issues of water and food insecurity.

About the Authors

Karen Salazar Ledezma is a field nutritionist for GlobalRise.

Paula Ibarra Flores is a researcher and graduate student at the Sorbonne.

Disclaimer

The results presented in this research were obtained with funding and logistical support from GlobalRise, which facilitated the household survey data collection. The findings, interpretations, and conclusions expressed in this paper are solely those of the authors and do not necessarily reflect the views of GlobalRise or its affiliates.

Acknowledgement

The authors thank Nelson Venegas Cordero from the University of Costa Rica for his collaboration as geographer to create the map in Figure 4.

Notes

1 United Nations, 'Sustainable Development Goal 6: Clean Water and Sanitation | United Nations in Uganda', *Clean Water and Sanitation* <https://uganda.un.org/en/sdgs/6> [accessed 4 April 2025]; Institute for Water, Environment and Health, *Water Security and the Global Water Agenda: A UN Water Analytical Brief* (UNU INWEH, 2013).
2 United Nations, UN-Water, 'Water Scarcity', *UN-Water* <https://www.unwater.org/water-facts/water-scarcity> [accessed 10 April 2025].
3 Institute for Water, Environment and Health, *Water Security*.
4 Food and Agriculture Organization (FAO) of the United Nations, 'Water Scarcity' <https://www.fao.org/land-water/water/water-scarcity/en/> [accessed 5 April 2025].
5 FAO, 'Water Scarcity'.
6 Carole D. Nounkeu and Dharod Jigna, 'Status on the Scale Development to Measure Water Insecurity Experiences at the Household Level: A Narrative Review', *American Society for Nutrition*, 10 (2019), pp. 864–75, doi:10.1093/advances/nmz008.
7 Nounkeu and Jigna.
8 Nounkeu and Jigna.
9 MAL-ED Network Investigators, 'Relationship between Growth and Illness, Enteropathogens and Dietary Intakes in the First 2 Years of Life: Findings from the MAL-ED Birth Cohort Study', *BMJ Global Health*, 2.4 (2017), doi:10.1136/bmjgh-2017-000370.
10 'What Are Food Systems?', World Food Programme <https://www.wfp.org/food-systems> [accessed 9 April 2025].
11 S. Galema and others, *An Overview of the Ugandan Food System: Outcomes, Drivers & Activities* (Foresight4Food, April 2024), doi:10.18174/657503.
12 Galema and others.
13 Galema and others.
14 Economist Impact, 'Global Food Security Index (GFSI)', 10 April 2025 <https://impact.economist.com/sustainability/project/food-security-index> [accessed 10 April 2025].
15 Galema and others, *Overview of the Ugandan Food System*.
16 Hilde Eggermont, Kay Van Damme, and James M. Russell, 'Rwenzori Mountains (Mountains of the Moon): Headwaters of the White Nile', in *The Nile*, ed. by Henri J. Dumont, Monographiae Biologicae 89 (Springer, 2009), pp. 243–61, <https://dx.doi.org/10.1007/978-1-4020-9726-3_13>.
17 Uganda Bureau of Statistics 2024: The National Population and Housing Census 2024 – Final Report – Volume 1 (Main), Kampala, Uganda <https://www.ubos.org/wp-content/uploads/2024/12/National-Population-and-Housing-Census-2024-Final-Report-Volume-1-Main.pdf> [accessed 25 January 2025].

18 Uganda Wildlife Authority (UWA), Rwenzori Mountains <https://ugandawildlife.org/> [accessed 15 February 2025].

19 Montserrat Lara Sutulov, Enrique Cruz, and Ashley Anderson, *Baseline Report: Rwenzori Region Case Study, Afromaison Project* (Mountains of the Moon University, 2013) <https://www.researchgate.net/publication/325158241_BASELINE_REPORT_RWENZORI_REGION_CASE_STUDY_AFROMAISON_PROJECT> [accessed 15 May 2025].

20 Uganda Bureau of Statistics (UBOS) and ICF, *Uganda Demographic and Health Survey 2016* (UBOS and ICF, 2018) <https://dhsprogram.com/pubs/pdf/FR333/FR333.pdf> [accessed 14 May 2025].

26

Food for Thought

The Story of Elements and the Brain

Anders Sandberg and Len Fisher

Introduction

Popular culture and the popular imagination are awash with beliefs, stories, and 'proofs' about the value of particular foods for brain function, as exemplified in P. G. Wodehouse's 1919 collection *My Man Jeeves*:

> 'It's brain', [Bertie Wooster] said; 'pure brain! What do you do to get like that, Jeeves? I believe you must eat a lot of fish, or something. Do you eat a lot of fish, Jeeves?' (2014: 194)

Fish and the brain both contain relatively high concentrations of phosphorus. But the brain only requires a certain amount of phosphorus for its phosphorus-containing molecules, and this is adequately provided by a normal diet. The idea that more will make you brighter is, well, fishy.

A similarly fishy story is the correlation between chocolate consumption *per capita* in different countries and the number of Nobel prizes awarded to scientists from that country (Messerli 2012) (Figure 1).

An attempt has been made to explain this correlation in terms of magnesium (a significant component of chocolate) as a brain food (Fehlhaber 2013). A more likely explanation is that scientists in well-off Western countries are more likely to have better access both to chocolate (a relatively expensive luxury commodity) and to the resources required to conduct significant, Nobel prize–winning research (Maurage, Heeren, and Pesenti 2013).

Belief in certain elements as 'brain foods' has even reached the marketplace. Marks & Spencer (2025) advertises a fruit and nut 'brain ball' 'proven' to support brain function because it contains iron and zinc (Figure 2).

Deficiency in these elements does indeed inhibit neural function, and in times or cultures where meat has been the major dietary source of iron, there has been a clear link between the amount and quality of meat and mental function (Gupta 2016). There is, however, an adequate amount of both elements in most modern balanced diets.

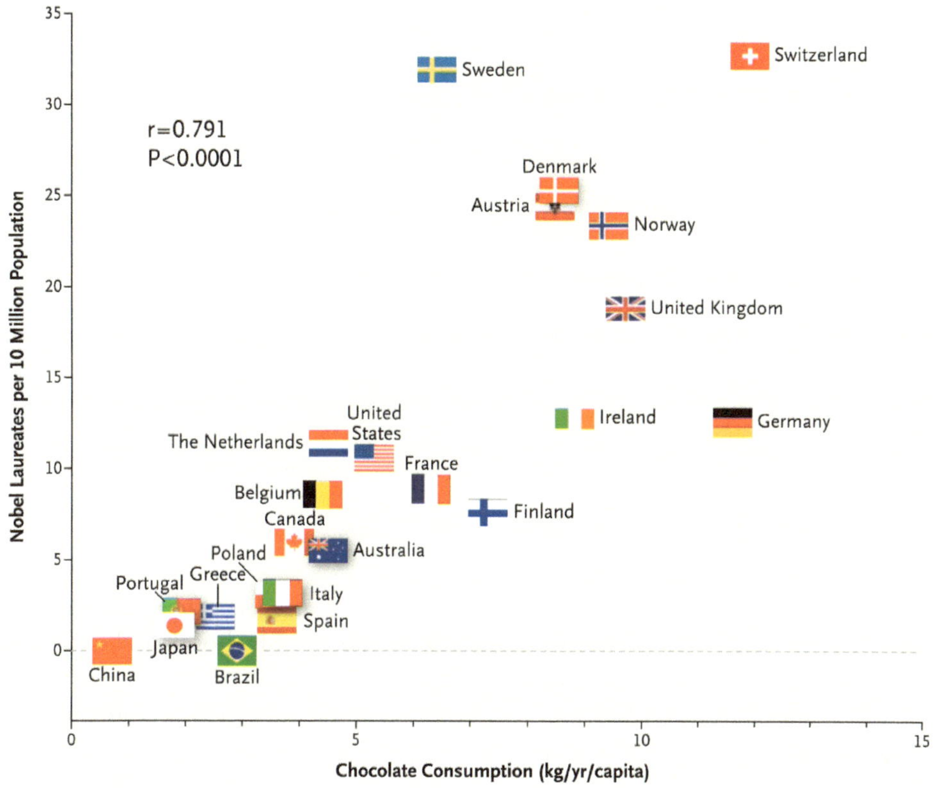

Figure 1. Number of Nobel laureates as a function of chocolate consumption. By permission from Franz Messerli.

Elements and the Brain

When it comes to brain function, we may conveniently divide the chemical elements into three categories: (1) those that link up to form the major physical structures of the brain; (2) those that otherwise contribute to brain function and which are present in significant amounts; and (3) those that are essential for brain function but which are present only in trace amounts. To these we might add a fourth category of elements that are not presently known to contribute to human brain function but which contribute to some biological function in other species. Here we describe those categories and the contribution of diet. As a general comment, we note that the role of diet can be very different between the early phases of brain development and the long-term functioning of the mature brain.

Figure 2. Marks & Spencer's 'brain ball'. Photo by the authors.

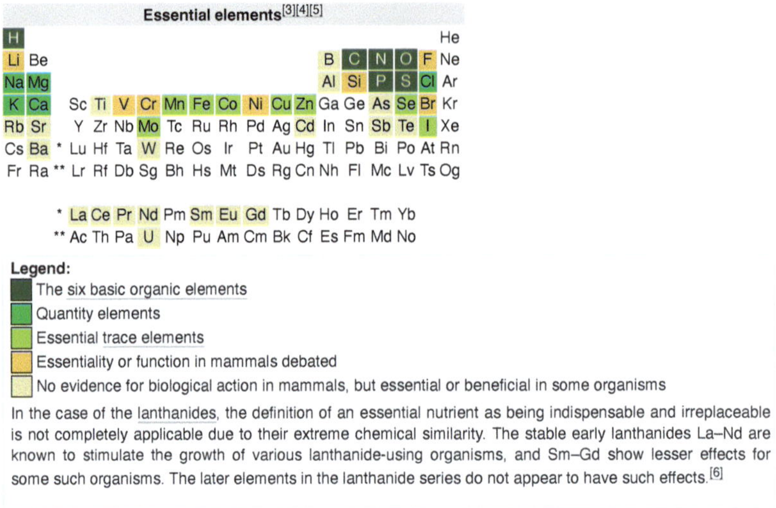

Figure 3. Elements and the brain. Creative Commons.

Brain tissue is constructed from four main elements – hydrogen, oxygen, carbon, and nitrogen. Some 80 percent of the mass is simply composed of hydrogen and oxygen in the form of water – a figure that needs to be closely maintained by dietary intake to balance loss through sweating, urination, etc. A short-term decrease of as little as 2 percent can lead to mental problems that include memory loss and failures in simple numerical computation (Fayaz n.d.), while a similar increase can lead to brain swelling, serious neurological failure, and death (Kozler, Maresova, and Pokorny 2022).

The remainder of the brain's mass consists mostly of organic molecules, such as proteins, lipids, and DNA, where the four main elements are linked together in various ways, sometimes together with minor components such as phosphorus (around 1.3 percent by weight) and sulphur (around 0.15 percent by weight).

Significant Functional Contributors

Six elements contribute significantly to brain functionality. These are (in increasing order of atomic weight) sodium, phosphorus, sulphur, chlorine, potassium, and calcium. All are exposed to some degree of turnover in the body (although not necessarily in the brain), and that loss must be provided from the diet.

Phosphorus has attracted the most public attention because of the fabled link with intelligence (Dronsfield and Ellis 2009). The French chemist Jean-Pierre Couerbe claimed in 1834 to have found scientific evidence for the link when he analyzed the brain tissue from deceased mental patients and found that the phosphorus content of normal brain tissue was '2.0–2.5 per cent, but in idiots [*sic*] it was 1.0–1.5 per cent whereas in madmen, 3.0–4.0 per cent or even 4.5 per cent'. He concluded:

> It would follow that the absence of phosphorus from the brain would reduce man to the sad state of the brute; that the great excess of this substance would irritate

> the nervous system, exalt the individual and drive him into that furious excitement which we call madness. (qtd. in Dronsfield and Ellis 2009)

Courbe's measurements, and his conclusions, were never taken very seriously by other scientists and soon became the butt of jokes. The French philosopher Paul Janet remarked that 'the brain of fishes, who do not pass for great thinkers, contains a large amount of phosphorus', while the humourist Mark Twain, replying to an aspiring author who wanted to know how he might improve his literary skills, suggested that

> if the specimen composition you send is about your fair usual average, I should judge that perhaps a couple of whales would be all you would want for the present. (qtd. in Bateman 1897: 54)

Phosphorus may not contribute directly to intelligence, but it does play a major role in many brain and other bodily functions. It is part of the structural backbone of DNA. It is an integral component of the phospholipids in cell membranes and of the energy-storage compounds ATP and ADP. It is also part of the molecular structure in many cell regulatory systems.

Folk beliefs remain strong, however, and the idea that phosphorus is somehow useful for cognition underlies the development of a number of brain 'tonics' containing phosphoric acid or other phosphorus compounds (Koch 1909), of which the most famous is surely Coca-Cola.

Of the other functional contributors, sulphur has the longest history. It was one of the three principal substances in the ancient Greek physician Paracelsus's theory of medicine – the combustible element, responsible for 'hot' states of matter. In the form of hydrogen sulphide, responsible for the foul smell of the many claimed 'healing' mineral waters around the world, it continues to be promoted for health benefits, especially for skin complaints.

In terms of factual chemistry, sulphur-containing compounds and amino acids (such as methionine and cysteine) have long been recognized as important biochemical components, and the sulphur-containing molecule glutathione is known to be a major brain antioxidant that plays an important neuroprotective role.

The remaining functional contributors – calcium (0.5–1.0 percent by weight), potassium (0.3–0.4 percent), sodium (0.1–0.2 percent), and chlorine (0.1–0.2 percent) – exist as electrolytes, involved in the electrical signalling of the brain. The mechanism is complex, involving the flow of different ionic species through channels in the cell membrane to produce a travelling 'action potential'.

These electrolytes are fairly evenly distributed over the brain, albeit with some minor differences (Ramos and others 2016). Maintaining the balance between them is important. Too much sodium can produce the confusion and jitteriness that are symptoms of salt intoxication. Overdosing on potassium is more nonspecific but can produce muscle weakness. More extreme cases of excess or deficiency can even be deadly. Maintaining the right intake of sodium and potassium is important for correct cognitive function (Na and others 2022). Both low and high calcium levels can give rise to tiredness and depression, but usually a healthy body manages this balance.

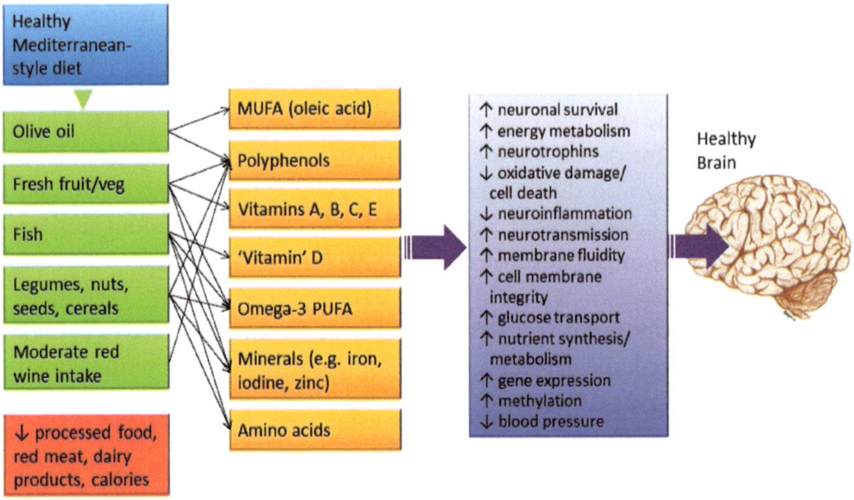

Figure 4. 'Overview of Links Between Mediterranean-Style Diet and Healthy Brain Function via Plant Compounds and Nutrients' (Parletta, Milte, and Meyer 2013: Medicine and Health Sciences Commons).

Micronutrients

There are many other elements whose presence in low concentrations is necessary for bodily function in general and brain function in particular, although higher concentrations can frequently be toxic. Many of these are metal ions that are good at accepting or delivering electrons to other molecules, facilitating (often as part of enzymes) various chemical reactions. They include magnesium (around 30 g in our bodies, or 0.05 percent of average adult body weight), iron (5 g), zinc (2 g), copper (0.1 g), manganese (0.016 g), molybdenum (0.0065 g), and cobalt (0.003 g) (Zoroddu and others 2019).

Magnesium

Magnesium in the brain is involved in neurotransmitter regulation, acts as a signal substance, and is involved in the transmission of nerve impulses. Low magnesium is linked to anxiety and cognitive issues and has been specifically implicated as an important factor in some forms of depression (Derom and others 2013), with many studies emphasizing the importance of magnesium in the diet (Li and others 2017).

Iron

Iron is well known for its role in transporting oxygen via haemoglobin in the blood and myoglobin in muscle. In the brain it participates in important functions that include making the neurotransmitters used to signal between nerve cells and synthesizing myelin, the 'insulator' between nerve cells. Its importance here can be gauged from the fact that, despite comprising about 2 percent of the total body weight, the brain contains 20 percent of all the iron. The distribution of iron in the brain is uneven, with more iron in the basal ganglia (deep parts of the brain involved in selecting and performing movements) than elsewhere. This is why the basal ganglia are more vulnerable to iron-related damage, making high levels associated with Parkinson's disease (Dexter and others 1989). Iron deficiency is also associated with impaired cognition and mood disorders.

Iron in spinach is associated in the popular imagination as a giver of strength, as in the popular Popeye cartoons. In fact, many other leafy greens contain more iron, as do many nuts and legumes.

The special claim for spinach is supposed to have come from a misplaced decimal point in the original published study (Hamblin 1981). In fact, this claim itself is a myth. The decimal point error was in a study about beans (Sutton 2010; Dagg 2016), which a cascade of publishing mistakes turned into spinach (Rekdal 2014)!

Zinc

Zinc plays an essential role in many enzymes and the immune system and can also act as a neuromodulator and a nerve growth signal (Choi and others 2020). The pituitary gland (found just under the brain) has the highest concentration of zinc of any organ, and the metal influences hormone regulation. Zinc deficiency in the diet stunts growth by reducing the production of growth hormone. As with many trace elements, zinc deficiency has also been linked to memory problems, depression, and anxiety.

Zinc is required for taste bud and olfactory epithelium health and is a key component of the saliva protein gustin that helps taste. Zinc deficiency can thus alter the perception of taste and smell (Henkin 1984; Mozaffar and others 2023). These in turn can affect appetite and nutrition.

An excess of zinc can, however, produce damaging oxidative stress. Zinc is also involved in the processes leading to neuron death after brain injury and/or oxygen depletion, where release of zinc from damaged neurons overexcites their neighbours, causing them to crash too.

Copper

The brain contains about 8–10 percent of the body's copper, reflecting the high demands of neural tissue. Copper is necessary for brain energy production, for producing protective antioxidants, and for making myelin. It also has a role in neurotransmitter synthesis.

Copper levels in the brain are maintained in a tight optimal range through copper chaperone proteins and efficient blood-brain barrier regulation. Genetic disorders like Menkes disease (too little copper) and Wilson's disease (copper buildup) have severe neurological consequences (Dooley 2024).

Copper and the other metals discussed here normally exist in the body as water-soluble ions. A somewhat surprising discovery has been the existence of metallic nanoparticles of iron and copper in the brain (Everett and others 2021). Their role – for example, in possible magnetic sensitivity or as signs of incipient dementia – is still unclear.

Manganese

Manganese is required by the brain as a component of enzymes that support neurotransmitter metabolism, energy production, and antioxidant defences (Takeda 2003). Rare genetic mutations affecting manganese transport can lead to severe developmental disability, which can be partially rescued by manganese supplementation. Careful monitoring is required, however, since manganese is neurotoxic in larger doses and has a narrow optimal range in the brain.

As with the chemically similar iron, excess manganese tends to accumulate in the basal ganglia – the part of the brain that is implicated in Parkinson's disease (Kwakye and others 2015), causing a Parkinson's-like disease called manganism that was observed in Chilean manganese miners as far back as 1830 (Couper 1837).

Cobalt

Cobalt is an integral component of vitamin B12, essential for brain health and cognitive function. B12 deficiency is linked to memory issues and mental fatigue.

Molybdenum

The essential nature of molybdenum for brain function was only discovered in the 1980s, when a man on long-term intravenous feeding developed sudden confusion, rapid heart rate, night blindness, and ultimately coma. Doctors were puzzled until they realized his intravenous diet lacked molybdenum. When they added it, he recovered. Molybdenum in trace amounts is now known to be essential because of its role in detoxifying brain-damaging substances (Abumrad and others 1981).

On the other hand, too much molybdenum in the diet can cause copper deficiency, since it interferes with copper absorption.

Iodine

The connection between dietary iodine and brain development is well established (Choudhry and Nasrullah 2018) and was one of the earliest recognized nutrient-brain relationships. In the eighteenth and nineteenth centuries, certain inland regions of England saw high rates of goitre and cognitive impairments. The condition was called 'Derbyshire neck', after one particularly affected region, and was eventually traced to the fact that these areas, far from the sea, had soil where the iodine was in a form not easily absorbed by plants.

'The effects of severe iodine deficiency during critical periods of brain development are well documented', according to Rosie Gordon and others (2009), whose double-blind study of 184 New Zealand children aged 10 to 12 with mild iodine deficiency showed a significant improvement in IQ test scores when iodine supplementation was added to the diet.

Iodine is an integral component of the thyroid hormone that is necessary for brain development. The mechanism through which iodine affects brain function is, however, still to be established.

Lithium

Lithium's mood-stabilizing effects were discovered by accident in 1947, when the Australian psychiatrist John Cade was testing the then-fashionable idea that some mental illnesses were due to blood toxins. Cade thought that uric acid might be the culprit and experimented by injecting urine from manic patients into guinea pigs. Uric acid is hard to dissolve in water (a cause of kidney stones!), so Cade used the more easily dissolved lithium salt. This made the guinea pigs more restful, and Cade (2000 [1949]) recognized that lithium was the important factor.

Lithium, which is chemically similar to sodium and potassium, also appears to support neurogenesis and neuroprotection at therapeutic doses (Greenblatt 2021).

The Tricky Cases

Less than a thousandth of the brain (about a gram) consists of elements other than those already mentioned. Their function – or whether they have any significant function at all – is often unclear. Here we offer a brief list.

Selenium

Selenium is both an essential element and toxic. As has been known since the time of Paracelsus, it is the dose that makes the poison.

Selenium is a component of some of the antioxidant enzymes that protect the brain from oxidative stress. A deficiency in the element has been linked to mood disorders and cognitive decline and some evidence of its involvement in multiple sclerosis, Parkinson's disease, and some forms of Alzheimer's disease. There is also evidence that food with higher selenium concentrations can be good for reducing cognitive impairment in older people (Cardoso and others 2016), perhaps by reversing deficiency.

High concentrations can, however, be toxic. Marco Polo described a region of China where the plants poisoned horses and cattle, causing hoof loss, now known to be due to their unusually high selenium content (Qin and others 2013). Brazil nuts are such a rich source of selenium that consuming more than eight per day can lead to selenium toxicity. Lemire and others (2012) note that the Indigenous Brazilians appear unharmed by a high-selenium diet; this might be due to adaptation or other environmental factors (Rayman 2020). Another, unfortunately more common, cause of selenosis is excessive supplementation (Morris and Crane 2013).

Strontium

Strontium is chemically similar to calcium and can easily replace it, e.g., in the bones. This is a cause for concern when the strontium comes in radioactive form as Sr^{90}, as occurs in nuclear fallout. In its nonradioactive form, however, strontium at suitable levels may even have some healthy effects – for example, by supporting synaptic transmission (Wigstrom and Swann 1980) – although, at high concentrations, strontium can cause neurotoxic and neuromuscular perturbations (ATSDR 2004).

Boron, Fluorine, and Ruthenium

Little is known about the role of these elements in human brain function. Boron may contribute to less drowsiness, higher mental alertness, and improved psychomotor skills (Penland 1998), but fluorine and ruthenium have no known function.

Ultratrace Elements

It has been suggested that there is a need for 'ultratrace elements' in exceedingly small amounts and often toxic at higher doses – arsenic, bromine, cadmium, nickel, silicon, tin, vanadium, aluminium, germanium, and rubidium have all been proposed (Nielsen 1998). Their biological roles in humans (if any) generally remain unclear. Lack of evidence, however, is of no concern for the supplement industry, which happily touts the entire periodic table as essentials. Microdeficiencies might happen and be bad for us, but that risk is more easily hedged by eating a wide variety of foods.

Toxic Elements

Some elements are just bad for us. Heavy metals such as lead, mercury, cadmium, arsenic, and thallium are bad because they typically bind very strongly to enzymes and other proteins, disrupting their structures and sabotaging their function. Often these metals can substitute for other elements, such as lead taking the place of zinc in some enzymes. Some heavy metals, like cadmium, can also generate reactive oxygen species that can damage DNA and other cell components.

Radionuclides such as radium are bad for us in a different way. Radioactive elements can enter the brain through similar pathways to lead, which is a frequent decay end-product. When radioactive atoms decay, the radiation and/or energetic particles that are released can break chemical bonds and lead to a cascade of damage.

Dark Histories

Lead has a dark history. The Roman architect Vitruvius noted that water from earthenware pipes was more wholesome than water from lead pipes (Prioreschi 1998). The lead intake in the upper classes was far beyond what is considered unhealthy today (Reddy and Braun 2010), including lead acetate that was used as a sweetener and preservative in wine (sometimes accidentally produced from the lead in pewter vessels). The use of lead acetate as a sweetener continued long beyond Roman times, and the idea that it caused the downfall of Rome is likely a myth.

Moving forward in history, an analysis of Beethoven's hair found high levels of lead, along with arsenic and mercury (Rifai and others 2024). The source may have been wine sweetened by lead acetate, lead-containing medications, and fish from the polluted Danube (Kuta 2024). Not everybody is convinced that this contributed to his deafness and gastrointestinal problems (Rzymski, Zarębska-Michaluk, and Flisiak 2024), but clearly environmental lead was (and is) dangerous. Two major sources in modern times have been lead-based paints and lead tetraethyl as an 'antiknock' agent in petrol. It is particularly bad for developing brains, causing almost 10 percent of intellectual disability of otherwise unknown causes worldwide. It can result in behavioural problems in children, delinquency in adolescents, and criminal behaviours in adults (Liu and others 2014).

Mercury also has a dark history. Damage from the metallic element is uncommon, although a man who injected himself intravenously with metallic mercury was found to have the element in most of his bodily organs when he died five months later (Lu, Liu, and Chen 2017).

Damage through exposure to the vapour is more common. An Elizabethan treatment for venereal disease was for the naked patient to sit over a vessel of heated mercury, which surely killed many of the patients. Mercury vapour has caused neural damage to mercury miners and to artisanal gold miners when liquid mercury was used for gold extraction.

Neural damage from water-soluble mercury compounds is, however, the most common form. Damage from mercuric nitrate, used by nineteenth-century hatters in the 'felting' process, was so common that it led to the popular expression 'mad as a hatter'. The highly toxic methylmercury is known to accumulate in aquatic food webs, building up to significant concentrations in top predators that form part of the human diet.

Toxic metals in food are often associated with industrial contamination but can also be a natural risk. Mining, and the erosion of rocks containing toxic metals, combined with hot, humid weather across a wide belt stretching from the Balkans to southern China, has released these metals into the soil (Hou and others 2025). This affects 14–17 percent of global cropland, which produces food for billions of people.

It would seem obvious that radium and other radionuclides should be avoided in food. There is, however, a long history of claimed health benefits, especially of radium in radioactive spa water, patent medicines, and radium-infused products. The English city of Bath even pronounced itself 'Britain's Radium Spa' and sold radium water at the pump room for two pence a glass (Santos 2020). The medicine Radithor (distilled water with radium) famously killed the business magnate Eben Byers in 1932, fairly decisively ending the worst radium fad (Macklis 1993). But to this day there remain spas offering mildly radioactive water, such as Jachymov in the Czech Republic (Vickery 2018).

It has been claimed that low doses of radiation are more harmful to precancerous and 'weak' cells than they are to normal cells (the 'hormetic effect'). Evidence for this effect is, however, inconsistent and scientifically controversial. As with many such borderline claims, a healthy dose of scepticism is likely to be the healthiest option in every sense.

Limits of Elemental Reductionism

Our conclusions from this whistle-stop tour are very simple: Yes, brains require certain elements and nutrients. Yes, some foods are good for us. But it is better to try to eat a healthy, varied diet than to shop around for brain food on Tuesdays and heart food on Wednesdays. Above all, don't eat too many Brazil nuts.

About the Authors

Anders Sandberg is a lapsed computational neuroscientist researching emerging technology, global disasters, and the very long-term future at the Institute of Futures Studies, Stockholm.

Len Fisher is a scientist, author, and broadcaster whose books range from *How to Dunk a Doughnut* to *Crashes, Crises and Calamities: How We Can Use Science to Read the Early-Warning Signs*. He won a spoof Ig Nobel prize for using physics to work out the best way to dunk a biscuit.

References

Abumrad, N. N., and others. 1981. 'Amino Acid Intolerance during Prolonged Total Parenteral Nutrition Reversed by Molybdate Therapy', *American Journal of Clinical Nutrition*, 34, pp. 2551–59.

Agency for Toxic Substances and Disease Registry (ATSDR). 2004. 'Toxicological Profile for Strontium', US Department of Health and Human Services, Public Health Service.

Bateman, Frederic. 1897. *The Idiot: His Place in Creation and His Claims on Society* (Jarold & Sons).

Cade, J. F. 2000 [1949]. 'Lithium Salts in the Treatment of Psychotic Excitement', *Bulletin of the World Health Organization*, 78, pp. 518–20.

Cardoso, Barbara Ritao, and others. 2016. 'Effects of Brazil Nut Consumption on Selenium Status and Cognitive Performance in Older Adults with Mild Cognitive Impairment: A Randomized Controlled Pilot Trial', *European Journal of Nutrition*, 55, pp. 107–16.

Choi, Seunghyuk, and others. 2020. 'Zinc in the Brain: Friend or Foe?', *International Journal of Molecular Sciences*, 21, p. 8941, doi:10.3390/ijms21238941.

Choudhry, Hani, and Md Nasrullah. 2018. 'Iodine Consumption and Cognitive Performance: Confirmation of Adequate Consumption', *Food Science & Nutrition*, 6, pp. 1341–51.

Couper, J. 1837. 'Sur les effets du peroxide de manganèse', *Journal de chimie médicale, de pharmacie et de toxicologie*, 3, pp. 223–25.

Dagg, Joachim. 2016. 'The Real Decimal Error That Transmogrified into the Spinach-Iron-Decimal-Error Myth', *Weltmurksbude* <https://historiesofecology.blogspot.com/2015/10/the-real-decimal-point-error-that.html> [accessed 16 April 2025].

Derom, Marie-Laure, and others. 2013. 'Magnesium and Depression: A Systematic Review', *Nutritional Neuroscience*, 16, pp. 191–206.

Dexter, D. T., and others. 1989. 'Increased Nigral Iron Content and Alterations in Other Metal Ions Occurring in Brain in Parkinson's Disease', *Journal of Neurochemistry*, 52, pp. 1830–36.

Dooley, James S. 2024. 'The History of Wilson Disease', *Clinical Liver Disease*, 23, e0238.

Dronsfield, Alan, and Peter Ellis. 2009. 'Phosphorus – Food for Thought', *Education in Chemistry*, 46 <https://edu.rsc.org/feature/phosphorus-food-for-thought/2020208.article> [accessed 16 March 2025].

Everett, James, and others. 2021. 'Biogenic Metallic Elements in the Human Brain?', *Science Advances*, 74.24 (June), doi:10.1126/sciadv.abf6707.

Fayaz, Imran. N.d. 'How Dehydration Affects the Brain', Brain and Spine Institute of North Houston <https://fayazneurosurgery.com/how-dehydration-affects-your-brain/> [accessed 16 March 2025].

Fehlhaber, Kate. 2013. 'Chocolate Consumption and Nobel Laureates', *Knowing Neurons* <https://knowingneurons.com/blog/2013/06/17/correlation-vs-causation-chocolate-consumption-and-nobel-laureates/> [accessed 16 March 2025].

Gordon, Rosie C., and others. 2009. 'Iodine Supplementation Improves Cognition in Mildly Iodine-Deficient Children', *American Journal of Clinical Nutrition*, 90, pp. 1264–71.

Greenblatt, James. 2021. 'Lithium as a Nutrient', Psychiatry Redefined <https://www.psychiatryredefined.org/lithium-as-a-nutrient/> [accessed 16 March 2025].

Gupta, Sujata. 2016. 'Clever Eating', *Nature*, 531, pp. S12–S13.

Hamblin, T. J. 1981. 'Fake', *British Medical Journal* (*Clinical Research Edn*), 283, p. 1671.

Henkin, Robert I. 1984. 'Zinc in Taste Function: A Critical Review', *Biological Trace Element Research*, 6, pp. 263–80.

Hou, Deyi, and others. 2025. 'Global Soil Pollution by Toxic Metals Threatens Agriculture and Human Health', *Science*, 388.6744 (April), pp. 316–21, doi:10.1126/science.adr5214.

Koch, W. 1909. 'Phosphorus Compounds as Brain Foods', *Journal of the American Medical Association*, 52, pp. 1381–83.

Kozler, Petr, Dana Maresova, and Jaroslav Pokorny. 2022. 'Determination of Brain Water Content by Dry/Wet Weight Measurement for the Detection of Experimental Brain Edema', *Physiol. Res.*, 71, pp. S277–S283.

Kuta, Sarah. 2024. 'Locks of Beethoven's Hair Are Unraveling the Mysteries of His Deafness and Illnesses', *Smithsonian Magazine* <https://www.smithsonianmag.com/smart-news/locks-of-beethovens-hair-are-unraveling-the-mysteries-of-his-deafness-and-illnesses-180984332/> [accessed 16 March 2025].

Kwakye, Gunnar F., and others. 2015. 'Manganese-Induced Parkinsonism and Parkinson's Disease: Shared and Distinguishable Features', *International Journal of Environmental Research and Public Health*, 12, pp. 7519–40.

Lemire, Melanie, and others. 2012. 'No Evidence of Selenosis from a Selenium-Rich Diet in the Brazilian Amazon', *Environment International*, 40, pp. 128–36.

Li, Bingrong, and others. 2017. 'Dietary Magnesium and Calcium Intake and Risk of Depression in the General Population: A Meta-Analysis', *Australian & New Zealand Journal of Psychiatry*, 51.3, pp. 219–29.

Liu, Jianghong, and others. 2014. 'Blood Lead Concentrations and Children's Behavioral and Emotional Problems: A Cohort Study', *JAMA Pediatrics*, 168, pp. 737–45.

Lu, Qiuying, Zilong Liu, and Xiaorui Chen. 2017. 'Mercury Poisoning through Intravenous Administration: Two Case Reports with Literature Review', *Medicine*, 96.46 (November), p. e8643, doi:10.1097/md.0000000000008643.

Macklis, Roger M. 1993. 'The Great Radium Scandal', *Scientific American*, 269, pp. 94–99.

Marks & Spencer. 2025. 'Brain Food Berry Ball' <https://www.marksandspencer.com/food/brain-food-brain-ball/p/fdp60713787> [accessed 16 March 2025].

Maurage, Pierre, Alexadre Heeren, and Mauro Pesenti. 2013. 'Does Chocolate Consumption Really Boost Nobel Award Chances? The Peril of Over-Interpreting Correlations in Health Studies', *Journal of Nutrition*, 143, pp. 931–33.

Messerli, Franz H. 2012. 'Chocolate Consumption, Cognitive Function, and Nobel Laureates', *New England Journal of Medicine*, 367, pp. 1562–64.

Morris, John Steven, and Stacy B. Crane. 2013. 'Selenium Toxicity from a Misformulated Dietary Supplement, Adverse Health Effects, and the Temporal Response in the Nail Biologic Monitor', *Nutrients*, 5, pp. 1024–57.

Mozaffar, Boshra, and others. 2023. 'The Effectiveness of Zinc Supplementation in Taste Disorder Treatment: A Systematic Review and Metaanalysis of Randomized Controlled Trials', *Journal of Nutrition and Metabolism*, 6711071.

Na, Xiana, and others. 2022. 'Association of Dietary Sodium, Potassium, Sodium/Potassium, and Salt with Objective and Subjective Cognitive Function among the Elderly in China: A Prospective Cohort Study', *Global Transitions*, 4, pp. 28–39.

Nielsen, Forrest H. 1998. 'Ultratrace Elements in Nutrition: Current Knowledge and Speculation. The Journal of Trace Elements in Experimental Medicine', *Official Publication of the International Society for Trace Element Research in Humans*, 11, pp. 251–74.

Parletta, Natalie, Catherine M. Milte, and Barbara J. Meyer. 2013. 'Nutritional Modulation of Cognitive Function and Mental Health', Faculty of Science, Medicine and Health – Papers: Part A, 778 <https://ro.uow.edu.au/smhpapers/778> [accessed 16 March 2025].

Penland, James G. 1998. 'The Importance of Boron Nutrition for Brain and Psychological Function', *Biological Trace Element Research*, 66, pp. 299–317.

Prioreschi, Plinio. 1998. 'Of Roman Medicine', *History of Medicine*, 3, p. 279.

Qin, Hai-bo, and others. 2013. 'The Bioavailability of Selenium and Risk Assessment for Human Selenium Poisoning in High-Se Areas', *Environment International*, 52, 66–74.

Ramos, Patricia, and others. 2016. 'Alkali Metals Levels in the Human Brain Tissue: Anatomical Region Differences and Age-Related Changes', *Journal of Trace Elements in Medicine and Biology*, 38, pp. 174–82.

Rayman, Margaret P. 2020. 'Selenium Intake, Status, and Health: A Complex Relationship', *Hormones*, 19, pp. 9–14.

Reddy, Aravind, and Charles L. Braun. 2010. 'Lead and the Romans', *Journal of Chemical Education*, 87, pp. 1052–55.

Rekdal, Ole Bjørn. 2014. 'Academic Urban Legends', *Soc. Stud. Sci.*, 44.4 (August), pp. 638–54, doi:10.1177/0306312714535679.

Rifai, Nader, and others. 2024. 'High Lead Levels in 2 Independent and Authenticated Locks of Beethoven's Hair', *Clinical Chemistry*, 70, pp. 878–79.

Rzymski, Piotr, Dorota Zarębska-Michaluk, and Robert Flisiak. 2024. 'Beethoven's Deafness More Likely Linked to Viral Infection Than Lead Poisoning', *Journal of Infection*, 88, p. 210.

Santos, Lucy Jane. 2020. *Half Lives: The Unlikely History of Radium* (Icon).

Sutton, Mike. 2010. 'Spinach, Iron and Popeye: Ironic Lessons from Biochemistry and History on the Importance of Healthy Eating, Healthy Scepticism and Adequate Citation', *Internet Journal of Criminology* <https://botanologia.gr/wp-content/uploads/2020/02/spinach.pdf> [accessed 16 April 2025].

Takeda, Atsushi. 2003. 'Manganese Action in Brain Function.', *Brain Research Reviews*, 41, pp. 79–87.

Vickery, Matthew. 2018. 'A Spa Where Patients Bathe in Radioactive Water', BBC Future <https://www.bbc.com/future/article/20180306-a-spa-where-patients-bathe-in-radiactive-water> [accessed 16 April 2025].

Wigstrom, H., and J. W. Swann. 1980. 'Strontium Supports Synaptic Transmission and Lasting Potentiation in the Hippocampus', *Brain Research*, 194, pp. 181–91.

Wodehouse, P. G. 2014 [1919]. 'The Aunt and the Sluggard', in *My Man Jeeves* (Bibliotech).

Zoroddu, Maria Antonietta, and others. 2019. 'The Essential Metals for Humans: A Brief Overview', *Journal of Inorganic Biochemistry*, 195, pp. 120–29.

27
The Death of Fire and the Rebirth of Magical Cooking

Richard Warren Shepro

The four classical elements – earth, air, fire, and water – are all essential to cookery, but fire, which requires air, produces water, and leaves earth as its residue, creates the most majestic display and has the only claim to having initiated cookery. There is beauty and grandeur and, I would say, magic in its display. Fire created cookery and may even have been the evolutionary spark that made us human. Despite its symbolic allure, cooking with live fire has come under siege for environmental and medical sins. The aesthetic pull of the flame has met regulatory and critical foes. Can it retain or recover its majesty, perhaps in a different way with the assistance of precision cookery?

Cooking over live fire is described in the earliest literary works to great effect. Richard Wrangham even argued that fire – the basis of cookery – provided the evolutionary spark for the transition from *Australopithecus* to the genus *Homo* that made us human.[1]

Bonfires are elemental and often connected to special events and ceremonies. Roasted and grilled meats are celebrated, especially when cooked over a wood fire or its embers, evoking a Promethean power (Figure 1). Candles create romance; birthday candles provide a special kind of magic (Figure 2). In Italy, the words *forno al legno* (wood-fire oven) are what bring in the crowd, not the word *pizza*. Fireworks illuminate holidays. The hearth is the spiritual centre of a home (Figure 3).[2]

Figure 1. Chickens on a clockwork spit roasting in front of embers and flames. Photo by the author.

The Trouble with Fire

Yet however primordial, the use of fire is coming under siege. Widely criticized as leading to health hazards and environmental disaster, its demise, or at least decline, seems inevitable. When the French chef Fernand Point, the dominant force in mid-twentieth-century cuisine in France, said that 'cooking demands the sacred fire [*le feu sacré*];

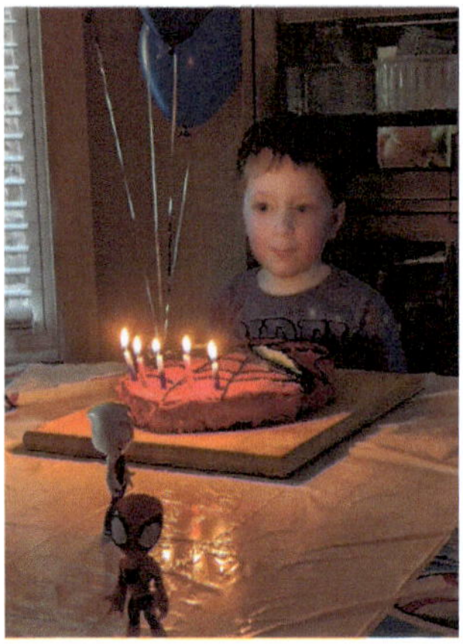

Figure 2. Simple birthday flames provide a special kind of magic. Photo by the author.

Figure 3. Roasted guinea fowl on a hearth. Photo by the author.

nothing can distract from the work', he had a double meaning in mind: the literal flames of the kitchen and the figurative fire that drives a great chef to success. But he also banned the attention-getting flames of *crèpes Suzette*: 'no pyrotechnics in the dining room'.[3] This was not for safety but to avoid being showy. Today, though, gas torches to brown meat or caramelize sugar are being banned even from some kitchens, not just dining rooms, because of worries about creating both pollutants and carcinogens.[4]

More dramatically, the World Health Organization has estimated that more than three million people die annually from lung problems induced by indoor air pollution from using wood and other solid-fuel fires to cook in poorly ventilated homes in sub-Saharan Africa, South Asia, and Southeast Asia. According to the WHO, about a third of the world's population's cooking is done with poorly ventilated indoor wood fires. This leads to chronic lung problems, lung cancer, strokes, and ischaemic heart disease, with women and children bearing the greatest health burden.[5] Live open fire use is so common and alternate energy resources so scarce that the problem cannot simply be solved by banning dangerous practices. The WHO's clean cooking initiatives have ambitious goals, but progress is slow. Specialized health advocacy groups have sprung up in many countries in an attempt to improve conditions incrementally, including an international coalition called Doctors and Scientists Against Wood Smoke Pollution (DSAWSP).

Cooking with fire in the developed world is criticized as well, where legislation is being increasingly enacted, most often at the local level. Fire and smoke have been regulated in some ways for centuries, but in 2000, the City Council of Berkeley, California, prohibited new wood fireplaces, pizza ovens, and other cooking devices fuelled with wood (a rule that was later revised in a nuanced way, as I will discuss). In Scotland, the New Build Heat Standard (NBHS) came into effect in April 2024, prohibiting the installation of 'direct emissions heating systems', including wood stoves, in new homes. A partial reversal of this directive, repealing the ban on wood stoves and peat burning and focusing instead on oil- and gas-based devices, has sparked new controversy and led to heated discussion on many sides. In the Netherlands, the city of Utrecht banned wood-burning pizza ovens, wood stoves, and fireplaces in January 2025, to be effective by 2030.[6]

These bans have been instituted because of environmental concerns about smoke and the medical risks of smoke inhalation. Smoke from burning wood releases toxic chemicals. These risks are well known and not controversial.[7] The difficult issues come in assessing the level of risk, principally in quantifying the amount of smoke in relation to the level of ventilation and ambient air quality.

The emergence of shockingly destructive wildfires in recent years has added to the concern. Some people have lost their taste for barbecue and other smoked meat because it reminds them of wildfires. Fire regulations are being broadened. In 2020, California enacted a law mandating the creation of an ember-resistant zone, called 'Zone 0', creating a five-foot buffer around structures in designated hazard areas where wildfires are deemed likely.[8] Unfortunately, the regulations that the California Board of Forestry and Fire Protection (CAL FIRE) needs to implement this law have not been finalized. A goal is to have them in place by the end of 2025, and localities may enact their own regulations earlier. The regulations will likely change residential landscaping dramatically. The Zone 0 buffers will need to be devoid of many kinds of plants and trees, wooden fences, and other combustible materials. Wood would not be able to be stored, and wood-fire cooking would be prohibited in those zones.

In legal terms, these restrictions focus attention on balancing individual rights to cook as one pleases against the right of the public to be free of danger from the polluting effects of neighbours and securing mutual protection against wildfires. These are similar to the issues raised in previous decades about secondhand smoke and vaccinations, pitting individual liberties against the collective good. In Berkeley, whose city council is trying to act faster than the state of California, some residents worry about aesthetic issues and the cost of compliance.

Natural gas stoves and butane burners have also been cited by health authorities as a serious health hazard. There is growing awareness that the harms of gas cooking stoves in terms of the release of combustion products in unvented homes have been underappreciated, much like the harms of smoking were in the 1950s. Although gas stoves do not have the romantic appeal of a wood fire (massive wok burners aside!), and poorly ventilated wood burning distributes more serious toxins, it is the spectre of gas stove regulation that in the United States has sparked the sharpest political controversy. Donald Trump's claim that '[Joe] Biden will take away your gas stove' became a signature political assertion in the 2024 US presidential election, even though the few ordinances enacted – local only – had to do with future construction, not confiscation. By contrast, prohibiting cooking with wood fire or wood-burning ovens has not become a national political issue. On 20 January 2025, the day of his inauguration, Trump signed an executive order 'to safe guard the American people's freedom to choose from a variety of goods and appliances, including but not limited to lightbulbs, dishwashers, washing machines, gas stoves, water heaters, toilets, and shower heads'.[9] (A US President does not actually have the authority to override local restrictions on cooking devices, though in some cases Congress may.)

The Rise of Precision Cookery

Who loses if cooking with wood fires is banned or regulated? Is live fire truly a superior method of cooking or, when replaced with modern cooking methods, is it unnecessary? A crucial related question is, Are the gastronomic benefits of a wood fire at least partially psychological, like many gustatory phenomena, unable to be identified in a blind taste test but very real to an appreciative diner who knows or can see the cooking method?

Manufacturers of gas and electric grills have argued for decades that results obtained with their products are identical to what is produced with wood or charcoal. Whether that is correct depends on what one means by 'identical'. One test would be whether an average person, given blind samples of similar products cooked with or without wood fire, could identify which is which. Another would be whether an expert or connoisseur could detect the difference.

However, blind comparisons and physics cannot end the inquiry. Taste is not objective. The romance and magic of fire and its nostalgic or even primordial appeal may lead to real pleasure whether or not differences in taste or smell can be detected by scientific analysis.

Differences can indeed be detected: wood fires produce detectable by-products including both wood smoke and a wide range of volatile compounds that are different from by-products produced by other forms of cooking. On the other hand, compounds created by Maillard reactions (often called caramelization) that are not unique to wood fires may often be what people remember after a meal and think of as the essence of wood-fire cooking, even though many of these flavours are not unique to the use of wood fire.[10] I cooked the guinea fowl shown in Figure 3 on a day when there was a fire in our fireplace. I liked the way it would look sitting on the hearth, so I placed it there for a photograph, but people who see this picture generally assume I cooked it over the embers.

Wood-fire cooking is an art. A true Neapolitan pizza (always cooked in a specialized wood-burning oven) is a handmade product demonstrating the work of a skilled artist-craftsman. At a time when all roasting was done using wood or wood embers, the nineteenth-century French culinary philosopher Jean Anthelme Brillat-Savarin wrote, 'One can learn to be a cook, but must be born knowing how to roast'. Like many memorable aphorisms, this is somewhat illogical – no one is born a great artist – but it emphasizes that Brillat-Savarin thought roasting over a wood fire, with all its vicissitudes, should be considered serious art, compared especially to routine cooking, where one might merely follow instructions.[11] Unpredictable factors can stymie a cook of limited experience cooking with fire: variations in logs or types of wood, varying airflow, different oven designs, and the possibility of flare-ups, among many other factors. Effective wood-fire cooking requires significant experience in balancing different types of thermal energy – fine results require a master chef of significant experience.

Both the difficulty of achieving consistent results with wood cooking and a growing awareness of the environmental issues have led to the use of special cooking techniques and high-tech precision cooking equipment to allow creation of wood-fire-esque dishes that can be similar to, as good as, or better than pure wood-fire ones and achieve traditional goals more consistently. Sometimes the risks of wood-fire cooking can be reduced while preserving its unique flavour by integrating modern high-precision techniques into portions of the cooking process. Thinking of Brillat-Savarin's comment on needing to be born a roaster, contemporary food writer Russ Parsons wrote, 'But that was before the invention of the instant-thermometer'.[12] To that we could add precise induction burners; remote thermometers that measure multiple points of temperature within a single dish; combi ovens that cook with any combination of radiant heat, steam, humidity, and convection; sous-vide cooking; and other developments.

In more than a few cases, these modern methods can provide an illusion of purely wood-fire cooking, appreciated by or even delighting customers. This could be viewed as dishonest, but cookery, like all art, encompasses many illusions. And the proof of the pudding is in the eating.

Here are some examples of these hybrid or nonfire approaches.

Pizza

In Italy, traditional pizza restaurants in Naples persuaded the Italian national government to codify legal rules that must be followed to use certain names (*pizza Napoletana, pizza Marinara, pizza Margherita*). Among the rules is that 'the woodfired oven is a key element' that must meet technical characteristics, including terra cotta construction and the ability to meet a floor temperature minimum of 485 degrees Celsius (905 degrees Fahrenheit). The European Union adopted the same regulations in 2010.[13] A private certifying body, the Associazione Verace Pizza Napoletana, awards its 'Vera Pizza Napoletana' status to compliant restaurants both in and outside the EU and licenses the restaurants to use the association's logo.[14]

One serious US pizza restaurant at the bottom of a high-rise building achieved both this certification and some critical acclaim. After a fire at the top of the oven's multistory flue and tired of dealing with environmental regulations, the owner engaged specialists from Naples this year to convert its massive oven, made in Naples, to gas fire. It continues nevertheless to post the 'Vera Pizza Napoletana' label, which requires wood fire. (Almost) no one seems to notice. The owner, an expert in the difficult art of balancing multiple sources of radiant and ambient heat in a wood fire, found that reducing the many variables involved in cooking with wood logs left the pizza 'just as good as ever' and more consistent. A purist might argue, however, that at least limited inconsistency, while maintaining quality, is a sign of an artist's handiwork and that some of the magic is now missing. It is not the wood smoke that makes the difference, though. Physicist-entrepreneur Nathan Myhrvold has shown that smoke from a Neapolitan wood-burning pizza oven cannot influence taste. It exits the dome too fast, and even aromas of particular kinds of wood can't affect the taste because any unique volatile compounds have been destroyed by the high heat.[15]

Less covertly, a serious chef in San Francisco, Thomas McNaughton, formerly devoted to the wood-burning oven, became a public convert at his Flour + Water Pizzeria to using an ultramodern Swedish-designed electric deck oven designed to re-create the results of the Neapolitan wood oven but made of metal with a high-tech ceramic floor that behaves much like terra cotta. Sceptical of this modern oven at first, McNaughton now touts its success as maintaining gastronomic excellence while promoting ecology, eliminating dangerous wood or gas emissions, and simplifying operations.[16] McNaughton claims the results match his previous wood-fire oven. I can say I've been similarly impressed with a small spot in Chicago called Munno that quietly uses the same Swedish oven. An Instagram user posted a picture of the impressive results at Munno, saying he had 'experienced' an 'electric oven magic show. [. . .] So impressive. Had someone sent me these pics before I arrived I would've asked them to take me in back to see the wood fired oven that isn't there'.[17] Also missing, though, because it also isn't there, is the opportunity to observe the skill of a master *pizzaiolo* at work, including the ballet as he introduces the pizza to the oven with the jerk of the traditional long-handled tool called a *palo* and the quick moves to rotate it the mandated 180 degrees with a different long-handled tool called a *palino*.

I really love and do prefer the pizza Napoletana I can make in the well-ventilated, terra cotta wood-burning oven we have in our kitchen, and I love following every detail in the Naples legislation, which I use as my recipe. I'm glad people can see magic in an electric oven, but for me there is a deeper magic I always feel, and my guests do as well, born of wood, smoke, and tradition (Figure 4).[18]

Figure 4. Firing of a pizza in a wood-burning oven, following the Neapolitan rules. Photo by the author.

'Traditional' Roasts

Michel Guérard (1933–2024), perhaps the greatest and most innovative chef in France of the last hundred years, began serving a remarkable roast lobster in 1981, gently cooked and lightly smoked in a grand ancient chimney over dying embers. The magical excitement of the dish came from the extreme precision and attention to the embers so that, as the chef said, the lobster would emerge supple and silky with a lovely, understated smokiness. But in the twenty-first century, the chimney became only the final, yet immensely important, step. The lobster meat is now cooked separately in an even gentler and more precise way, with electronically regulated heat. Then the shell alone is placed in the smoking chimney for just seconds before the lobster meat is restored to the shell and the dish is plated, sauced, and served. The risk of even slight overcooking from the unpredictable embers is eliminated. The revision created a more succulent dish, still with the deep, impressive flavour from the chimney, yet more complex, subtle, and memorable. The cooking went beyond what would have been possible with fire alone.[19]

Paella and Arròs

The prominent Spanish chef Quique Dacosta, well known for many things, is also a paella specialist. A wood fire is as *de rigueur* for paella in Valencia as it is for pizza in Naples. In Spain and in his London restaurant, Dacosta and his team meticulously prepare traditional paella over a wood fire that can be seen by the guests. But he has other, similar dishes where there is no live fire. Modern temperature sensors monitor the rice cooking in traditional paella pans as it simmers and can adjust the heat, while robotic arms stir according to an algorithm that considers measurements of the viscosity of the cooking liquid and determines when the rice can be declared a finished dish with a crispy *socarrat* (the prized thin crispy layer of rice at the bottom of the pan, which can develop unpredictably over a wood fire). Guests seated at tables away from the cooking may marvel at the quality and authenticity; guests sitting at a counter where the cooking is done marvel also at the technology – and at the fully traditional cooking with fire also being done in the same kitchen. The chef follows tradition precisely as far as nomenclature is concerned: only the wood-fire rice can be called *paella*. All the other dishes, however much the results may resemble a particularly successful version of the wood-fire dish, are merely called *arròs*.

Binchotan Charcoal

Not all wood fires are the same, and even charcoals vary greatly. There is a special type of Japanese charcoal, called *binchotan* or *shiro-zumi*, that produces heat at unusually high temperatures yet produces so little smoke that it can safely be used indoors. This presents a real opportunity for chefs who want to reduce harmful emissions without giving up the benefits of high-temperature cooking with charcoal. Also called white charcoal, *binchotan* was developed in Japan hundreds of years ago and has long been a fixture in specialized grilling restaurants in Japan. It is produced by charring wood at a relatively low temperature, then putting it in a kiln at 1000° Celsius (1832° Fahrenheit) and then cooling it (Figure 5).

Recently, this kind of charcoal has become an item of interest outside of Japanese cuisine because, when skilfully used on an appropriate grill, it can allow precise application of wood-fire-like radiant heat to small amounts of food. An acclaimed, highly experimental

Figure 5. Binchotan charcoal. Creative Commons/Wikimedia.

restaurant, vegetable garden, and mushroom centre called Locavore NXT in Indonesia has an almost obsessive focus on ecology and uses this charcoal just as the finishing step in cooking a duck breast in its spacious dining room located in a wide agricultural field in the centre of Bali. The duck breast has been preliminarily cooked in a high-tech oven and then finished using the *binchotan* in view of the guests on a tiny Japanese *shichirin* or *konro* grill. In Paris, this same technique is used for precise, smoke-free searing at the restaurant Hakuba, a collaboration between two prominent chefs, one French and one Japanese.

(American) Southern Barbecue

The traditions of Southern barbecue are myriad, deeply held, and often deeply divergent from one small geographic area to another, but all traditionally depend on long, slow smoking of meat at relatively low temperatures over charcoal.[20] Traditionally, expert pitmasters produce the charcoal themselves by controlled burning of logs. Pitmasters can control temperature, smoke, and seasoning with precision because – like *pizzaioli* or *rôtisseurs* – they have learned from long experience.[21] Pitmasters work seemingly by intuition, gauging doneness by feel, texture, or how the meat jiggles when pressed. One might say a pitmaster must be born to the task.

Modernist equipment and concern about smoke are both impinging on the role of the revered pitmaster, a gruelling profession whose demands were always a reason many of the best Southern barbecue places were open only a few days a week. A combination of scientific knowledge and high-tech ovens and smokers has changed the barbecue world, expanding it to larger parts of the United States and beyond and making it possible for large restaurants to produce at least decent barbecue on a consistent basis, seven days a week. Nathan Myhrvold and his large team of chefs and scientists – the same group who studied smoke and pizza – have done detailed explorations of the physics, benefits, and drawbacks of many approaches to creating and applying radiant energy to foods found in both modern and traditional cooking methods and devices.[22]

They and a number of like-minded scientist-cooks and cook-scientists, including Aaron Franklin and a man who calls himself Meathead, have applied this same detailed approach to barbecue, explaining Southern barbecue methods in a scientific manner, correcting or corroborating long-held beliefs, and introducing techniques such as long sous-vide cooking in a water bath, cold-smoking meat to reduce emissions, and using liquid nitrogen and electric deep-frying to create – consistently – the kind of crispy outside layer ('bark') that those using wood fire alone seek to achieve but manage only on their best days.[23] Some of the romance may be gone, but it may be replaced by other forms of romanticism: Franklin is a barbecue hero in Austin, Texas, and Myhrvold has been part of winning teams in barbecue contests.[24]

Hybrid Approaches: The Future?

With nuanced legislation, the romance of wood fire can be preserved without artifice, although precision techniques may in time become so successful that wood fire begins to die out. Fire adds to excitement and appreciation, but it's not always the most effective way to cook. Restaurant Troisgros in France, which has held the top Michelin star rating for longer than any other in the world, shocked the restaurant world in the late 1970s by converting from traditional French wood-burning graded ranges to electric cooktops and switched again in 2006 to induction cooktops, a more powerful and yet dramatically more supple and precise technology that can also maintain precisely controlled low temperatures.

Meanwhile, in some areas, banning of wood burning has been also replaced by more carefully measured approaches. Berkeley, for example, has rescinded its ban on wood burning as far as cooking devices are concerned while continuing to prohibit non-EPA-certified wood-burning fireplaces and heating stoves. The relaxation of rules was generally considered to be a sound ecological compromise based on the limited number of wood-burning cooking devices, the importance of gastronomy to Berkeley, and the fact that Berkeley is also governed by a regulatory body covering a larger area, the Bay Area Air Quality Management District, that will still require temporary suspensions of wood and many other types of fire when airborne smoke particulates trigger air quality alerts.

Despite the primordial attraction many people (including myself) have to the flame, one person's majestic display of cooking fire may be another person's poison. A favourite argument used by anti-wood-fire groups in England is to deride wood fires as dangerous pollution caused by thoughtless people who are willing to endanger others because they want to be 'cozy'. It is my hope that improved ventilation, thoughtful air quality regulation, judicious use of modern precision cooking techniques, and appreciation of cooking with fire as a special event may allow this magical method of cooking to survive and even flourish in new ways.

About the Author

Richard Warren Shepro is an international lawyer and a food scholar who teaches at the University of Chicago. He is the author of ten Oxford Symposium papers, the chair of the Symposium's Trustees, the *président* of the Alliance Française de Chicago, and a former editor of the *Harvard Law Review*.

Notes

1 Richard Wrangham, *Catching Fire: How Cooking Made Us Human* (Basic Books, 2009). Wrangham's argument was novel at the time and well supported but somewhat speculative. His colleagues Katherine Zink and Daniel Lieberman have since argued that Wrangham's timeline was off, claiming that the relevant change was not when our distant forebears began to eat cooked meat but an earlier time when they prepared sliced, or sliced and pounded, raw meat, which they then dried for later consumption ('Impact of Meat and Lower Paleolithic Food Processing Techniques on Chewing in Humans', *Nature*, 531 (2016), pp. 500–03 <https://doi.org/10.1038/nature16990>).

2 Many exceptional books on cooking with live fire have been published over the years. Among the most useful are Hélène and Philippe Boisseau Béharn, *La Cuisine de Braise: L'art de cuuisiner dans la cheminée et au barbecue* (Hachette, 1996); William Rubel, *The Magic of Fire: Hearth Cooking* (10 Speed Press, 2002); and Jacqueline Queneau, *La Cuisine au Feu de Bois* (Aubanel, 2004).

3 'La cuisine exige le feu sacré. Il ne faut penser qu'à son travail' (*Ma Gastronomie* (Flammarion, 1969), p. 46). There does not seem to be a clear record of his often-quoted saying about pyrotechnics, 'Pas de pyrotechnie dans la salle', but it did fit his style.

4 'L'album du Chef Jean-François Piège', *Thuries: Le Magazine de la Gastronomie*, December 2024, p. 15. When asked how he had chosen the location for one of his restaurants, Jean-François Piège cited the pre-existing ventilation system, noting the difficulty of properly ventilating kitchens in old structures in Paris.

5 World Health Organization, *Household Air Pollution*, 16 October 2024 <https://www.who.int/news-room/fact-sheets/detail/household-air-pollution-and-health> [accessed 12 March 2026].

6 'City to Ban Wood Heating', *Doctors + Scientists Against Wood Smoke Pollution* <https://www.dsawsp.org/blog/city-to-ban-wood-heating> [accessed 12 March 2026].

7 See, for example, Harold McGee, *On Food and Cooking: The Science and Lore of the Kitchen*, rev. edn (Scribner, 2004), pp. 448–50.

8 Assembly Bill 3074.

9 'Unleashing American Energy', Executive Order 14154, US Office of the Federal Register, 20 January 2025 <https://www.whitehouse.gov/presidential-actions/2025/01/unleashing-american-energy/> [accessed 12 March 2026].

10 There are separate possible health issues relating to the ingestion of charred meats, but these issues, where research is ongoing, are common to all methods of roasting and broiling foods and are not unique to wood fires. Maillard reactions can produce compounds such as pyrazines and advanced glycation end products (AGEs) that some researchers feel contribute to chronic diseases and ageing. There is also some evidence that these health effects may be ameliorated by carefully controlling time and temperature or may be repaired by the body itself. Effects of the accumulation of AGEs, for example, may be able to be monitored through tests like the Haemoglobin A1c test. Foods produced without browning produce far fewer AGEs. See, for example, Aleksandra Twarda-Clapa and others, 'Advanced Glycation End-Products (AGEs): Formation, Chemistry, Classification, Receptors, and Diseases Related to AGEs', *Cells*, 12 April 2022 <https://www.mdpi.com/2073-4409/11/8/1312> [accessed 12 March 2026].

11 Jean Anthelme Brillat-Savarin, 'On devient cuisinier, mais on naît rôtisseur', *La physiologie du goût* (1825), in *Aphorismes du Professeur Pour Servir de Prolégomènes a Son Ouvrage et de Base Éternelle A La Science*, 15 <https://www.gutenberg.org/ebooks/22741> [accessed 12 March 2026]. I am not the first to note this illogic. In 1952 in *Le Monde*, the critic Robert Courtine (writing as La Reynière), one of the best chroniclers of cooking as art, commented on this arresting aphorism, making a plea for both cooks and roasters to be considered artists who learn deep skills over a long period of time ('Un cuisinier', *Le Monde*, 13 November 1952 <https://www.lemonde.fr/archives/article/1952/11/13/uncuisinier_2001783_1819218.html> [accessed 12 March 2026]).

12 Russ Parsons, 'Roasting Beef and Lamb Just Right Is an Art Form', *Los Angeles Times*, 15 February 1995, available at *The Spokesman Review* <https://www.spokesman.com/stories/1995/feb/15/roasting-beef-and-lamb-just-right-is-an-artform/#:~:text=%E2%80%9CWe%20can%20learn%20to%20be,Physiology%20of%20Taste%E2%80%9D%20in%201825> [accessed 12 March 2026].

13 European Commission Regulation No. 97/2010, 4 February 2010.

14 See Richard Warren Shepro, 'True Bread, *Pizza Napoletana*, and Wedding Cakes: The Changing Ways Legal Power Has Shaped Bakers' Lives', in *Food & Power: Proceedings of the Oxford Symposium on Food and Cookery 2019*, ed. by Mark McWilliams (Prospect Books, 2020), pp. 295–301.

15 Nathan Myhrvold and Francisco Migoya, *Modernist Pizza* (The Cooking Lab, 2021), vol. 1, pp. 375–89.

16 Mario Cortex, 'This Hot New Oven Is Fueling a Pizza Boom in the Bay Area', *San Francisco Chronicle*, 22 April 2024 <https://www.sfchronicle.com/food/restaurants/article/electric-pizza-ovens-19410392.php> [accessed 12 March 2026].

17 Forzapizza, 'Munno', Instagram, 24 June 2019 <https://www.instagram.com/forzapizza/p/BzGI5kcFp2e/> [accessed 12 March 2026].

18 At roasting temperatures, one can taste the effects of the wood, unlike at the high temperatures in a Neapolitan pizza oven.

19 These methods have been preserved and refined by M. Guérard's successor, Hugo Souchet, and this magical dish is still available at the restaurant.

20 I refer to American Southern barbecue because the word barbecue means so many different things in different parts of the world. In the American Southeast, it is a particular smoked meat dish, called barbecue, and not a verb, a cookout, or a cooking device.

21 While the stereotypical pitmaster is a grizzled, older man, younger folks like Teresa 'Chuck' Charnichart are receiving high praise these days (Tejal Rao, 'The Best New Reason to Visit the Barbecue Capital of Texas', *New York Times*, 24 February 2026 <https://www.nytimes.com/2026/02/24/dining/restaurant-review-barbs-b-q-texas.html> [accessed 12 March 2026]).

22 See particularly Nathan Myhrvold, Chris Young, and Maxime Bilet, *Modernist Cuisine: The Art and Science of Cooking* (The Cooking Lab, 2011), vol. 2.

23 Myhrvold, Young, and Bilet, pp. 132–53, and see also vol. 3; Aaron Franklin and Jordan Mackay, *Franklin Barbecue: A Meat-Smoking Manifesto* (Ten Speed Press 2015); and Meathead Goldwyn and Greg Blonder, *Meathead: The Science of Great Barbecue and Grilling* (Houghton Mifflin Harcourt, 2016).

24 Charlie Burton, 'How to Barbecue Perfectly according to a Scientist (Turned Barbecue World Champion)', *GQ*, 24 April 2020 <https://www.gq-magazine.co.uk/food-drink/article/how-tobarbecue-champion-nathan-myhrvold-science> [accessed 12 March 2026]; 'Nathan Myhrvold', *Steven Raichlen's Barbecue Bible*, 17 August 2012 <https://barbecuebible.com/2012/08/17/nathan-myhrvold/> [accessed 12 March 2026].

28
Ashes without Borders
Culinary Histories and Practices

Ozoz Sokoh and Danielle Chike-Ogwo

Introduction
Salt is essential to life and to our taste buds. Historically, it played a key role in shaping civilization, trade, and religion. Through time, it has also featured in wars, conflicts, protests, and shortages, from the American Revolution to the 1930s Salt March in India, paving the way for the end of British rule.[1]

The majority of salts in use today originate from water bodies – sea and lake salts which are transformed to table and other salts – but potassium-rich salts from plant matter, commonly extracted from ashes, were and remain significant across different cultures. These plant-based salts reflect the ingenuity of sustainability and have been mostly unaffected by the economic and political power struggles that have shaped the adoption of water-based and extracted salts.

We're curious about how ashes became a useful culinary ingredient, why culinary ashes are salts (and the definition(s) of salts), the various functions of culinary ashes, the role of ashes beyond cooking and eating, and the future of culinary ashes. Did the reverence of fire lead to ashes being treated as treasure? What are some ways ashes are used, and where do they come from?

We travel around the world through culinary ashes as salts, preservatives, additives, and leavening. Along the way, we examine beliefs, rites, and rituals around culinary ashes, from Nigeria and New Mexico to Italy and India, places where ashes serve not only as functional additives but as cultural signifiers.

Methodology
This research draws on a blend of personal experience, ethnographic exploration, and textual analysis. Our familiarity with culinary ashes in Nigerian cuisine began through personal interactions on social media, informal networks, and visits to Oyingbo Market in Lagos, where culinary ashes are sold alongside traditional condiments.

We conducted a broad literature review spanning food blogs, academic papers, historical accounts, and cultural studies from various regions. Sources were chosen to reflect a range of geographies, voices, and levels of formality – from journals to Indigenous-led platforms and oral histories.

As part of our approach, we also engaged in corrective reading. For instance, one published paper conflated *akanwu* (limestone) with ash-derived potash; our field knowledge clarified that *akanwu* is geological, not plant based. Another paper shared that ash usage was found on almost every continent except Oceania, yet our research demonstrates many examples in Australia using ash – as leavening and in the making of damper bread.[2] We relied on both dictionary definitions and lived culinary knowledge to interpret and contextualize such claims.

Our framework is narrative and comparative – guided by lived experience, critical reading, and a desire to bring to the fore and situate understudied culinary traditions in a global context.

On Culinary Ashes and Salts

Where Do Culinary Ashes Come From?

Culinary ashes, also known as vegetable salts, plant-based salts, and traditional alkaline salts, are the detritus of burning plant materials – husks, pods, cobs, vegetable skins and peels, stalks, tree bark, and wood, including select softwoods like juniper and hardwoods like birch and oak – until light grey-white ashes form.[3] Once the burning process is complete, the ashes are left to cool and, afterward, are typically sieved to separate hard, not-completely burned bits from the fine ashes.

Ashes intended for culinary use are usually made from specific kinds of plant matter, deliberately chosen for properties such as salinity and flavour – because the salt properties of the ashes are inherent in their sources. Pia Sorensen describes the chemical process that underpins the creation of culinary ashes as incomplete combustion, where a lack of oxygen results in the production of carbon monoxide, carbon, flavour molecules (such as smoky flavours, depending on the original plant matter), and mineral salts like calcium carbonate.[4]

The resulting ashes are useful in culinary applications because, depending on the source material, they are rich in various types of salts – potassium hydroxides, potassium carbonates, calcium carbonates, calcium bicarbonates, etc. – which help with flavouring food, tenderizing meats and legumes, emulsifying oils, and so on.

How Do Culinary Ashes Become Salts?

A salt is an ionic network of metals and nonmetals, often the product of a reaction between an acid and a base. While sea salt ($NaCl$) is formed by the reaction between hydrochloric acid (HCl) and sodium hydroxide ($NaOH$), our research shows that in many areas around the world, ash salts result from combining ashes with water. In some other communities, once plant matter is dried and burned, the resulting ashes are stored in an airtight container and used as needed in cooking, preserving foodstuffs, and so on.

Combining ashes with water produces two primary outputs: cleaned ashes and a liquid filtrate formed by leaching. In the first case, the ashes are soaked in water, stirred, and allowed to settle, then the water is discarded. This is repeated once or twice, and then the cleaned ashes are ready to be dried and stored. In South India, the cleaned ashes of *pirandai* (veld grape or *Cissus quadrangularis*) are combined with honey, ghee, or milk – after several washes – to make a thick paste which is then portioned into small peppercorn sizes, known as *pirandai uppu*, used to treat a variety of ailments.[5]

In the case of liquid filtrates, the alkali salts are leached from the ashes into water, yielding a strained salt-rich liquid as the primary product, while the remaining ashes are discarded.[6] Sometimes, there is further processing, in which the liquid filtrate is heated over an extended period until the water evaporates, leaving behind a residue of black-coloured cakes and solid 'salt' pellets, known as pot ashes. Where processing facilities allowed, these salt cake residues were further baked until white; thus they were known as pearl ash. Ashes in both filtrate and cake forms are used in cooking, leavening, and emulsifying oils, as well as in nonculinary applications including agriculture, soap-making, fibre processing (of wool and cotton), and glass-making.

Historically, while much of the production of ashes was on a small scale (usually at home), things were different in Lower Canada (present-day southern Quebec). There, black salts became the raw material for over five hundred asheries and potasheries in Canada in the mid-nineteenth century, representing the burning of 2.8 million pounds of hardwood, which – even if it wasn't a consideration at the time – must have had a longer-term environmental impact.

Today, potash, a rock salt with similar properties to historical pot ashes, is also mined.

Where Do We Find Culinary Ashes?

Culinary ashes are made and used all over the world. As an ingredient, culinary ash is a powerful link across global foodways, as it was and continues to be important in many Indigenous cultures stretching back centuries.

Across Nigeria, one finds regional variation in the source and usage of ashes and by-products. There is *ita ngu* in the southeast, ash commonly made by burning palm fruit (*Elaeis guineensis*) husks, as well as *odo* from the salty mangroves of the Edo region, used as both thickener and medicine. In the north, there is *toka*, made of sorghum (*Sorghum bicolor*) stalks. Other sources of ashes include skins of vegetables like plantains (*Musa × paradisiaca*), the pods of legumes like beans (*Vigna unguiculata*), corn (*Zea mays*), straws and stalks of millet (*Pennisetum glaucum*), sesame seeds (*Sesamum orientale*), the bark of the baobab tree (*Adansonia digitata*), and the shells of groundnuts (*Arachis hypogaea*) and bambara beans (*Vigna subterranean*).

In Papua New Guinea, culinary ash from a diverse range of plants – including ferns, grasses, herbs, and palms – is used as both condiment and additive.[7]

The Assamese of northeastern India are known as *Khar Khuwa Axomiya*, meaning '*khar* eaters'.[8] *Khar*, in Assamese, refers to many things related to culinary ashes – the ashes themselves, liquid decants made from the ashes, associated cooking techniques, and the dishes themselves.[9] For example, *kola khar*, a dark liquid filtrate, is made from the ashes of ripe peels of a species of seeded bananas known locally as *bhim kol* or *athiya kol* (*Musa balbisiana*). Coconut husks are sometimes added as fire starters in the process. In the same region, *dokhora khar* is an extract from the ashes of water hyacinth (*Eichhornia crassipes*), and *sajji khar* is made from a variety of ashes from different plants in the Amaranth family.

In the Loire Valley in central France, grape vine clippings were a rich source of ashes used to coat cheeses like *valençay*, a pyramid-shaped goat's cheese; today a combination of vegetable ashes and salt is used to form the coating.[10] Double Dutch is a cheese that combines a creamy goat's cheese with cow's milk cheese from Meuse-Rhine-IJssel (MRIJ) cattle – in which the cheeses are separated by a layer of grape leaf ash.[11]

In Mexico, Central America, and South America, a common practice in processing corn is nixtamalization – a derivative of the words *nextli*, meaning ashes, and *tamal(l)i*, referring to unformed maize dough in Nahuatl, the language of the Aztecs. The ashes are made from juniper, onions, and fourwing saltbushes, among others.[12] Some ash sources are better than others: poplar, juniper, hickory, and oak are some recommended sources. The Mayans made salt from a variety of palms, including the *escoba* palm (*Crysophila sp.*) and the *botán* palm (*Sabal yapa*).[13] The Hopi people of northeastern Arizona make Hopi *piki*, which is a thin, papery flatbread made of blue cornmeal and ashes from a variety of woods, including juniper, greasewood, saltbush, and chamisa.

How Are Culinary Ashes Used?

The use and effectiveness of culinary ashes for different purposes are largely dependent on properties of the ash components.

As Salts and Flavouring

Salt-loving plants, also known as halophytes, were the source material for early plant salts. These plants have specially designed root networks adapted to high-saline environments, and their leaves collect beads of crystalized salt which would historically have been used for diverse culinary purposes; sometimes plants of this nature – like the fourwing saltbush (*Atriplex canescens*) – were burned to create culinary ashes.

In Kenya, people of the West Pokot region – in the villages of Tarsoi, Tartur, Lition, and Chaunet – historically made an ash yogurt known as *mala ya kienyeji* or *kamabele kambou*, which was a staple for pastoral herdsmen.[14] To make this, ashes of the *Cromwo* or *Kromwo* (*Ozoroa insignis*) tree are combined with cow or goat milk yogurt (processed separately) and fermented in smoked gourds. This blend results in a smoky, light-grey-coloured yogurt prized both for its unique aroma and its antiseptic qualities.

Among the Hausa, we see a more direct relationship between ashes and salt because the word for ashes is *toka*, but it is likewise known as *gishiri*, a word also used to describe (water-derived) salt.[15]

As Additives

Culinary ashes often boost the flavour and nutritional value of food and drinks, depending on the salt composition, taste profile, and elemental properties of the ash sources – i.e. plantain peel ashes likely provide a different flavour and nutrition boost than corn cob ashes.

In Mexico, ashes are one option for alkalizing corn during nixtamalization, which changes the chemistry of the corn, improves its nutritive qualities, and reduces toxicities; the resulting corn is used in all sorts of dishes – hominy, tamales, etc. The Indigenous peoples of Mesoamerica avoided pellagra, a disease associated with a niacin or vitamin B3 deficiency, thanks to nixtamalization with ashes, slaked lime, or other potassium-rich salts. In Uganda, boiling beans with banana-peel ash extract improves the cooking time and texture and likely makes lysine bioavailable, improving the overall quality of the dish.[16] In Yorubaland – in the southwest of Nigeria – in a manner similar to Mexican nixtamalization, dried corn kernels are soaked in a wooden mortar with wood ashes and water to remove the skins in the preparation of *egbo*, a dish of boiled and mashed corn, a bit like Mexican hominy. Snacks of parched corn kernels, known as *guguru* (popcorn), *gbugburu*, or *eyan igbado*, are made by toasting dry or fresh kernels in ashes, fine sand, or a pan over the fire.

The addition of ashes also has an effect on the visual and aesthetic qualities of food. In Nigeria, ash decants are used to fix the green, vibrant colours of vegetables in dishes such as cooked jute leaves (*Corchorus olitorius*) and okra soup. In New Mexico, among the Pueblo people, juniper ash deepens the blue in a number of blue corn dishes, from mush to tortillas.

Okinawa *soba* noodles are made with versions of alkaline solutions which impart a unique texture; traditionally, these noodles would have been made with wood ash filtrates. Today, noodles made this way are referred to as *mokkai soba*.

For Preservation

Different cultures have historically used ashes as preservatives to halt the decay of fruits and vegetables; examples include laying fruits and vegetables on ashes – just like some farmers and the Amish do to preserve tomatoes for months – and coating cuttings (such as pieces of yam) before planting.[17]

Across West Africa, there are many indigenous alkaline fermented condiments, including Senegalese *nététou* and Beninois *afiti*.[18] In Nigeria, ashes are incorporated into mixes of nuts and seeds in the process of making condiments like *ogiri* Igbo (also known as *ogili* Igbo), beloved in the southeast. To make *ogiri*, steamed castor oil seeds are combined with ashes of palm nut (*Elaeis guineensis*) rachis (the central stem of the oil palm tree); the ashes turn the mix grey, encourage fermentation while decreasing the microbial load, and create a rich and stable alkaline environment.[19] The result is a deeply flavourful, incredibly pungent grey paste with distinctive notes.

These processes also feature in Japanese cuisine. Wood ash from camellia, Japanese zelkova, and oak (*Akita Konno*) is used in one method of making *kōji*, steamed rice cultured with *Aspergillus oryzae* mould which lends food and beverages depth of flavour.[20] In this process, steamed rice is inoculated with a mixture of wood ash and *kōji* spores; the ash creates an alkaline environment with minerals (such as potassium, phosphorus, magnesium), sterilizes the mixture, prevents the growth of bacteria, acts as a fertilizer for the mould, and increases activity and growth in the rice while keeping the *Aspergillus oryzae* mould stable and free from aflatoxins.[21]

In South Africa, *biltong* – a snack similar to beef jerky – is made by curing meat in some kind of salt or vinegar before air drying, and culinary ash is one of the curing substances, useful to deter insects and to enhance rapid dehydration.[22] In Italy, some prized hams are aged for many months in ash.[23] *Lutefisk*, cured fish from Scandinavia (and popular at Christmas), was historically made from dried, salted cod and rehydrated in an extract of ashes until a textural transformation resulted, turning the fish gelatinous.

There's a long history of ash-covered cheeses, from Canada to France. Grey Owl goat's cheese, produced by Fromagerie Le Detour in Notre-Dame-du-Lac, Quebec, is dusted in vegetable ashes which enhance rind development. In France, cheeses like *morbier*, *valençay*, *selles-sur-cher*, and *sainte-maure de touraine* have ash on the inside and/or on the rind for flavour and preservation. In Veneto, the northern Italian region, *sottocenere al tartufo* cheeses are wheels of black truffle cheese buried in ashes and flavoured with warming spices – cinnamon, cloves, nutmeg, and more. The ashes neutralize the pH, protect the rind, lengthen ageing, and enhance flavour development.

From roasted eggs to the Chinese delicacy century eggs, ashes have been used as a cooking cocoon (ash-roasted eggs have featured in lore and literature stretching back hundreds of years) and as a curing component.[24] Today, century eggs are cured in a mixture of table salt, calcium hydroxide, and sodium carbonate, but historically (and still artisanally today), the combination of oak ashes, salt, hot-brewed tea leaves, and calcium oxide created an alkaline environment in which eggs were steeped to transform in colour, translucence, flavour, and aroma.

For Tenderization

All over the world, people employ various methods to tenderize food – including adding alkaline substances such as baking soda or potash or acidic ones like bromelain (from pineapples) to their pots, in addition to methods like pressure cooking used to reduce preparation times and conserve cooking fuel. In different parts of Africa and across the world, alkaline substances such as culinary ash filtrates have historically been used to lower cooking times of foodstuffs like *ukwa* (African breadfruit), different kinds of legumes, and meat.[25]

For Emulsification

In parts of southeastern Nigeria, ash filtrate (*ntu ngu* in Igbo) is combined with oil to create a creamy, orange sauce with the qualities of salad cream or mayonnaise – a unique case of oil and water mixing. This emulsion is used in a variety of ways – as a sauce for vegetables, to dress salads of rehydrated cassava shreds in *abacha ncha*, to add to cooked specialty meats like *nkwobi* (cow foot), *isi ewu* (goat head), and *piom piom* (goat foot and periwinkles). One might think of emulsification as liquid saponification, where some form of hydrolysis produces what are technically salts of fatty acids thanks to the ash filtrate being rich in potassium hydroxide, which is similar to lye, or sodium hydroxide.

For Detoxification

The Icelanders and Iroquois washed lichen in water and ashes to temper toxicity and remove bitterness prior to cooking. In Sardinia, acorns were boiled in a mixture of water, ashes, and clay before consumption; this process also helped reduce toxicity and bitterness by lowering the tannin content of the acorns.[26]

In Japanese cuisine, there's a process referred to as *akunuki* through which bitterness – of ferns such as fiddleheads, *zenmai* (flowering or royal ferns), and other mountain and foraged vegetables – is tempered, sweetness highlighted, and flavour and colour preserved, often through the use of culinary ash.[27] According to Elizabeth Andoh, the word *aku* is written with the characters 灰 (ash) and 汁 (broth), describing the ancient practice of mixing alkaline hearth ash with water to create a solution that tames enzymes which would otherwise make certain foods inedible.[28]

As Leavening

The alkali-rich nature of culinary ashes makes them great leaveners, like baking soda, which is a basic salt. For example, the process of burning saltbush leaves converts sodium carbonate into sodium bicarbonate, similar to baking soda, and works well for leavening in baked goods when combined with an acid.[29]

As Condiments

In parts of Papua New Guinea, ashes are used as an accompaniment to fruit, in which the fruit is repeatedly dipped in ash before eating.[30]

To Reduce Acidity

In Nigeria, ashes and decants of ashes are added to food to reduce acidity. In southwestern Nigeria, dried white or yellow hibiscus *sabdariffa* calyces – known as *ishapa* in Yoruba – are rehydrated in ash decants until they soften and are then used in soups and stews.

In Filtration

On St John Island in the US Virgin Islands, ashes were an important part of clarification in sugar processing in the early eighteenth century, during the Danish colonial period. Ashes of select woods and sources – along with other alkali powders – were utilized as tempers or rafts; the tempers worked by pulling impurities to the surface of the sugar cane juice, where the scum could then be skimmed away with ease.[31]

How Did Humans Learn about Culinary Ashes and Salts?

Our ancestors may have deduced things about culinary ashes by observation, hypothesis, and experimentation – just like scientists do today – but they also learned things by accident, as is postulated by Tom Standage, who links the discovery of beer to the process of fermenting porridge.[32]

Extending the Use of Halophytes

In some places, the use of ash as a culinary ingredient evolved from already-existing practices. In regions without direct access to natural salt deposits, many Indigenous communities figured out how to extract minerals from plants through fire. For example, in the Sonoran Desert, juniper, saltbush, and even dried beanstalks were burned to produce ash used as a seasoning and mineral supplement – a practice echoing ancient Egyptian techniques, where halophytes like saltbush were burned and the resulting ash combined with water to yield salt-rich solutions, which were processed further to produce salt residues used in cooking.[33] Ash from hickory nuts – and other nuts and roots traditionally used by the Cherokee – has long served as both seasoning and mineral supplement in dishes like hominy and cornmeal; archaeological evidence in the Eastern Woodlands suggests that such practices date back as far as 8700 BCE.[34] Additionally, the Bukusu community – originally from Congo – settled in what is now western Kenya sometime in the seventeenth century and have been making ash salts by burning and processing *muchua* reeds from the Nzoia River.[35]

Putting Residual Heat to Good Use

People in other regions likely made similar discoveries about the culinary value of ash via ash cakes, also known as ash breads, ash pones, or fire cakes. These dough-based bakes were cooked in ashes, which may have led to the discovery that the breads had certain distinguishing qualities related to preservability, taste, texture, and nutrition.

But how did they come to cook with ashes to begin with? Were ash cakes an easy way to use residual heat as fires died down after cooking? Were they perfect for slow cooking such that our ancestors could prepare them in advance, leaving dough discs to bake as they went about their lives or slept?

Mentions of ash cakes stretch back to the fourth century in Cyprus, as Epiphanius, the bishop of Salamis, describes:

> Now the preparation of the ash cake is in this manner. When the bread has been kneaded and has afterward fermented, it is kneaded again. They bake this bread not in an oven but on a rock. Collecting smooth stones and piling them upon the ground, by means of much brushwood they heat them until they make of the smooth (stones) glowing embers. Then they remove the ashes from them, cover them with dough, and again spread the ashes over all the dough, spreading it out as one loaf; and hence it is called 'hidden,' because [it is] concealed in the ashes.[36]

For enslaved Africans in America, ash cakes were a mainstay consumed out in the fields, according to Frederick Douglass, American abolitionist and writer:

> The dinner of the slaves consisted of a huge piece of ash cake, and a small piece of pork, or two salt herrings. Not having ovens, nor any suitable cooking utensils, the slaves mixed their meal with a little water, to such thickness that a spoon would stand erect in it; and, after the wood had burned away to coals and ashes, they would place the dough between oak leaves and lay it carefully in the ashes, completely covering it; hence, the bread is called ash cake. The surface of this peculiar bread is covered with ashes, to the depth of a sixteenth part of an inch, and the ashes, certainly, do not make it very grateful to the teeth, nor render it very palatable. The bran, or coarse part of the meal, is baked with the fine, and bright scales run through the bread.[37]

In the Modena Apennines, crescentines have a long history. These twice-cooked flatbreads are first baked between terracotta or refractory stone discs (embossed with flower patterns and known as *tigella*), then removed and covered with ashes to cook through.[38] Once puffed up and partly bronzed, the cooked breads are dusted to remove any ash residue before serving with *pesto modenese*.

Root-to-Shoot Consumption

Our ancestors likely developed waste-not-want-not practices out of sheer necessity. Burning inedible vegetable matter, such as corn cobs in the Americas or palm fruit husks in West Africa, may have been one way our ancestors extended the usefulness of food sources and by-products, perhaps by introducing ashes into meals and then observing how different ashes acted on their bodies and affected their health.

Culinary Ashes and Culture

From food to fire to the sacred, societies around the world have rites, rituals, and beliefs about many aspects of their ways of life – and ash and salt are essential aspects.

In their ethnomedical system, the Cherokee had numerous restrictions on the use of salt, some of which extended to culinary ashes. There were also taboos associated with the type of wood ashes used in cooking and ramifications for actions that appeared to desecrate the sacred fire and its ashes. These constraints suggest a complex relationship to salt and the minerals left behind when plant materials are burned.

Some of these practices have been around for such a long time that the original intentions are difficult to decipher. For example, in Benin City, in the south of Nigeria, it is forbidden to use the ashes from the day before – only fresh ashes are permitted. The idea is that one must clean the fireplace every day and begin anew the next day, which is likely connected to cultural beliefs about purity, freshness, and cleanliness.

There are many other cultural practices involving the use of culinary ashes. For example, in the Kinaaldá, a traditional Navajo coming-of-age ceremony, girls learn to make corn cakes prepared with ashes, making a large batch for their community as part of the celebration. Ashes feature differently in other areas. The Huitoto of the Colombian Amazon process ashes into salts for ritual use and to prepare *ambil*, a tobacco paste also employed in ritual contexts.[39]

The Future of Culinary Ashes

In tracing the global story of culinary ash – from ancient salt substitutes to contemporary cheesemaking – this paper shows how culinary innovation often begins at the margins, in what's burned, buried, stumbled upon, and overlooked: sometimes, therein lies treasure.

While the research began from a personal curiosity anchored in the authors' Nigerian roots, the scope includes global practices related to the use of culinary ash. What unites these diverse foodways is not geography but the ingenuity that led different cultures to similar conclusions about transforming the elemental – fire, wood, earth – into sustenance, flavour, and meaning. Future research is an opportunity to expand our collective ideas of how to use plant-based 'waste' sustainably.

About the Authors

Ozoz Sokoh is an ex-exploration geologist, food and tourism studies educator, and author of *Chop Chop: Cooking the Food of Nigeria* (2025). She researches and interrogates connected foodways across the world, highlighting similarities between popular, little-known cultures and cuisines.

Danielle Chike-Ogwo works in tech as a product manager. In her spare time, she's a cultural anthropology researcher focused on how the past informs the present, exploring the incentives and choices that shaped human history and continue to impact our lives today.

Notes

1 Mark Kurlansky, *Salt: A World History* (Walker, 2002); Georgiana Andra Liciu, 'Salt: An Addiction that Engineered the World – The Study of Salt through the History of the British Empire', *Dawson English Journal*, 9 (Winter 2018) <https://www.dawsonenglishjournal.ca/article/salt-an-addiction-that-engineered-the-world-the-study-of-salt-through-the-history-of-the-british-empire/> [accessed 18 April 2025].

2 Samara Vercoe, 'Saltbush – The Unlikely Delicacy', *Warndu*, 1 December 2020 <https://warndu.com/blogs/first-nations-food-guide/saltbush-the-unlikely-delicacy> [accessed 16 October 2025]; 'Victorian Campfire Damper Recipe – Historical Food Fortnightly', *Sew Historically* <https://www.sewhistorically.com/victorian-campfire-damper-recipe-historical-food-fortnightly/> [accessed 16 October 2025]; 'Food Adventure – Down Under: Grubs!', *Unlikely Lady of Leisure*, 13 July 2016 <https://unlikelyladyofleisure.wordpress.com/2016/07/13/food-adventure-grubs/> [accessed 16 October 2025].

3 Many soft woods don't make good-tasting ashes because they contain a lot of resin, which is bitter; see Mike Yarbrough, 'How to Make Blue Corn Ash Pone', *Wolf & Iron* <https://wolfandiron.com/blogs/feedthewolf/how-to-make-blue-corn-ash-pone> [accessed 16 October 2025].

4 Pia Sorensen, 'Culinary Ash in Contemporary Native American Cuisine', YouTube, 25 October 2020 <https://www.youtube.com/watch?v=oWD_f_Lm5Go> [accessed 18 April 2025]. Calcium carbonate is a basic or alkali compound with pH values from nine to twelve and enriches food like corn during the nixtamalization process. It is also used as a source of calcium. See Kim Smith, 'Cooking Blue Corn: Indigenous Science', *Indigenous Goddess Gang*, 22 March 2021 <https://www.indigenousgoddessgang.com/self-care-medicine/2021/3/22/cooking-blue-corn-indigenous-sciencenbsp> [accessed 18 April 2025].

5 Deepa S. Reddy, 'Pirandai Uppu', *Paticheri*, 23 March 2023 <https://www.paticheri.com/2023/03/23/pirandai-uppu/> [accessed 18 April 2025].

6 When ashes are soaked in water and mixed to combine, the resulting liquid becomes rich in potassium, carbonates, and bicarbonate salts over time. This liquid is akin to food-grade lye water, like *lihia* in Filipino cuisine, made from burnt coconut husks.

7 In their study of sodium-rich tree ash as a native salt source in lowland Papua, Ryutaro Ohtsuka, Tsuguyoshi Suzuki, and Masatoshi Morita document the use of a wide variety of plants – from ferns and grasses to palms and herbs – as ash sources ('Sodium-Rich Tree Ash as a Native Salt Source in Lowland Papua', *Economic Botany*, 41.1 (1987), pp. 55–59).

8 Jyotirmay Choudhury, Mridu Pawan Rabha, and Bhola Chourasia, *The Great Assamese Cuisine*, vol. 1 (Rudra, 2021).

9 Aysha Tanya, 'Khar: A Cornerstone of Assamese Cooking', *Goya Media*, 7 September 2021 <https://www.goya.in/blog/khar-a-cornerstone-of-assamese-cooking> [accessed 18 April 2025].

10 Jennifer Meier, 'Why Vegetable Ash Is Used in Cheesemaking', *The Spruce Eats*, 19 September 2019 <https://www.thespruceeats.com/why-is-ash-used-in-cheesemaking-591201> [accessed 18 April 2025].

11 'Double Dutch Bio', *Meester-Affineurs* <https://www.meester-affineurs.com/c/our-cheeses/p/double-dutch-bio/> [accessed 16 October 2025].

12 'Cooking with Ashes', *Onondaga Nation*, 2011 <https://www.onondaganation.org/blog/2011/cooking-with-ashes/> [accessed 18 April 2025].

13 Dennis V. Johnson, 'Palm Ash: A Humble Product of Many Uses', *Palms*, 55.3 (2011), pp. 117–21 <https://palms.org/wp-content/uploads/2016/05/vol55n3p117-121.pdf> [accessed 18 April 2025].

14 'Pokot Ash Yogurt – Slow Food Presidia', *Google Arts & Culture* <https://artsandculture.google.com/story/pokot-ash-yogurt-slow-food/cAVBARuYOhMA8A?hl=en> [accessed 18 April 2025]; 'Pokot Ash Yogurt – Slow Food Presidia', *Fondazione Slow Food* <https://www.fondazioneslowfood.com/en/slow-food-presidia/pokot-ash-yogurt/> [accessed 18 April 2025].

15 Neil Skinner, *Hausa Comparative Dictionary*, ed. by Herrmann Jungraithmayr and Norbert Cyffer (Rüdiger Köppe Verlag, 1996).

16 R. C. J. Mallemo and others, 'Effects of Banana Peel-Ash Extract on Cooking Time and Acceptability of Hard-to-Cook Beans (*Phaseolus vulgaris L*)', *International Journal of Tropical Agriculture and Food Systems*, 3.1 (January 2009), doi:10.4314/ijotafs.v3i1.50014.

17 Jean de Dieu Ininahazwe, 'Burundi: Farmer Finds New Technique for Preserving Tomatoes', *Barza Wire*, 28 November 2016 <https://wire.farmradio.fm/farmer-stories/burundi-farmer-finds-new-technique-for-preserving-tomatoes/> [accessed 18 April 2025].

18 @eatwithafia, 'Demonstration of the Preparation of Fermented Locust Beans', Instagram, 2025 <https://www.instagram.com/reel/DHHCrYkJ81O/> [accessed 18 April 2025].

19 William J. Henderson and others, 'Oil Palm (*Elaeis guineensis* Jacq.) Bunch Structure Variation and Limitations', *Scientific Research Journal*, 3.1 (January 2015), pp. 5–10 <https://www.scirj.org/papers-0115/scirj-P0115223.pdf> [accessed 18 April 2025].

20 Josh Evans, 'Koji: History and Process', *Nordic Food Lab*, 8 August 2013 <https://nordicfoodlab.org/blog/2013/08/koji-history-and-process/> [accessed 18 April 2025].

21 Stephanie Houng, 'What Is Koji? A Guide to Making Koji', *Made with Ine* <https://madewithine.com/blogs/news/what-is-koji-a-guide-to-making-koji> [accessed 18 April 2025]; Meredith Leigh and Kirsten K. Shockey, 'Koji: An Ancient Mold and Its Modern Renaissance', *My Fermentation*, 13 August 2019 <https://www.myfermentation.com/grains/koji-an-ancient-mold-and-its-modern-renaissance-zmoz19fzwoo/> [accessed 18 April 2025].

22 Eben van Tonder, 'Salt and the Ancient People of Southern Africa', *Earthworm Express*, 5 August 2018 <https://earthwormexpress.com/salt-salz-sul-盐-yan-sol-αλασ-usawoti-ملح-letsoai-sel-sal-نمک-muoi-鹽-sout-塩-shio-소/salt-and-the-ancient-people-of-southern-africa/#Ash_in_Food> [accessed 18 April 2025].

23 'Il Prosciutto sotto Cenere', *Macelleria Falorni* <https://falorni.it/en/prodotti/il-prosciutto-sotto-cenerer/> [accessed 18 April 2025].

24 Wei Guo, 'Century Eggs, the Myths and Recipe (Pi Dan, 皮蛋)', *Red House Spice*, 16 January 2021 <https://redhousespice.com/century-eggs/> [accessed 18 April 2025]; Paula Marcoux, 'The Lost Art of Roasting Eggs in Ash', *Atlas Obscura*, 1 September 2023 <https://www.atlasobscura.com/articles/how-to-roast-eggs-in-ash> [accessed 18 April 2025].

25 Chinwe's Creative Kitchen, 'How to Cook Party Standard Ukwa (Anambra Style) | Breadfruit Recipe', YouTube, 18 March 2022 <https://www.youtube.com/watch?v=RK7dAqA_9rI> [accessed 18 April 2025]; 'Mushelekha', *Fondazione Slow Food* <https://www.fondazioneslowfood.com/en/ark-of-taste-slow-food/mushelekha/> [accessed 18 April 2025].

26 Dauro Mattia Zocchi and others, 'Food Security beyond Cereals: A Cross-Geographical Comparative Study on Acorn Bread Heritage in the Mediterranean and the Middle East', *Foods*, 11.23 (2022), 3898, doi:10.3390/foods11233898.

27 'Aku Nuki: Removing the Bitterness from Vegetables', *Umamitopia*, 18 May 2010 <https://umamitopia.com/blog/article_12.html> [accessed 18 April 2025].

28 Elizabeth Andoh, 'A Taste of Culture – Aku Nuki', *Japan Living Arts*, 15 March 2021 <https://japanlivingarts.com/a-taste-of-culture-aku-nuki/> [accessed 18 April 2025].

29 'Saltbush', *Native Memory Project* <https://nativememoryproject.org/plant/saltbush/> [accessed 18 April 2025].

30 Ohtsuka, Suzuki, and Morita, 'Sodium-Rich Tree Ash.'

31 David W. Knight, 'How Sugar Was Made on St John during the Danish Colonial Period', *St John Historical Society*, 2001 <https://stjohnhistoricalsociety.org/how-sugar-was-made-on-st-john-during-the-danish-colonial-period-david-w-knight2001/> [accessed 16 October 2025].

32 Tom Standage, *A History of the World in Six Glasses* (Walker, 2006).

33 Arlo Starr, 'How Salt Built Civilization', *ICT News*, 24 December 2016 <https://ictnews.org/archive/how-salt-built-civilization> [accessed 18 April 2025].

34 Kyra Gibson, 'Blue Corn Ash', *Bittersweet: The Ozark Quarterly*, Winter 1976, archived by Springfield-Greene County (MO) Library <https://sgcld.thelibrary.org/lochist/periodicals/bittersweet/wi76h.htm> [accessed 18 April 2025].

35 'Nzoia River Reed Salt', *Slow Food Foundation for Biodiversity* <https://www.fondazioneslowfood.com/en/slow-food-presidia/nzoia-river-reed-salt/> [accessed 16 October 2025]; Arlo Starr, 'How Salt Built Civilization', *ICT News*, 24 December 2016 <https://ictnews.org/archive/how-salt-built-civilization> [accessed 18 April 2025].

36 Epiphanius of Salamis, *Treatise on Weights and Measures*, trans. by James Elmer Dean (University of Chicago Press, 1935) <https://archive.org/details/DeanSprengling1935EpiphaniusTreatiseOnWeightsAndMeasures/page/n3/mode/2up> [accessed 18 April 2025].

37 Frederick Douglass, *My Bondage and My Freedom* (Miller, Orton & Mulligan, 1855) <https://archive.org/details/mybondagemyfreedooodoug_y3h4/mode/1up> [accessed 18 April 2025].

38 TasteAtlas, 'The History of Crescentine, a Particular Type of Bread Typical of the Modena Apennines, Italy', Instagram, 2023 <https://www.instagram.com/reel/CyDw41coBq2/> [accessed 18 April 2025].

39 Jimmy Weiskopf, 'Quarantine Chronicle from Choachí: How to Make Salt from Grass (While Making Smokeless Tobacco)', *Mundo Amazónico*, 11.2 (2020), pp. 169–78, doi:10.15446/mav11n2.87661; O. Enokakuiodo and others, 'Vegetable Salt: An Ignored Resource', in *Des sources du savoir aux médicaments du futur*, ed. by Jacques Fleurentin and others (IRD, 2002), doi:10.4000/books.irdeditions.7255.

29
Elements of an Amazonian 'Uber Superfood'
The Case of the Aguaje (*Mauritia flexuosa*)

Diana Tung

What makes a superfood? Despite the lack of a scientific or medical definition – and proven benefits in many cases – the international superfood market is estimated to be worth over USD 162 billion annually.[1] While the strongest association with superfoods is health-boosting properties, those of us interested in the politics of food know that it is never just about the food. Indeed, as this paper highlights, superfood consumption, particularly in wealthy Western societies, is fraught with tensions and contradictions reflecting broader concerns. As such, the hypervisibility and pervasiveness of superfoods in the public imagination and global market demand that we engage critically with this phenomenon.

Despite the lack of an official definition for superfoods, there are particular elements that render a foodstuff identifiable as a superfood. Tina Sikka points out that eating superfoods is fraught with notions of morality for the consumer and that the consumption of superfoods is a 'normative food trend among affluent groups in the global North' which is 'mobilized using neoliberal, postfeminist, and food justice discourses'.[2] Sikka argues that superfood consumption relies on a distinction between 'clean' and 'dirty' foods, which in turn extends to the consumer in an embodied manner. In other words, when choosing between a fattening, greasy, and highly processed burger or a blueberry acai smoothie with guaraná, consumers are forced to deliberate on what kind of person they want to be and what kinds of ethics they want to practice through their consumption. A closer examination of such dynamics in superfood consumption also uncovers predominant ideas and imagery perpetuating neocolonial notions of the producer-consumer relationship, masked in relations of benevolence and ethical consumption.[3] Altogether, these elements reaffirm how the consumption of superfoods is not a singular act but a lifestyle with moralistic aspirations.[4]

This paper takes 'superfoods' to be a category of foods composed of three interrelated and co-constituting elements: firstly, its perceived benefits for individual consumer health; secondly, its production by racialized 'Others' from the Global South; and thirdly, its environmentally sustainable production.[5] This paper examines how a new category of superfoods, what I call the 'uber superfood', promises to assuage contemporary anxieties of living in the Anthropocene, a distinct geological era of anthropogenic climate change.[6] The uber superfood not only is good for producers, consumers, and the environment but also contributes to the viability of our planetary future. This paper is informed by original data gathered from almost twenty-four months of ethnographic fieldwork in the Peruvian Amazon and utilizes the *aguaje* palm (*Mauritia flexuosa*) in Peru as a case study.

This paper has three main parts. I first provide some historical context on the notion of superfoods and introduce the key elements that constitute a superfood. Here I also lay out my core argument, which is that in an age of climate anxiety, the uber superfood has emerged as a new category of superfood, with explicit benefits for our collective planetary future. The second section introduces the case of the aguaje palm and situates it within its social and economic context in the Peruvian Amazon. The third section demonstrates how the aguaje fulfils aptly traditional notions of what constitutes a superfood but, in addition, has carbon-capturing capacities as a palm growing in tropical peatland swamps that extend health benefits to the planet.

Elements of a Superfood

Without needing to have ever read an academic paper on the topic, most people will have heard of the term 'superfoods' or consumed them in the form of acai bowls, goji berries, chia seeds, maca powder, quinoa, and so on. All these foods are well known for their supposed health benefits, whether it be their energy-boosting properties or their richness in fibre, protein, antioxidants, vitamins and minerals, and so on. Even the term '*super*food' already denotes that this product is something out of the ordinary. Yet superfoods have no clear definition and 'should be understood as more of a marketing term', and they can encapsulate such broad meanings that they 'lie on the boundary between nature and culture, food and medicine'.[7] At the same time, the term 'superfood' has been around for at least a century and is so well integrated into the popular lexicon that it appears in the *Oxford English Dictionary* as a 'food considered especially nutritious or otherwise beneficial to health and well-being'.[8] Of the many elements that can constitute superfoods, three key elements can be easily identified.

A Boon to Individual Consumer Health

Jeffrey P. Miller posits the persistent urge to find superfoods is 'tied to the human fascination with immortality' and part and parcel of a larger quest to defy mortality.[9] As such, a direct benefit to the health of the consumer is a crucial element of superfoods, even if this is expressed in vastly differing ways. Superfoods have been claimed to prevent disease, induce weight loss, and improve mental health, as well as 'improve overall health, boosting the immune system, increasing the production of serotonin and other hormones, and promoting the smooth operation of organic systems of the human body'.[10] Superfoods may also contain higher concentrations of a 'desired nutrient and/or a chemopreventive or chemotherapeutic effect'.[11] In turn, the broad and all-encompassing promise of superfood consumption has also led to far-fetched health claims in labelling, prompting calls for better policing of food fraud.[12] Additional categories of distinction such as 'organic' or 'raw' place emphasis on the purity of these foods and their benefits for the individual consumer, even when they may be heavily processed.[13]

Production by Racialized Others from the Global South

A second element that emerges in the literature on superfoods is how producers are portrayed as being exotic, far away, and located in the Global South. Producers are seen as engaging in '"traditional" agricultural practices' that are close to nature and therefore assumed to be better for human health.[14] This image of superfoods as being purer and more authentic – both culturally and ecologically – is important to consumers trying to counter the vicissitudes and excessiveness of the 'modern Western diet', as symbolized by refined carbohydrates.[15]

At the same time, the marketing of superfood products often promotes a narrow vision of 'nutritional primitivism', which Christine Knight describes as the 'pursuit of supposedly simpler, more natural and authentic ways of eating, as part of a quest for health'.[16] In their article on the maca (*Lepidium meyenii*) market in Australia, Jessica Loyer and Christine Knight illustrate how indigeneity is utilized and portrayed in the packaging. As they describe, 'The women – uniformly plump, dark-skinned, black-braided, felt-hatted, and traditionally dressed in Andean skirts and mantas (shawls) – labor against a serene backdrop of snow-capped peaks and blue sky dotted with fluffy clouds'.[17] This image, while idyllic and romantic, utilizes and perpetuates notions of the 'racialized Other' even when it bears little resemblance to who actually harvests maca and how.[18] Particularly in the context of superfoods, essentializing portrayals of Indigenous producers are compelling because they build upon long-standing notions of Indigenous peoples as being 'deeply connected to nature and spirituality' and as peoples who have maintained their traditional foodways in harmony with Mother Earth.[19] The consumption of superfoods by mostly wealthy Western individuals reflects a desire to become and perform a certain type of moral being, one who supports the livelihoods and traditions of Indigenous producers. In addition to caring about their own health, individuals can therefore consume superfoods produced in faraway places with the reassurance of their benevolent and ethical engagements with the racialized Other.

Environmental Sustainability

As Ana Fernández-Ríos and others argue, the environmental dimension of superfoods is a key but 'under-represented' element informing consumers' perceptions and consumption of superfoods, regardless of the actual environmental impacts on the ground.[20] Indeed, the elements of consumer health, racialized producers, and environmental benefits are so tightly intertwined that many of the same dynamics introduced earlier inform the widespread perception of superfoods as being environmentally friendly.

As superfood consumers are predisposed to engage in 'nutritional primitivism', there are inbuilt assumptions about the producers being 'closer to nature' and therefore utilizing traditional methods that are small-scale, organic, less harmful to the environment, and so on.[21] In addition, as superfoods are often automatically equated with plants, they are perceived to have a lower carbon footprint than that of meat products. As Aneta Popova and others argue, 'The high demand for "*superfoods*" arises not only from their beneficial nutritional profile but also because of the widespread belief in their environmental benignancy and climate neutrality'.[22] However, similar to how the actual production of superfoods can depart sharply from what is depicted in idyllic marketing campaigns, the production of superfoods may incur significant food miles, directly contribute to the degradation of soil and loss of biodiversity, and be heavily processed even when portrayed otherwise.[23]

A New Category of Superfoods

So far, we have established superfoods as being, in theory, good for the health of the consumer, produced by racialized Others, and environmentally friendly products. Eating superfoods is not just about ingesting singular nutritional building blocks but deeply entangled with contemporary concerns about morality, authenticity and exoticism, identity politics, food colonialism, and more. Indeed, 'charismatic nutrient[s]' can 'go in and out of fashion'.[24] At the same time, the amplification of health concerns during and after the Covid-19 pandemic has led to a marked increase in the consumption of superfoods.[25] In the Anthropocene, concerns about human survival on a warming planet are only likely to intensify. Superfood production, marketing, and consumption have already begun to reflect these changing societal priorities, with the introduction of new food products that promise to address concerns on a planetary scale. While superfoods denote a category of foods that are extraordinary in their healthful qualities, production, and environmental sustainability, 'uber superfoods' can be seen as being fundamental to our existence due to their beneficial impact on the planet as a whole. As much as the term is clunky and slightly absurd, arguably so is the notion that eating one particular foodstuff will save the planet, even though that has not dampened attempts to portray it as such. The aguaje palm fruit is one such superfood that has been marketed as a solution to urgent planetary concerns.

Aguaje as Amazonian Fruit

Among the 150 palm species in the Amazon known to science, the aguaje palm (*Mauritia flexuosa*) is one of the most commonly occurring palms and one of the most valuable both for its food and for its economic importance (Figure 1).[26] In Iquitos, the largest city in the Peruvian Amazon – and the largest city in the world that cannot be reached by car – the fruit of the aguaje palm reigns supreme. The aguaje palm is known as 'el arbol de la vida' ('the tree of life'), and its consumption is synonymous with Amazonian identity. As I was told by a prominent forestry engineer in Iquitos, 'si tú no comes aguaje, no eres amazónico' ('if you don't eat aguaje, you're not Amazonian').[27] In a city of about half a million people, a conservative estimate of 5 tonnes of aguaje fruits are consumed daily (Figure 2), or about twenty-nine kilograms per person per year.[28]

Iquiteños are partial to the aguaje for its *agridulce* (sweet and sour) flavour, and it is consumed in a dizzying array of forms. It is commercialized as *aguajina*, *curichi*, *adoquín*, *marciano*, liquor, *cremolada*, ice creams, popsicles, beverages, cakes, sweets, and jams and in its most popular form, as the fruit itself (*maduro*).[29] The local aguaje economy supports over five hundred aguaje street vendors and their households.[30] As such, the aguaje is significant not only as a widespread object of consumption but also as the centre around which significant social, economic, and ecological interactions occur. Sensationalist mythologies accompany the consumption of aguaje: it can act as an aphrodisiac, enhance

Figure 1. A female fruit-bearing aguaje palm in Iquitos. Photo by the author.

Figure 2. Matured aguaje fruits for sale in Iquitos. Photo by the author.

women's curves, and stave off the worst effects of menopause. If consumed in excess, the aguaje can be queering and turn straight men gay, though street vendors responded acerbically that 'el aguaje se come por la boca, no por el culo' ('the aguaje is eaten through the mouth, not through the ass').[31]

The aguaje's reputation as a superfood is largely built upon it being the 'richest natural source of beta carotene known' to science, in addition to its high levels of calcium, phosphorus, iron, and vitamin C.[32] At 260 milligrams per 100 grams, the beta carotene levels in aguaje pulp surpass 'any other known fruit of fruit trees up to now', and its consumption would help meet the daily requirements of children and half the daily requirement for adults to successfully treat vitamin A deficiencies.[33] In comparison, carrots contain concentrations of beta carotene at 8.3 milligrams per 100 grams.[34] Due to its high carotenoid content, consuming aguaje could have photoprotection potential while providing a multitude of antioxidant and antimicrobial benefits.[35] Aguaje could also treat 'protein energy malnutrition, hidden hunger and chronic diseases related to nutrition' – particularly in the Amazon, where these issues are widespread – in addition to cardiovascular diseases.[36] As a fruit that is wild harvested by Amazonian residents, the aguaje duly fulfils the three superfood elements presented earlier: it is beneficial for individual health, produced by racialized Others from the Global South, and environmentally sustainable when harvested by climbing techniques.

The Aguaje as Uber Superfood

If carbon emissions are one of the root causes of anthropogenic climate change and a major cause of irreversible damage to the planet, one way of solving it would be to capture the carbon through photosynthesis, effectively removing it from the atmosphere and storing it in biomass. Following this logic, for decades the practice of establishing new forest areas, known as afforestation, has been seen as a sure solution to climate change precisely for its ability to capture carbon emissions.[37] However, the widespread adoption of carbon-offsetting strategies such as tree-planting schemes has largely led to ineffective or perverse outcomes, such as the logging of old-growth forests to clear land for new plantations.[38] Carbon-offsetting activities have come under fire for a lack of accountability, implementation, and monitoring, even when certification schemes are in place.[39] As such, critics have labelled these programmes as largely corporate 'transition washing' and green fraud, whereby the purported aims and the actual impact on reducing carbon emissions diverge greatly.[40]

As the hype surrounding carbon offsetting through afforestation has slowed down in recent years, carbon sequestration, either capturing or preserving what carbon already exists in the ground, is gaining new interest. In turn, peatlands have emerged as an unlikely yet promising bulwark against climate change due to their high carbon sequestration capacities. Peatlands, as a subcategory of wetlands categorized by the compact buildup of organic material over time, cover only about 3 percent of the planet's surface but store approximately two-thirds of the entire atmospheric carbon pool.[41] Peatlands store twice the amount of carbon as all the world's forests and are 'the most efficient carbon stores of all terrestrial ecosystems'.[42] As only 17 percent of peatlands lie within protected areas, peatland conservation is gaining rapid traction among environmentalists and policymakers alike for their utility as significant carbon 'sinks'.[43]

Peatlands in tropical regions are estimated to store about half of the global peatland-emitted carbon, the bulk of these located in South America and specifically in the Amazon and in areas where aguaje palms grow.[44] In fact, 'due to the high level of conservation and the expected high capacity of carbon accumulation, *the Amazon region peatlands are particularly important* in the context of climate change mitigation'.[45] This sense of urgency is compounded by the extent of peatland devastation elsewhere – for example, in Indonesia as well as in the Brazilian Amazon, particularly under the 2019–2023 Bolsonaro administration.[46] In addition, due to their vast carbon stores, rising temperatures and drier climates could unleash a disastrous 'carbon bomb' in peatlands with no way to recuperate the large volumes of emitted carbon.[47] As such, Amazonian peatlands have increasingly gained international attention not only as landscapes of localized environmental importance but as areas of planetary significance that require urgent conservation.[48]

Within the Peruvian Amazon, peatlands account for approximately 11 percent of the area, but their carbon storage capacity is equal to approximately 75 percent of mature forest biomass carbon storage in the entire region.[49] Carbon stocks in *aguajales*, or aguaje palm swamps, in particular are so significant that researchers argue that they warrant a recalculation of national carbon stocks.[50] Within this context, the consumption of aguaje is not just about boosting individual health, exotic producers, or low-environmental impact foods but about the potential to store extraordinary amounts of carbon underground and to prevent global temperatures from rising irreversibly.

Current attempts to commercialize aguaje are based on a logic that if wealthy, far-off consumers develop an appetite for aguaje, Amazonian harvesters will see the economic value of caring for the palm and will climb it to harvest aguaje instead of cutting the palms down, as is customarily done. In turn, this will keep the carbon underground and the rainforest intact, and harvesters will earn a better income. Like other nontimber forest products (NTFPs), it's 'a win-win situation for poverty reduction and bio-diversity conservation'.[51] The logic presented is that of a closed and self-sustaining loop (Figure 3), imbued with even greater urgency and significance given the broader implications of conserving the aguaje. Thus, saving the aguaje palm is not just about preventing deforestation, improving harvester's livelihoods, or preserving cultural values but about saving the planet.

Within the past 5 years, Grupo Aje, one of the largest multinational beverage corporations in Latin America, has seized upon the opportunity to commercialize aguaje as

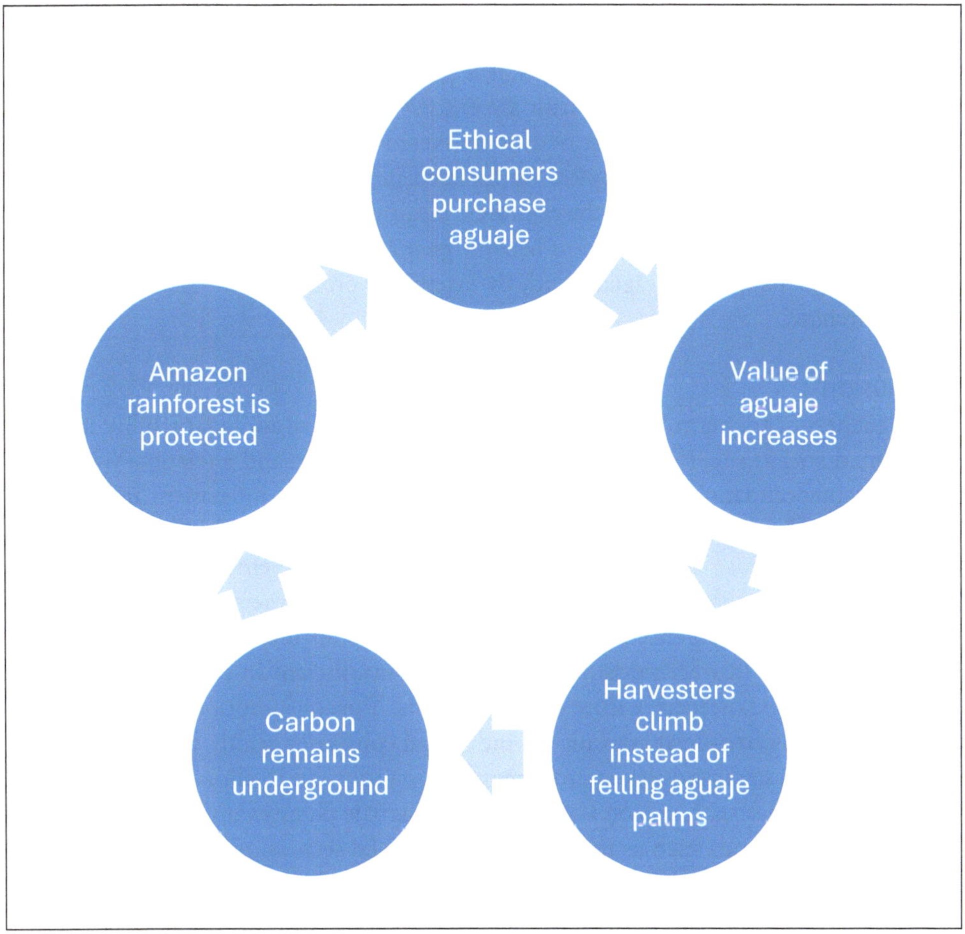

Figure 3. The circular logic of aguaje commercialization. Diagram by the author.

a superfood for wider markets.[52] Through its subsidiary Bio Amayu, it has executed a comprehensive social media strategy emphasizing the company's commitment to sustainability as well as the importance of aguaje for planetary health. The company's sustainably harvested aguaje from the Pacaya Samiria National Reserve is certified by the Peruvian National Service of Protected Natural Areas by the State (SERNANP) and proudly displayed on the aguaje beverage label (Figure 4).[53] The Pacaya Samiria National Reserve is located within the Pastaza-Marañón Foreland Basin, one of the largest tracts of intact tropical peatlands in the world and a site of tremendous interest to conservationists.[54]

Figure 4. An image from the 'Nuestra Historia' ('Our History') page from Grupo Aje's website. The green SERNANP seal shows that the product is certified under the Aliado por la Conservación (Allied for Conservation) programme. Source: https://www.ajegroup.com/nuestra-historia.

In Bio Amayu's marketing campaigns, it is perhaps unsurprising that aguaje's locally valued characteristics as a women's fruit are entirely omitted. Instead, the aguaje, in its transformation into an uber superfood, is marketed for its high vitamin content and its harvesting by Indigenous peoples. As depicted in Bio Amayu's promotional video, these 'guardianes' ('guardians') of the rainforest in turn protect the Amazon, the lungs of the planet.[55] Consumers are invited to become *guardianes* themselves by consuming the Amazonian 'uber superfood' aguaje and, by extension, protecting the planet with their ethical purchase.

Conclusion

The vagueness of the term 'superfoods' facilitates malleability and adaptability in what can be marketed as superfoods to meet contemporary desires and anxieties. With contemporary preoccupations of anthropogenic climate change, the judgement and scrutiny placed upon individual consumption choices have only intensified. This paper highlights how 'uber superfoods', as a new category of superfoods, have emerged in response to anxieties about anthropogenic climate change and allow consumers to partake in ethical consumption. The emergence of the 'uber superfood' is an indication of the anxieties of our current era, in which existing categories are insufficient in capturing the depth and breadth of the existential nature of our concerns. The consumption of aguaje is thus marketed as being beneficial not only for individual consumer health, the livelihoods of Amazonian harvesters, and the environment but as a solution to climate change in preserving the immense carbon stores of *aguajales*, or aguaje palm swamps in the Amazon. Eating uber superfoods such as the aguaje, then, can be seen as one attempt to salvage not only individual health, the exotic Other, and the environment but the very planet that we live on.

About the Author

Diana Tung conducted two years of original ethnographic fieldwork in the Peruvian Amazon on the commercialization of aguaje as part of her doctoral studies in anthropology. She is currently a postdoctoral researcher at the University of Barcelona, examining migrant experiences in citrus production in Spain.

Acknowledgements

I would like to thank the American Friends of the Oxford Symposium, the Julia Child Foundation, and the Student Extracurricular Enrichment Fund of the Australian National University Students' Association for their support of my attendance at the 2025 Oxford Food Symposium.

Notes

1 Ana Fernández-Ríos and others, 'A Critical Review of Superfoods from a Holistic Nutritional and Environmental Approach', *Journal of Cleaner Production*, 379 (December 2022), doi:10.1016/j.jclepro.2022.134491; Cognitive Market Research, 'Superfoods Market Report 2025 (Global Edition)', March 2025 <https://www.cognitivemarketresearch.com/superfoods-market-report> [accessed 1 October 2025].

2 Tina Sikka, 'The Contradictions of a Superfood Consumerism in a Postfeminist, Neoliberal World', *Food, Culture & Society*, 22.3 (2019), p. 354, doi:10.1080/15528014.2019.1580534.

3 Lisa M. Heldke, *Exotic Appetites: Ruminations of a Food Adventurer* (Routledge, 2003).

4 Sikka, 'Contradictions.'

5 Heldke, *Exotic Appetites.*

6 Paul J. Crutzen and Eugene F. Stoermer, 'The "Anthropocene"', *International Geosphere-Biosphere Programme Newsletter*, 41 (2000), pp. 17–18.

7 Sikka, 'Contradictions', p. 355; Richard R. Wilk and Emma McDonell, 'Tracking Superfoods: An Introduction', in *Critical Approaches to Superfoods*, ed. by Richard R. Wilk and Emma McDonell (Bloomsbury, 2020), p. 2, doi:10.5040/9781350123908.

8 Jeffrey P. Miller, 'Introduction to Superfoods', in *Superfoods: Cultural and Scientific Perspectives*, ed. by Jeffrey P. Miller and Charlene Van Buiten, Food and Health (Springer International, 2022), doi:10.1007/978-3-030-93240-4; entry 'superfood', OED <https://www.oed.com> [accessed 1 October 2025].

9 Miller, 'Introduction', p. 1.

10 Miller; Fernández-Ríos and others, 'Critical Review', p. 1.

11 Miller, 'Introduction', p. 1.

12 Janine Curll and others, 'Unlocking the Energy of the Amazon? The Need for a Food Fraud Policy Approach to the Regulation of Anti-ageing Health Claims on Superfood Labelling', *Federal Law Review*, 44.3 (2016), pp. 419–49, doi:10.1177/0067205X1604400304.

13 Richard R. Wilk and Emma McDonell, eds., *Critical Approaches to Superfoods* (Bloomsbury, 2020), doi:10.5040/9781350123908; Monika Prakash Singh and others, 'Superfood: Value and Need', *Current Nutrition & Food Science*, 18.1 (2022), pp. 65–68, doi:10.2174/1573401317666210420123013.

14 Jessica Loyer and Christine Knight, 'Selling the "Inca Superfood": Nutritional Primitivism in Superfoods Books and Maca Marketing', *Food, Culture & Society*, 21.4 (2018), p. 449, doi:10.1080/15528014.2018.1480645.

15 Loyer and Knight, p. 452.

16 Christine Knight, '"We Can't Go Back a Hundred Million Years": Low-Carbohydrate Dieters' Responses to Nutritional Primitivism', *Food, Culture & Society*, 18.3 (2015), p. 441, doi:10.1080/15528014.2015.1043107.

17 Loyer and Knight, 'Selling', p. 457.

18 Michael Omi and Howard Winant, *Racial Formation in the United States* (Routledge, 2014); Loyer and Knight.

19 Loyer and Knight, 'Selling', p. 462.

20 Ana Fernández-Ríos and others, 'Superfoods: A Super Impact on Health and the Environment?', *Current Opinion in Environmental Science & Health*, 31 (February 2023), p. 1, doi:10.1016/j.coesh.2022.100410.

21 Heldke, *Exotic Appetites*.

22 Aneta Popova and others, 'Promoting Plant-Based "*Superfoods*" – A Delicate Balance between Health Benefits and Sustainability Challenges', *Food and Bioprocess Technology*, 17 (2024), p. 1387.

23 Popova and others; Ainhoa Magrach and María José Sanz, 'Environmental and Social Consequences of the Increase in the Demand for "Superfoods" World-wide', *People and Nature*, 2.2 (2020), pp. 267–78, doi:10.1002/pan3.10085; Wilk and McDonell, 'Tracking Superfoods', p. 3.

24 Wilk and McDonell, *Critical Approaches to Superfoods*, p. 7.

25 Fernández-Ríos and others, 'Superfoods'; Abdo Hassoun and others, 'Truths and Myths about Superfoods in the Era of the COVID-19 Pandemic', *Critical Reviews in Food Science and Nutrition*, 64.3 (2024), pp. 585–602, doi:10.1080/10408398.2022.2106939.

26 Nigel Smith, *Palms and People in the Amazon*, Geobotany Studies (Springer International, 2015), doi:10.1007/978-3-319-05509-1; Jennifer A. Holm and others, 'Population Dynamics of the Dioecious Amazonian Palm *Mauritia flexuosa*: Simulation Analysis of Sustainable Harvesting', *Biotropica*, 40.5 (2008), pp. 550–58, doi:10.1111/j.1744-7429.2008.00412.x.

27 Dennis del Castillo Torres, personal communication with the author, 16 January 2020.

28 Cesar Delgado and others, '*Mauritia flexuosa* (Arecaceae: Calamoideae), an Amazonian Palm with Cultivation Purposes in Peru', *Fruits*, 62.3 (2007), pp. 157–69, doi:10.1051/fruits:2007011.

29 Delgado and others.

30 Christine Padoch, 'Aguaje (*Mauritia flexuosa* L. f.) in the Economy of Iquitos, Peru', *Advances in Economic Botany*, 6 (1988), pp. 214–24.

31 Padoch.

32 Leonor Maria Pacheco Santos, 'Nutritional and Ecological Aspects of Buriti or Aguaje (*Mauritia flexuosa* Linnaeus filius): A Carotene-Rich Palm Fruit from Latin America', *Ecology of Food and Nutrition*, 44.5 (2005), pp. 345–58 (p. 345), doi:10.1080/03670240500253369.

33 Delgado and others, '*Mauritia flexuosa*', p. 158; Jaime Paiva Lopes Aguiar and Francisca Das Chagas Do Amaral Souza, 'Dehydration and Spraying of Buriti Pulp (*Mauritia flexuosa* L.): Shelf-Life Evaluation', *Revista Brasileira de Fruticultura*, 39.0-34 (2017), doi:10.1590/0100-29452017034.

34 Ali Ikram and others, 'Exploring the Health Benefits and Utility of Carrots and Carrot Pomace: A Systematic Review', *International Journal of Food Properties*, 27.1 (2024), p. 182, doi:10.1080/10942912.2023.2301569.

35 Hector H. F. Koolen and others, 'Antioxidant, Antimicrobial Activities and Characterization of Phenolic Compounds from Buriti (*Mauritia flexuosa* L. f.) by UPLC–ESI-MS/MS', *Food Research International*, 51.2 (2013), pp. 467–73, doi:10.1016/j.foodres.2013.01.039; Aguiar and Souza, 'Dehydration.'

36 Aguiar and Souza, 'Dehydration', p. 2; Hiroshi Shimoda and others, 'Hydroxypterocarpans with Estrogenic Activity in Aguaje, the Fruit of Mauritia flexuosa (Peruvian Moriche Palm)', *Bioactive Compounds in Health and Disease*, 2.4 (2019), pp. 64–76, doi:10.31989/bchd.v2i4.585.

37 Dong Kwan Kim, 'Afforestation Can Help to Tackle Climate Change. Here's How', *World Economic Forum*, 3 November 2021 <https://www.weforum.org/stories/2021/11/afforestation-can-help-tackle-climate-change-heres-how/> [accessed 1 October 2025]; Eilidh J. Forster and others, 'Commercial Afforestation Can Deliver Effective Climate Change Mitigation under Multiple Decarbonisation Pathways', *Nature Communications*, 12.1 (2021), p. 3831, doi:10.1038/s41467-021-24084-x.

38 Falk Huettmann and Brian D. Young, 'The So-Called Modern "Sustainable Forestry" Destroys Wilderness, Old-Growth Forest Landscapes and Ecological Services Worldwide: A Short First-Hand Review and Global Narrative on the Use of "Growth-and-Yield" as a Destructive and Even Impossible Goal', in *Forest Dynamics and Conservation*, ed. by Manoj Kumar and others (Springer Nature, 2022), doi:10.1007/978-981-19-0071-6_3.

39 Thales A. P. West and others, 'Action Needed to Make Carbon Offsets from Forest Conservation Work for Climate Change Mitigation', *Science*, 381.6660 (2023), pp. 873–77, doi:10.1126/science.ade3535.

40 Sehoon Kim and others, 'Carbon Offsets: Decarbonization or Transition-Washing?', *SSRN*, 31 January 2024, doi:10.2139/ssrn.4762799; Peter Martin and Reece Walters, 'Fraud Risk and the Visibility of

Carbon', *International Journal for Crime, Justice and Social Democracy*, 2.2 (2013), pp. 27–42, doi:10.5204/ijcjsd.v2i2.95.

41 Christian Dunn and Chris Freeman, 'Peatlands: Our Greatest Source of Carbon Credits?', *Carbon Management*, 2.3 (2011), pp. 289–301, doi:10.4155/cmt.11.23.

42 UNEP, *Global Peatlands Assessment: The State of the World's Peatlands – Evidence for Action toward the Conservation, Restoration, and Sustainable Management of Peatlands* (United Nations Environment Programme, 2022), doi:10.59117/20.500.11822/41222; Dunn and Freeman, p. 289.

43 Kemen G. Austin and others, 'Mismatch between Global Importance of Peatlands and the Extent of Their Protection', *Conservation Letters*, 18.1 (2025), e13080, doi:10.1111/conl.13080.

44 Kelly Ribeiro and others, 'Tropical Peatlands and Their Contribution to the Global Carbon Cycle and Climate Change', *Global Change Biology*, 27.3 (2021), pp. 489–505, doi:10.1111/gcb.15408.

45 Ribeiro and others, p. 499 (emphasis added).

46 Erik Lilleskov and others, 'Is Indonesian Peatland Loss a Cautionary Tale for Peru? A Two-Country Comparison of the Magnitude and Causes of Tropical Peatland Degradation', *Mitigation and Adaptation Strategies for Global Change*, 24.4 (2019), pp. 591–623, doi:10.1007/s11027-018-9790-3; Rosa Maria Dias and others, 'Amazon Peatlands in Peril: A Warning for Global Warming', *Oecologia Australis*, 25.01 (2021), pp. 1–6, doi:10.4257/oeco.2021.2501.01.

47 Damian Carrington, 'World's Largely Unprotected Peatlands Are Ticking "Carbon Bomb", Warns Study', *The Guardian*, 13 February 2025 <https://www.theguardian.com/environment/2025/feb/13/worlds-largely-unprotected-peatlands-are-ticking-carbon-bomb-warns-study> [accessed 1 October 2025].

48 K. H. Roucoux and others, 'Threats to Intact Tropical Peatlands and Opportunities for Their Conservation', *Conservation Biology*, 31.10 (2017), doi:10.1111/cobi.12925.

49 Lilleskov and others, 'Is Indonesian Peatland', p. 593.

50 Frederick C. Draper and others, 'The Distribution and Amount of Carbon in the Largest Peatland Complex in Amazonia', *Environmental Research Letters*, 9.12 (2014), 124017, doi:10.1088/1748-9326/9/12/124017.

51 Albert Ahenkan and Emmanuel Boon, 'Non-Timber Forest Products (NTFPs): Clearing the Confusion in Semantics', *Journal of Human Ecology*, 33.1 (2011), p. 1, doi:10.1080/09709274.2011.11906342.

52 Récord, 'Grupo AJE sube en ranking Merco Empresas y Líderes 2024', 25 November 2024 <https://www.record.com.mx/contra/grupo-aje-sube-en-ranking-merco-empresas-y-lideres-2024> [accessed 1 October 2025]; Romina Landeo and others, 'Natural Revolution: Sharing Value with the First Owners: The Case of AJE in Peru', *360: Revista de Ciencias de La Gestión*, ahead of print, 28 October 2023, doi:10.18800/360gestion.202308.012.

53 Roucoux and others, 'Threats to Intact Tropical Peatlands'.

54 Roucoux and others.

55 Bio Amayu, 'Join the AMARUMAYU Movement with BIO AMAYU', YouTube, 4 December 2020 <https://www.youtube.com/watch?v=WURjo3Sh_Uo> [accessed 1 October 2025].

30
H*Au*te Cuisine
Considerations on Edible Gold

Anne Urbancic

When, many years ago, I was gifted with a bottle of Bols Bride's Tears Liqueur (now renamed Gold Strike), with its specks of gold floating like an expensive snow globe on the table in front of me, I was both flattered and bewildered. How would I deal with the contents? Would I ingest both liqueur and flakes? Would I spit the flecks out? Finally, I did neither. I put the bottle away at the back of a cupboard; it followed me through several changes of address until years later my family insisted on opening and tasting it because it had evaporated somewhat. What did I find? Drinking the liqueur only increased in me a curious interest in learning more about edible gold (also known as alimentary gold or edible gold leaf). I soon discovered that serious consideration of this topic was decidedly limited. Among the few sources I did find is an excellent article by Carole Bloom entitled 'Decorating Pastries and Confections with Gold' published in the *1993 Proceedings of the Oxford Food Symposium*. I use this as a point of departure for my own brief study here. In her work, Bloom describes the typology of edible gold, the tools and techniques required by confectioners and bakers to manipulate it, and its use on pastry and confections. She also gives a nod to silver, reviewing it similarly as an edible product. While I cannot overlook referring to Bloom's seminal work in my own study, I will add to her information with a discussion about the often-invented history of edible gold in Western Europe and describe its current use. Edible gold was known in Egypt, Greece, and Rome from ancient times, but I will focus on the western hemisphere from the late Middle Ages until today and leave the traditions and customs for edible gold in other areas of the world for future scholars.

Gold, *Au* on the periodic table of elements, has always been a highly prized, highly coveted commodity, the stuff of luxury, of wealth and ostentation, the stuff that seemed reserved for the most haute of cuisines. In the past, edible gold has been the stereotypical purview only of the wealthy class; today, however, it has a wider presence even in popular television shows like the *Great British Bake Off*, where two judges – but as far as I am aware, no contestants – have published recipes for challenging cakes decorated with gold leaf. This indicates to me that edible gold bears semiotic, sociological, and political markers that change its associations with food.

First, some basic information about edible gold. While gold was and is very well known for its traditional use by jewellers, artists, and artisans over the centuries and for its importance as proof of fiscal resiliency, it has also found not-so-traditional service in the kitchen beyond table and service ware. For culinary use, it is generally available in booklets of very fine sheets measuring about 8 cm × 8 cm or as ribbon that comes in various widths. It should be between 22 and 24 carats pure gold. It is also available as a fine powder and in flakes.[1] The artisans at Italy's Giusto Manetti Battiloro Gold Chef company, masters in beaten gold in Florence since 1600, also offer it as an ingestible spray and edible pills, all in 23 karat edible grade (free of dangerous copper additives).[2] Edible gold today is 90 percent pure gold and 10 percent silver to give it a heftier structure for better workability.[3] While alimentary gold leaf is not regulated in North America, when used as an additive for food, it is assigned as E-175 by the worldwide standards of the Codex Alimentarius, valid in the EU, the UK, Canada, and the US.[4] Prices, of course, depend on the market value at the time of purchase. There are some expenses for tools to manipulate the edible gold in sheets and ribbons: tweezers, tamping quills, and gilders' knives, all available, as Bloom tells us, wherever art supplies are sold, although such stores do not sell alimentary-grade gold. Close attention must be paid to applying the gold in an environment of complete stillness. Gold ribbon must be applied carefully through a dispenser, as touching it will break the ribbon. For excellent adherence, the surface where the gold will be applied must be slightly greasy or buttery.[5] You cannot touch it with your hands because it will adhere; even breathing directly on it will make it wrinkle.[6]

Bloom does not make mention of this, but today, manufacture of edible gold must follow strict regulations, including GMP (Good Manufacturing Practice), sanitation rules, and quality standards, which include absence of allergens and indications of production traceability.[7]

The history of edible gold is not always fully or reliably recorded. Several sources I consulted for my brief overview below describe that it was used in both China and Egypt by the second millennium BCE.[8] In the latter case, gold powder was often baked into the bread of the wealthy. Sources also point to recipes for fish enveloped in gold described in the work of Marcus Gabius Apicius. However, checking the manuscript of the *De re coquinaria*, I did not find such a recipe, although I did find two sauces for a prized fish called golden dory.[9] It appears that in Roman times, gold ware offered an ostentatious presentation for various dishes but may not have been used as an edible. It was, however, a horrific beverage for those who were punished by having molten gold poured into their throat.[10]

There is no further information about edible gold until mentions appear in fairly current sources stating that the practice was noted in the twelfth century in *Il Milione*, the book describing the travels of Marco Polo.[11] Again, however, in my reading of the manuscript about Marco Polo's voyages, I could not find any reference to comestible gold, although there were many instances of gold plates, platters, cutlery, and decorations, as well as clothing and jewellery.

That gold was ingestible and not harmful to the body was certainly known by the fourteenth century. In the *Historia di Milano*, Bernardino Corio recounts that in 1368, when 13-year-old Violante, daughter of the lord of Pavia and Milano, Galeazzo II Visconti, married Lionel, Duke of Clarence, son of Edward III of England, her father offered an extravagant banquet with eighteen courses. Corio describes how the meat and fish courses, including boar, hare, heron, partridge, peacock, quail, trout, and veal, were all gilded.[12] Five months later, Lionel was dead of suspected food poisoning, although not, it seems, from having ingested gold leaf.[13]

Later, in the fifteenth century, the recipe manuscript *An Ordinance of Pottage* offers a more modest suggestion for using edible gold: a pie made with dried fruits decorated with walnut halves encased in gold foil.[14] Barbara Ketcham Wheaton describes how, in this time period, gold leaf was used mostly on pastry for visual effect, including on the elaborate *tourte parmerienne*, which was shaped like a castle with turrets made from chicken drumsticks, each bearing coats of arms in edible gold.[15] She also describes stuffed roast chicken garnished with gold leaf–wrapped meatballs found among the recipes of *Le Viandier* also produced around the same time.[16] There, roast peacock was presented with gilded beak and feet.[17] Moreover, gold and other precious stones were used as restoratives in broths; the stones and gold nuggets were encased in cloth packets while they simmered. Since the packet contents were not ingested, they could be recooked many times. *Du fait de cuisine*, the 1420 cookbook of Maistre Chiquart, has one such recipe.[18]

Wheaton's assertion that gold foil was used predominantly to make an unforgettable impression on banquet guests is also confirmed later by Italian cooks of the fifteenth century. An anonymous Neapolitan cook, perhaps the renowned Maestro Martino of Como himself in an early version of what would later become the *Libro de arte coquinaria*, offers sweets called *Zonchate* enveloped in gold leaf.[19] Like *Le Viandier*, his manuscript also has a recipe for a spectacular presentation of a peacock covered in gold leaf.[20]

Following in Maestro Martino's footsteps is Bartolomeo Sacchi, best known as Il Platina. Il Platina was acquainted with Maestro Martino, and in a tribute that would today constitute plagiarism, he incorporated the latter's work into his own *De honesta voluptate et valetudine* (*On Right Pleasure and Health*, 1474). Written in Latin, this work is generally considered to be the first cookbook to appear in print. Il Platina repeats Maestro Martino's treatment for peacock.[21] He also has pastilles of sugared pine nuts wrapped in gold leaf to begin or end Lenten meals of 'the nobles and rich'.[22] Ken Albala has pointed out that, in this period, powdered gold was used on foods described in the fictional *Hypnerotomachia Poliphili*, perhaps by Francesco Colonna (1499).[23] Half a century later, in the middle of the sixteenth century, Cristoforo da Messisbugo, in his posthumous book on banquet preparation (1549), mentions nut pastries that are '*dorato*'; the term is ambiguous, however, as it could mean 'covered with gold leaf' or 'baked until golden brown'.[24] Bartolomeo Scappi, arguably the most famous cook of the sixteenth century, includes two recipes requiring gold in his encyclopaedic *Opera dell'arte del cucinare* (1570). For a soft cheese, raisin, ginger, and cinnamon torte, he advises presenting the dish with gold or silver foil on its top. In his section on preparations for invalids, he outlines a distillation made from capon interspersed either with peeled lemon slices or with sorrel and ground gold.[25] Similarly, the Spanish court chef Francisco Martínez Montiño reports that chicken broth intended for patients could be enhanced by the addition of gold, or pumpkin or beans, according to the request of the attending doctor.[26]

It seems that the history of alimentary gold becomes attenuated with the waning of the Renaissance. Since sumptuary laws which prohibited extravagance in the social interactions of the ultrawealthy had already been in effect, and often ignored, across Europe since 1294, when they were first decreed by Philip IV of France, their presence seems not to be a factor in creating the paucity of information about how edible gold was used after the sixteenth century.[27] Sumptuary laws were often intended to keep the classes in their place so that the conspicuous and ostentatious display of material opulence was the purview only of the wealthy. Thus gold was still in high demand to accentuate and show off the elaborate tables of the French court in the seventeenth century, but I have found no mention of it as a comestible.[28] The extensive and detailed correspondence of Mme de Graffigny, for example, mentions the food served, but there is no reference to dishes that included edible gold.[29] Later still, in the nineteenth century, Lady Judith Montefiore included a cake-like dessert, *Bola d'amor*, 'ornamented with myrtle and gold and silver leaf' in her *Practical Information in Jewish and Modern Cookery* (1846).[30] Alexis Soyer used gold flakes only to enhance jellies, as did Mrs Beeton, who noted that '[a] little gold leaf in tiny flakes makes [orange] jellies look very pretty'.[31] Victorian British chef Charles Elmé Francatelli did not appear to use it, nor could I find it in the book of his French mentor, the renowned Parisian pâtissier Marie-Antoine Carême.[32] Similarly, Italian cookbook writer Pellegrino Artusi did not mention edible gold in his groundbreaking work of 1891.[33] Early in the twentieth century, French chef Auguste Escoffier offered two recipes using gold flakes, both named in honour of the Rothschilds.[34] In the latter half of the twentieth century, Julia Child is silent on the matter, as far as I can discover.[35]

I found it fascinating to speculate on why, after the Renaissance, the use of edible gold was mostly relegated to medical distillations. Perhaps the Columbian effect of foods which began to arrive from the newly discovered Americas during the Renaissance played a role here; now there were other comestibles from the Americas, and elsewhere, that became more prestigious and exotic and more desired and recognized as a marker of high social status: cane sugar, coffee, and chocolate come to mind. Even for medicinal purposes, the rarer and therefore more prestigious and expensive ambergris (a whale duct secretion) caught the attention and recommendations of pharmacists of the wealthy, whereas previously decoctions containing gold had been used widely as restoratives.[36]

Furthermore, with the industrial age and the rise of the various levels of the middle class, tastes seem to have changed, as did table aesthetics and etiquette; alimentary gold was used in smaller quantities, as flecks or powder, rather than sheets. Food scholar Daniel Bender suggests that in the late nineteenth century, and in the twentieth, the knowledge of and ability to procure and offer fine and expensive wines became the fundamental marker of class and discerning taste.[37] I would also posit that the industrialized production of food with its concomitant focus on speed and efficiency of preparation contributed to the shift of attention away from edible gold for making an impression.

Generally, by the twentieth century, alimentary gold was perhaps relegated mostly to liqueurs. Taking his cue from other curative drinks, Ambrosien Vermellon, a Dutchman residing in Danzig (today Gdansk), had invented *Goldwasser* at the end of the sixteenth century. The combination of gold flakes floating in a distillation of roots and herbs soon caught the eye and taste of the rich and powerful, among them Louis XIV, Catherine the Great, and Peter the Great. The drink continues to be made and sold today, a consistent use of alimentary gold now over 400 years old.[38]

Edible gold on food has not disappeared completely, however. To the contrary, it seems to have had a prominent resurgence which began in Milan in 1981, when Chef Gualtiero Marchesi was asked to impress the uncle of one of his regular clients, a producer of gold leaf. His dish, inspired by the iconic *risotto alla milanese*, was a last-minute brain flash: a single square sheet of gold leaf placed on a plate of traditional risotto flavoured with saffron, a spice that is already expensive in and of itself. Marchesi forgot about the dish until 1985, when photographer Riccardo Marcialis wanted to do an article on the colour yellow.[39] Here was the stimulus for our most recent use of gold leaf. Today you will easily find gold, especially on baked goods, as ribbon, sprinkles, flecks, or wrapping. You will also find it in many drinks, of course, not only in the traditional *Goldwasser* but also on meat, vegetable dishes, sushi, and so on.[40]

Unlike cooks and gourmands of the past, we recognize today that this element is biologically inert; while ingestible and permitted by kosher and halal traditions (although since 2023, the Fatwa Center of the Muslim Board of Uzbekistan has forbidden it for its arrogant ostentation[41]), edible gold has no taste, no aroma, and no nutritional benefit for the body. Once on the tongue, it slides quickly down your throat, leaving no trace in your mouth.

Is it used, then, only because 'this shimmering, metallic sheet has the power to transform plates of food into works of art'?[42]

If so, it is not unproblematic. There is a thorough and concerning body of work describing the negative impacts of gold mining on people, communities, and the environment. Like blood diamonds, gold mining, especially on a small scale, has been described as 'dirty' because of its harmful environmental practices.[43] Admittedly, most of the gold extracted from the mines currently is not intended for food. Some companies that make and sell edible gold acknowledge this with a nod to the fact that they provide guarantees for the manner in which their alimentary gold has been sourced and worked.[44] Undeniably, however, as in the past, edible gold still flouts alimentary fairness and justice and loudly insists on flaunting its overtly exclusionary aspects; while no longer offered only at royal and aristocratic tables, it boldly proclaims its partiality for wealthy clients as we see in the reactions to the creations of Chef Salt Bae of Türkiye, such as the 2000 USD gold leaf pizza and the tomahawk steak, about a kilogram of meat enveloped in gold leaf with its unfortunate allusion to the tools used by North American Indigenous peoples, or the bread roll wrapped in edible gold leaf by Chef Carlo Cracco.[45] Another paradoxical example comes to us from the multi-Michelin winner Massimo Bottura, whose website promotes food equality and inclusion for all and cautions against food waste but offers a recipe for a dessert crisp made of leftover bread and gold powder.[46] And of course, there is the gold cappuccino from Abu Dhabi.[47]

How has this come about? Undeniably, a fundamental impetus is social media, for both chefs and consumers. The idea of the celebrity chef has grown with the proliferation of websites and apps that encourage chefs like those mentioned above to work with exotic ingredients and presentations. Theirs is the spectacle of excess as theorized by Roland Barthes, Jean Baudrillard, Guy Debord, and others.[48] However, even amateur cooks can participate in this phenomenon through numerous DIY websites describing how to produce pastries and cakes using edible gold. Social media have allowed and 'grown' increasing numbers of viewers of spectacle, a growth that to me also seems supported by the fact that we are attracted to 'little treat culture', sometimes called the 'lipstick effect'. We practice dream scrolling on the internet, which allows us to view, consider, and perhaps order more expensive small items we might not usually purchase.[49] Studies have shown that 'little treat culture', where 'people attempt to boost their mood with small treats in difficult times', decreases stress hormones like cortisol and adrenaline and boosts endorphins.[50] 'Being short of money is psychologically daunting for people and the way to make yourself feel better, even if it's ever so little, is to purchase something that you think will cheer you up', according to Dr Jansson-Boyd, an associate professor in consumer psychology at Anglia Ruskin University.[51] A cursory internet survey of products using alimentary gold shows how well these products fit within such a category; a single piece of cake, some cookies, a few luxurious chocolates, beverages, a celebrity chef bread roll or sweet bread crisps, or cappuccino embellished with gold leaf are treats that ordinary consumers can allow themselves occasionally even in times of economic and social insecurity. They may not be able to afford a house, ever, but the 25 USD gold cappuccino is within their means.

Edible gold has been a marker of exclusionary sociopolitics for millennia. In the past, it was always the exclusive purview of the rich. Today, through 'little treat culture', it is a readily available treat for those who may feel that their well-being, confidence, and self-esteem are threatened by large political and economic forces beyond their control. Edible gold is no longer simply an inert mineral used to show off wealth; to me it seems that, for those who indulge, it has evolved to become a marker of self-well-being, and although without any nutritional or health value, aroma, or taste, it has also returned to signal gustatory sophistication at a relatively affordable price.

About the Author

Anne Urbancic, the Mary Rowell Jackman Professor in the Humanities at Victoria College, University of Toronto, specializes in nineteenth- and twentieth-century Italian literature, which, to her delight, has led her to the fascinating cookbooks and foodways of Italy in that period.

Notes

1 Carole Bloom, 'Decorating Cakes and Pastries', *Look and Feel: Studies in Texture, Appearance and Incidental Characteristics of Food. Proceedings of the Oxford Symposium on Food and Cookery 1993*, ed. by Harlan Walker (Prospect Books, 1994), pp. 36–40 (p. 36).

2 'Giusto Manetti Battiloro, dal 1600 produttore di foglia d'oro a Firenze', n.d. <https://www.manetti.com/> [accessed 3 March 2025].

3 'What Is Gold Leaf and Why Can We Eat It?', n.d. <https://www.americastestkitchen.com/articles/7774-all-about-edible-gold-leaf> [accessed 3 March 2025].

4 'What Is Gold Leaf and Why Can We Eat It?' and Government of Canada, 'Food Colours', n.d. <https://inspection.canada.ca/en/food-labels/labelling/industry/food-additives/food-colours> [accessed 18 March 2025].

5 Bloom, 'Decorating', pp. 37–38. Edible gold today is beaten by machine and sold loose or layered between waxy transfer sheets, which make it more easily manipulated. See also 'What Is Gold Leaf and Why Can We Eat It?'

6 'Sweet Art', n.d. <https://www.sweetartcreations.ca/supply/24k-gold-leafloose-sheets/5674> [accessed 18 March 2025].

7 Gold Chef, 'Certificazioni', n.d. <https://www.goldchef.shop/certificazioni/> [accessed 20 March 2025].

8 Sandra Ianni, 'L'oro a tavola', *Epulae News. Food and Wine Web Magazine*, 1 November 2023 <https://epulaenews.it/loro-a-tavola/> [accessed 20 March 2025].

9 'L'oro a tavola'. For Apicius, see Recipes 461 and 462, in Apicius, *Cookery and Dining in Imperial Rome* (Project Gutenberg, 19 August 2009) <https://www.gutenberg.org/files/29728/29728-h/29728-h.htm> [accessed 3 March 2025]. Who Apicius was remains unresolved. There may have been several gourmands by this pseudonym in between the first and fourth centuries CE. *De re coquinaria* is usually attributed to the earliest of these; the manuscript consulted here was likely expanded in the fourth century. See also Robert Marchese, *Storia della cucina medievale* (Youcanprint, n.d.) <https://www.everand.com/read/286732747/Storia-della-cucina-La-cucina-medievale> [accessed 3 March 2025].

10 Rachel Nuwer, 'Here's What Actually Happens during an Execution by Molten Gold', *Smithsonian Magazine*, 10 June 2014 <https://www.smithsonianmag.com/smart-news/heres-what-happened-people-who-were-executed-having-molten-gold-poured-down-their-throat-180951695/?no-ist> [accessed 3 March 2025].

11 See Gold Chef, 'The History of Edible Gold' <https://www.goldchef.shop/en/magazine-the-history-of-edible-gold/> [accessed 20 March 2025]. I examined the edition coordinated by Eugenio Burgio, Marina Buzzoni, and Antonella Ghersetti, *I Viaggi di Messer Marco Polo gentiluomo veneziano*, ed. by Samuela Simion and Eugenio Burgio. This is a version of *Il Milione* ed. by Giovanni Battista Ramusio (1485–1557) for his own *Navigationi et viaggi* (Venezia: Giunti, 1559, reprinted from the first edition of *Il Milione*, n.d.) <https://risorse-esterne.edizionicafoscari.it/main/index.html> [accessed 25 February 2025].

12 Bernardino Corio, *L'Historia di Milano* (John Adams Library, Boston Public Library), pp. 239–40 <https://archive.org/details/lhistoriadimilanoocori/page/n543/mode/2up> [accessed 18 March 2025].

13 Mortimer History Society, 'Lionel of Antwerp', 29 November 2016 <https://themortimersblog.wordpress.com/2016/11/29/lionel-of-antwerp/> [accessed 11 October 2025].

14 *An Ordinance of Pottage. An Edition of the Fifteenth Century Culinary Recipes in Yale University. Ms Beinecke 163*, ed. by Constance B. Hieatt (Prospect Books, 1988), p. 90. The MS is at Yale University's Beineke Library.

15 Barbara Ketcham Wheaton, *Savoring the Past: The French Kitchen and Table from 1300 to 1789* (University of Pennsylvania Press, 1983), p. 15 <https://archive.org/details/unsetooooounse_d5z8/page/16/mode/2up> [accessed 10 March 2025].

16 Wheaton, p. 21. Recipes for 'parma tortes' are also described in Maistre Chiquart's cookbook. See D. Eleanor Scully and Terence Scully, *Early French Cookery. Sources, History, Original Recipes and Modern Adaptations* (University of Michigan Press, 2002), p. 177.

17 Wheaton, *Savoring the Past*, p. 13.

18 Scully and Scully, *Early French Cookery*, p. 287.

19 Terence Scully, *Cuoco napoletano: The Neapolitan Recipe Collection* (Pierpont Morgan Library, MS Bühler 19), a critical edition and English translation (University of Michigan Press, 2000), p. 19 <https://archive.org/details/neapolitanrecipeootere/page/100/mode/2up> [accessed 10 March 2025]; since this cook offers recipes based on the tastes of the Spanish rulers in Naples at the time (Scully, p. 105), these *Zonchate* may well be versions of what will become shell-shaped concha, famous in Mexico today: see 'Concha', Wikipedia <https://en.wikipedia.org/wiki/Concha> [accessed 10 March 2025].

20 Scully, *Cuoco napoletano*, p. 187: 'To make it breathe fire through its mouth, get a little camphor with a little fine cottonwool around it and put this into the peacock's beak and soak it with a little aquavita or else with a little fumey old wine that is volatile; when you want to serve it, set fire to the cotton-wool: in this way it will breathe fire for a long time. To make it more magnificent you can cover the peacock with gold leaf and then cover it with its skin'.

21 Il Platina (Bartolomeo Sacchi), *On Right Pleasure and Good Health: A Critical Edition and Translation of* De Honesta Voluptate et Valetudine *by Mary Ella Milham* (Medieval and Renaissance Texts and Studies, 1988), Book VI, Recipe 14, p. 277.

22 Platina, *On Right Pleasure*, Book III, Recipe 4, p. 177.

23 Ken Albala, 'Poliphilo's Dream of Divine Feeding', *CEA Critic*, 69.1/2 (Fall 2006–Winter 2007), pp. 77–85 <https://www.jstor.org/stable/44377637> [accessed 15 March 2025].

24 Cristoforo da Messisbugo, *Banchetti, composizioni di vivande e apparecchio generale* (Giovanni de Buglhat et Antonio Hucher Compagni, 1549) <https://archive.org/details/bub_gb_9yA8AAAAcAAJ> [accessed 15 March 2025]. Messisbugo lists 'marzapane in parte dorato' and 'pignocato grande dorato', without recipes.

25 Bartolomeo Scappi, *Opera di M. Bartolomeo Scappi, cuoco secreto di Papa Paolo V* (Michele Tramezzino, 1570), Book V, Recipe 81 p. 357 and Book VI, Recipe 21, p. 395, respectively <https://archive.org/details/operavenetiascapooscap/mode/2up> [accessed 15 March 2025].

26 Francisco Martínez Montiño, *Arte de cocina, pastelería, vizcochería, y conservería* (Juan Francisco Piferrer, 1750[?]), p. 354 <https://archive.org/details/gri_33125016924801/page/354/mode/2up?q=oro> [accessed 20 March 2025].

27 Editors of the Encyclopaedia Britannica, 'Sumptuary Law', n.d. <https://www.britannica.com/topic/sumptuary-law> [accessed 18 March 2025].

28 Carolin C. Young, *Apples of Gold in Settings of Silver* (Simon & Shuster, 2002).

29 'La correspondance de Mme de Graffigny', Voltaire Foundation, 23 June 2022 <https://graffigny.artsci.utoronto.ca/> [accessed 10 March 2025].

30 By a Lady [Lady Judith Cohen Montefiore], '*Bola d'amor*', in *Practical Information in Jewish and Modern Cookery* (Boone, 1846) <https://archive.org/details/b22021371/page/n3/mode/2up> [accessed 12 October 2025]. Hers is considered the first cookbook of Jewish recipes in the English language, and the inclusion of the *Bola d'amor*, using its Spanish name, suggests a dish of a much longer tradition, preceding the arrival of the recipe in England. I am grateful to Dr Susan Weingarten for bringing this book to my attention.

31 Alexis Soyer, 'Recipe 742', in *The Modern Housewife* (Appleton, 1850) <https://www.gutenberg.org/files/41899/41899-h/41899-h.htm> [accessed 20 March 2025]; Isabella Beeton, 'Recipe 626', in *Mrs Beeton's Cookery Book. New and Enlarged Edition* (Ward, Lock & Co., 1898), p. 179 <https://www.google.ca/books/edition/Mrs_Beeton_s_Cookery_Book/vsBQAQAAIAAJ?hl=en&gbpv=1&pg=PR3&printsec=frontcover> [accessed 22 March 2025].

32 Charles Elmé Francatelli, *The Cook's Guide and Housekeeper's and Butler's Assistant* (Richard Bentley, 1861) <https://archive.org/details/cooksguidehousekoofran> [accessed 22 March 2025]; Marie Antonin Carême, *French Cookery: Comprising* L'art de la cuisine française, Le pâtissier royal, Le cuisinier parisien *by A. M. Carême*, trans. by W. Hall (John Murray, 1836) <https://wellcomecollection.org/works/vzqkcs65/items> [accessed 22 March 2025].

33 Pellegrino Artusi, *La Scienza In Cucina e L'Arte Di Mangiar Bene* (Landi, 1891) <https://archive.org/details/artusi-1891> [accessed 22 March 2025].

34 Auguste Escoffier, *Le Guide Culinaire* (Flammarion, 1903).

35 Julia Child, *Mastering the Art of French Cooking* (Knopf, vol. 1, 1961; vol. 2, 1970).

36 Ashley Lynn Buchanan, 'The Politics of Medicine at the Late Medici Court: The Recipe Collection of Anna Maria Luisa de' Medici (1667–1743)' (unpublished dissertation, University of South Florida, 2018), pp. 71–75. For medical recipes made with ambergris, Buchanan cites the collected recipes of both Anna Maria Luisa de' Medici (in Archivio di Stato di Firenze, Miscellanea Medicea 1, ins. 2, fols. 1–218) and Pierre Pomet (*A Complete History of Drugs. Written in French by Monsieur Pomet, Chief Druggist to the Late French King Lewis 14* (BNCF, 1748), pp. 437–39) <https://digitalcommons.usf.edu/cgi/viewcontent.cgi?article=9303&context=etd> [accessed 10 March 2025].

37 Daniel Bender, '"À Bas Le Sommelier": Stewards, Drinkers, and the Jokes They Told', Lecture for Victoria College Fellows' Colloquium, Victoria College, University of Toronto, 6 March 2025.

38 James Bass Relish, 'Ten Great Drinks of Europe', *New European*, 11 July 2017 <https://web.archive.org/web/20180210062355/http://www.theneweuropean.co.uk/culture/ten-great-drinks-of-europe-1-5101522> [accessed 22 March 2025].

39 Marco Bolasco, 'Gualtiero Marchesi e il segreto del piatto icona Riso, oro e zafferano', *Cucina italiana*, 10 January 2025 <https://www.lacucinaitaliana.it/article/gualtiero-marchesi-riso-oro-zafferano-ricetta/> [accessed 22 March 2025]; see also Gabriele Zanatta, 'Riso oro e zafferano, 40 Years of Magic', *Identità golose*, 17 December 2021 <https://www.identitagolose.com/sito/en/44/29708/dallitalia/riso-oro-e-zafferano-40-years-of-magic.html> [accessed 22 March 2025].

40 'Why Do People Eat Gold Today', *Global Bullion Suppliers*, 13 February 2019 <https://globalbullionsuppliers.com/blogs/blog/why-do-people-eat-gold> [accessed 20 March 2025].

41 'Gold and Silver Cannot Be Consumed as Food-Fatwa Center', *Aktualno.UZ*, 18 September 2023 <https://aktualno.uz/ru/a/10600-zoloto-i-serebro-potreblyat-v-pishhu-nelzya-centr-fetvy> [accessed 22 March 2025].

42 See 'What Is Gold Leaf and Why Can We Eat It?'

43 'No Dirty Gold', *Earthworks*, 23 June 2013 <https://earthworks.org/blog/no-dirty-gold-retailer-profiles/> [accessed 20 March 2025].

44 'Edible Gold Powder', Gold Chef, 2026 <https://www.goldchef.shop/en/gold-powder/> [accessed 20 March 2026].

45 '7 Extremely Expensive Dishes that Have Edible Gold in Them', *GQ*, 31 March 2024 <https://www.gqindia.com/content/7-extremely-expensive-dishes-that-have-edible-gold-in-them> [accessed 23 March 2025]; Vincenzo Pagano, 'Salt Bae, *la bistecca d'oro a* 650 €', 6 May 2023 <https://www.scattidigusto.it/salt-bae-e-la-bistecca-in-foglia-doro-a-650-euro> [accessed 23 March 2025]; Sebastiano Spinelli, 'Dopo il tagliere sul Frecciarossa, tocca al "pane d'oro"', 15 June 2023 <https://www.step1.it/dopo-il-tagliere-sul-frecciarossa-tocca-al-pane-doro-non-ce-pace-per-carlo-cracco-scoppia-la-polemica/> [accessed 23 March 2025].

46 'Bread, Time, and the Power of Transformation', *Food for Soul*, n.d. <https://www.foodforsoul.it/about-us/news-stories/news/bread-time-and-the-power-of-transformation/> [accessed 24 March 2025].

47 'De l'or alimentaire dans vos assiettes!', *Or en cash*, 11 August 2021 <https://www.orencash.fr/blog/or-alimentaire/> [accessed 24 March 2025].

48 Roland Barthes, *Mythologies* (Hill & Wang, 1972); Jean Baudrillard, *Simulacra and Simulation* (University of Michigan Press, 1994 [1981]); Guy Debord, *The Society of the Spectacle*, trans. by Fredy Perlman and Jon Supak (Black & Red, 1970; rev. ed. 1977).

49 SWNS, 'Americans Spend 2 ½ Hours "Dreamscrolling" Each Day', *New York Post*, 20 May 2024 <https://nypost.com/2024/05/20/us-news/average-american-spends-nearly-2-and-a-half-hours-dreamscrolling-each-day-poll/> [accessed 25 March 2025].

50 Kokho Jason Sit, 'Treat Culture: Why Indulging in Small, Affordable Pleasures Can Help You Cope with Tough Times', *The Conversation*, 10 July 2023 <https://theconversation.com/treat-culture-why-indulging-in-small-affordable-pleasures-can-help-you-cope-with-tough-times-207957> [accessed 25 March 2025].

51 Nicola Bryan, 'Lipstick Index: Do Shoppers Turn to Low-Cost Luxuries in a Recession?', *BBC.com*, 1 October 2022 <https://www.bbc.com/news/uk-wales-63047913> [accessed 25 March 2025].

31
The Creation of Adam
From the Elements to the Edible

Susan Weingarten

> Nature, that framed us of four elements
> Warring within our breasts for regiment,
> Doth teach us all to have aspiring minds.
>
> —Christopher Marlowe, *Tamburlaine* (*c.* 1588)

In his play *Tamburlaine*, Marlowe reflects the commonly held ancient concept of the four elements: earth, water, air, and fire. Everything in the world is made up of them, but the proportions differ, resulting in people looking different and behaving differently. Only in gold are the proportions exactly balanced so that this substance never changes. This concept descends from common widely held ideas in the ancient and not-so-ancient world. In this paper I look at some late antique Jewish concepts of these elements and how they morph into the edible.

Accounts of Creation

I begin with the most elemental of all scenes, the Biblical account of creation in the book of Genesis. In fact, there are two accounts of creation in chapters 1 and 2 of Genesis, and they do not agree on all points. One of the most common characteristics of Jewish literature through the ages is the widespread presence of *maḥloqet*, or disagreement: it is usual to record even the minority opinion of the losing side in the argument. It is thus interesting to see disagreement right from the beginning, in the account of creation itself.

The first chapter of Genesis relates how God created heaven and earth and all their components in six days. We are told what is created day by day – the first day begins with the creation of heaven and earth, by the third day the earth sprouts plants, and we end on the sixth day with the creation of humankind, in God's own image, specified as male and female. On the seventh day, God rests. So far so good.

But then in Chapter 2 of Genesis we get another, rather different account of creation. Here we are told that there were no plants sprouting, because there was no rain, but then a mist went up from the earth and watered the ground. From this wet dust, God formed *adam*, man or the human, and breathed life into him, and man became a 'living soul'. God then put Adam into Eden and made woman out of one of his ribs.[1]

Creation of *Adam/Adamah*

Thus in Genesis 2, the human (*adam* in Hebrew) is formed by God from the dust of the earth (*adamah* in Hebrew), together with water. The verb used here (*y-tz-r*, to form) is one often used to describe the work of a potter who creates his pots with clay. Later in the Hebrew Bible, the prophets Isaiah and Jeremiah talk of the relationship between a potter and his clay as an image for God and His human creation:

Isaiah writes:

> Yet you, Lord, are our father.
> We are the clay and you are our potter (*yotzrenu*);
> We are all the work of your hand.[2]

The prophet Jeremiah uses the image of clay in the hands of a potter (*yotzer*) as the people of Israel in God's hands. The wet clay may be spoiled, but the potter can choose to rebuild on his wheel:

> Then the word of the Lord came to me saying: Can I not do with you, Israel, as this potter does? See, as the clay in the potter's hand, so are you in my hand, Israel.[3]

This image of us humans as clay in the hands of the Divine potter still forms a very moving part of the sung liturgy of the Day of Atonement.[4]

Returning to the Genesis account, Adam is formed by God from earth and water and brought to life by God's breathing the breath of life into his nostrils – he is made up of three of the four elements. Only fire is missing from the Genesis accounts. It belongs to heaven.[5] Fire, indeed, according to rabbinic ideas in the late antique *midrashim*, did not exist before the expulsion from Eden.[6]

Adam: Origins of the Name and the Elements

I noted the likeness between the Hebrew words *adam* (man or the human) and *adamah* (earth). In *Midrash* Pirqei deRabbi Eliezer (PRE), we see rabbis disagreeing on the origin of the name *Adam*.[7] R Yehudah says the first human was called *Adam* after *adamah*, the earth from which he was taken. R Joshua b Karḥa, however, says his name was derived from (flesh and) blood, which is *dam* in Hebrew. We shall see a number of further exegeses which support these two different possibilities. The *midrash* does not decide.

The Genesis text continues:

> But there went up a mist from the land and watered the whole face of the earth, *adamah*. Then the Lord God formed the human, *adam*, of the dust of the earth, *adamah*, and breathed into his nostrils the breath of life and the human, *adam*, became a living soul.[8]

Adam, man or the human, is thus made by God of clay and animated by God's breath. There are also other Jewish traditions related to his name and the elements that he is made of. The so-called Sibylline oracles, from the second century BCE passed on in Greek, show God creating using the *logos*, the word. Here the created *Adam* is said to be a *tetragrammaton*, a four-lettered word. Adam in Greek does indeed have four letters, Αδαμ. In Hebrew, אדם *Adam* only has three letters, although אדמה *adamah*, earth, does have four letters in Hebrew. The Greek term *tetragrammaton* is usually used to denote God's four-lettered Hebrew name: the four-lettered name of *Adam* is thus an image of the Divine name. The first creation account in Genesis 1 told us that Adam, the human, was made in the Divine image. The oracles also note that the Greek *Adam* is an acrostic of the initials of all the directions of the compass in Greek. Later the Babylonian Talmud tells us that God took dust to make the human from all different parts of the world.[9] Midrash PRE has the dust for Adam's body coming from the four corners of the world in four different colours: red, black, white, and green/yellow.[10] This *midrash* also notes that Adam was formed in 'a pure place', later specified as the navel of the world, while the fourth–fifth century Midrash Genesis Rabbah has him deriving from 'the place of atonement' – that is, the site of the future Temple.[11]

Mesopotamian Myth: *Atrahasis*

We shall return to talmudic texts later. Now I turn to look at other ancient accounts of the creation of humankind, where we can see certain parallels. The Mesopotamian *Atrahasis* epic (early second millennium) tells that mankind was created to do the hard work the gods did not want to do.[12] Here too the first man is made of earth and water. But Mesopotamian man is animated by the slaughter of one of the gods, whose blood and flesh give him life.

Hesiod and Ovid

Greek myth, as seen in Hesiod (eighth century BCE), has Prometheus, a Titan, part god and part human, together with the goddess Athene creating man from earth and water mixed with blood of slain Titans.[13] Here again the elements of creation do not include fire: indeed, Zeus forbids it. But Prometheus brings fire to humans, together with reason, and is punished for it. Fire is necessary for civilization, for cooking and for firing pots. Later the first-century Latin poet Ovid, in his *Metamorphoses*, hedges his bets: either the creator made man with Divine seed or Prometheus used earth, which was still partly Divine because it was only recently separated from heaven, and mixed it with water to form man in the image of the gods.[14]

In all these cases, mere clay and water are not enough to create a human: some part of the Divine, be it breath or seed, blood and flesh, or Divine essence, is needed as well. And fire has an ambivalent status.

Fishbane on Myth

In writing about Biblical mythology and rabbinic mythmaking, the modern scholar Michael Fishbane has analyzed rabbinic myth as developing along two tracks. One track has thematic continuity with mythic material found in earlier Near Eastern and Biblical sources and traditions. The second track, however, is characterized by thematic innovation: new mythic material is developed.[15]

I turn now to look at what Fishbane would presumably see as new mythic themes in talmudic concepts of the creation of the first man and woman. We begin with the *golem*.

The *Golem*

In the Biblical and some of the talmudic accounts, the verbs used for the creation of man come from the work of a potter – Adam was initially made of clay. Potters shape the lump of clay referred to as a *golem*. A *golem* is a shapeless lump with future potential. This term appears as a *hapax* in the Bible: the book of Psalms uses the word *golmi*, 'my *golem*' (RV, 'my unshaped flesh') to connote the unfinished human being before God's eyes.[16] The third-century Mishnah uses the term *golem* to refer to someone who is not learned: 'Seven characteristics are in a *golem*, an uncultivated person, and seven in a learned one'.[17] Today the word *golem* is used for a pupa or chrysalis, the brown lump where caterpillar eventually morphs to butterfly.

We shall see later that the fifth–seventh century CE Babylonian Talmud, describing the stages of creation, specifies that Adam was a *golem*, a shapeless lump, before God animated him.[18] The late antique rabbis explain the *golem* in another *midrash* which deals with the creation of Adam.[19] They base their exposition on a verse from Deuteronomy 4.32 writing of the

> first days [. . .] from the day that God created Adam/a human on the land and from one end of the heavens to the other end of the heavens.

They explain this verse as meaning that when Adam was first created, he was a *golem*, a shapeless mass, which filled the whole space between heaven and earth. In this state, God showed him all of history passing like sheep before his eyes. The Palestinian Talmud also tells us that the human embryo is a shapeless *golem* before birth.[20]

There are also stories or myths of rabbis creating living beings in the Babylonian Talmud. Tractate Sanhedrin describes the rabbi called Rava creating a *gavra*, a male, whom he then sends to Rav Ze'ira:

> R. Ze'ira spoke to him, but got no answer. So he said to him: 'you are a creature of the magicians. Return to your dust.'[21]

We noted above how the same tractate, Sanhedrin, shows God forming Adam as a *golem* before He animates him.[22] He takes his dust from different parts of the world to make different parts of his body: his chest derives from Babylonia, but his head is from the Land of Israel.[23]

Later *Golems*

Moving forward, early modern rabbinic literature tells stories of *golems* made by rabbis of clay and water, animated by magic formulas such as a secret name of God placed in their mouths. The most famous of these was the *golem* of Prague, attributed to the fifteenth-century Maharal, Rabbi Judah Leib b Betzalel Loewe, by a nineteenth-century legend. This robot-like creature was made to work for its creator, as we saw in the early Mesopotamian myths about the creation of mankind. The Maharal would remove God's name every Sabbath so the *golem* could rest. One week he forgot, and the *golem* went on a rampage. He was finally defeated by forcible removal of the name of God.[24] His remains are said to be kept in the attic of the Alt-Neu synagogue in Prague to this day, but entry is forbidden.[25]

From Divine Potter to Divine Baker

Returning to late antiquity and talmudic rabbis, I have already noted that descriptions of God making Adam began in Genesis with the verb *y-tz-r* (formed), associated with a potter, *yotzer*. Later talmudic exegesis uses terms that belong to making bread and baking rather than making pots and pottery. When I make Sabbath bread and roll out the leavened dough, which changes shape before my eyes, I feel the dough is a living thing. So it is perhaps not surprising that accounts of the clay and water of Divine creation morph to dough – indeed, to human flesh. Bread formed between 50 and 80 percent of the diet in antiquity. It was the basic food, converted into flesh every time we eat. So the move from clay made of dust and water to dough made of flour and water is immediately comprehensible. The Greek writer Athenaeus in his *Deipnosophistae* metaphorized women as dough, white and desirable, moving around the kitchen – and, of course, eventually to be consumed by men.[26] Grinding flour and baking were traditionally women's tasks. Hesiod (whose *Theogony* I cited earlier) describes Pandora, the woman created by Prometheus and Athene who brings troubles to the world, as a 'woeful gift to bread-eating men'.[27]

The Twelve Stages of the Sixth Day of Creation

We have seen talmudic exegesis of some creation accounts, but most late antique Jewish exegesis of creation is contained in narrative *midrashim*. There are at least a dozen *midrashim* where rabbis analyze the process of creation. Humans, in both Biblical accounts, were created on the sixth day, and the rabbis visualize the process in stages, matching each of the twelve hours of the day. But they do not agree on which stage was which and which stage belonged to which hour.[28] Thus one of the minor tractates of the Talmud, which is midrashic in essence, Avot deRabbi Natan, version A (AdRN A), writes,

> How was Adam created? In the first hour his dust was collected. Second his shape was created. Third he was made a *golem*, a soulless lump. Fourth his parts were connected. Fifth his tubes were opened. Sixth he was given the spirit of life. Seventh he stood up on his legs. Eighth he was coupled with Eve. Ninth he was put in the Garden of Eden. Tenth he was commanded [sc. not to eat from the tree]. Eleventh he transgressed. Twelfth he was driven out.[29]

From *Golem* to Bread Dough

Adam as *golem* filled the whole space between earth and heaven.[30] After he sinned, God put His hands on him and shrunk him.[31] There are also a number of *midrashim* that tell us that God *g-v-l*, worked or kneaded, Adam's body with His two hands, just as women work or knead bread.[32] We have moved from working clay to working bread.

Adam as Dough-Offering, God as Woman

We see this change from clay to dough in action in Midrash Genesis Rabbah, commenting on the verb in Genesis *vayyitzer*, [God] formed, which we have already seen as the term used in the Bible for the work of a potter. Now in the *midrash* it also refers to dough:

> Adam, the first man, was the perfect finishing touch, *ḥallah*, the dough-offering of the world. [. . .] Rabbi Yossi b Ketzarta said: Like this woman who mixes her dough with water and lifts up her dough offering from the very centre, so at first 'a mist went up from the land and watered all the face of the *adamah*, earth,' and then 'the Lord God formed, *vayyitzer* man, *adam*, of the dust of the earth, *adamah*'.[33]

Midrash Exodus Rabbah clarifies this:

> What is the way of woman? When she wishes to separate her *ḥallah*, dough offering, she works, *g-v-l*, the flour [sc. with water] and after this she takes the dough-offering. God did the same: He worked, *g-v-l*, the world [sc. with water] and after this he took Adam [sc. as the dough offering], as it says (Genesis 2.6) 'a mist went up from the land,' and after this (Gen 2.7) 'He formed, *vayyitzer*'.[34]

Thus the rabbis tell us that in the process of creation, God behaved like a woman making bread dough: He first worked the dust of the earth with water, like a woman working flour and water till it becomes dough, then just as she then takes a portion of the dough and sets it aside for the dough-offering, so He took Adam from the earth.

What Is *Ḥallah*, the Dough-Offering?

The origins of the dough-offering are Biblical, elaborated in the Mishnah, where there is a whole tractate called *Ḥallah*, the dough-offering.[35] Both a woman baking at home and the public baker are required by the Mishnah to separate a piece of the dough they make as the priests' portion. In the days when the Temple stood in Jerusalem, this portion would actually have been given to the priests. This ritual is called 'taking *ḥallah*' (or *terumah*, a donation), and as such predates the habit of calling Sabbath bread *ḥallah*, as is the general custom today. The handful of *ḥallah* is taken from the unbaked dough, and since the verb for making an offering is the same as the verb for lifting something up, the *ḥallah* is often lifted up. This process is accompanied by a blessing. Now that there is no Temple, the *ḥallah*, the priests' portion, is burned.

So Adam as the '*ḥallah* of the world' is either a priest's portion or a burned piece of sacrificial dough.

The Dough-Offering of the World

The talmudic picture has a further take on the human as bread. Adam, we noted, is metaphorized as the 'dough-offering of the world'. The *midrashim* add that he is also the 'blood of the world' and the 'lamp/light of the world'. In the Garden of Eden, Eve is tempted by the serpent and manages to tempt Adam with the fruit of the forbidden tree and 'spoil' him. Christians call this the 'Fall' and see humankind as henceforth tainted from birth, only to be redeemed by Divine intervention. But the rabbis' attitude is different. Because woman was responsible for spoiling man, she is endangered, but she is also given the opportunity of atoning for her sins and mending the fractured world. She is thus given three special commandments aimed at her as woman. If she fails to observe these regulations, she is in danger: 'For three transgressions women die in childbirth', says Mishnah Shabbat, 'for not heeding the laws of *niddah*, the purity laws, *ḥallah*, the dough-offering, and lighting the Sabbath lamp'.

If she does observe them, however, then she can 'correct' the world and resanctify it. The list of three transgressions and their correction appears again and again in many *midrashim*.[36] We noted the shift from clay to dough. Adam, we saw, say the rabbis, is the *ḥallah*, the 'dough-offering of the world', or even the perfect, or the purest, dough-offering. By tempting him with the fruit of 'that forbidden tree whose mortal taste brought death into the world', Eve is regarded as contaminating or spoiling the dough-offering, which is why she and her female descendants become responsible for resanctifying it by setting it aside for the priests (or in the absence of the Temple, burning it).[37] Because of the threat of dying in childbirth, various apotropaic ceremonies have grown up around this regulation. It is the fashion today for groups of women to meet together before a wedding or an expected birth to remind God, as it were, that this woman has not forgotten her commandment.

Secondly, Adam is also metaphorized as the 'light/lamp of the world' (which incidentally solves the problem of the earlier absence of the element of fire). Eve by her actions has 'extinguished' the light. She and her descendants are therefore now given the commandment of lighting the Sabbath lamp or candles every week to atone for this.

Finally, by bringing mortality into the world, Eve has metaphorically shed the blood of the world. She is therefore given the laws of *niddah*, whereby she can purify herself after menstruation. Thus she atones for Adam as the dough-offering of the world, the light of the world, and the blood of the world.

Blood of the World – and of God

Most of the *midrashim* which talk of Eve's atonement are similarly worded. But there is one exception. Avot deRabbi Natan, version B (AdRN B) begins like the Mishnah just quoted with the warning that women who do not observe these regulations are in danger of dying in childbirth. It then continues like all the other midrashim, metaphorizing Adam as the blood, dough, and light of the world. But it also includes a surprising alternative version: Adam, it says, is (also) metaphorized as the blood, dough-offering, and light – of God Himself.[38] The light is explained using a quotation from Proverbs: for the soul of man/*Adam* is the light/lamp of God.[39] Eve extinguished the light (by causing humans to become mortal), so she is given the commandment of lighting the Sabbath lamp. AdRN B then continues:

> And why was the commandment about *ḥallah*, the dough-offering, given to the woman and not the man? Because Adam, the first human, was the pure *ḥallah* of the Holy One Blessed be He and she contaminated it, therefore they gave her the commandment of *ḥallah* so that she can atone for the *ḥallah* which she contaminated.

But perhaps the most surprising expression in this *midrash* is when Eve is described as shedding Adam's blood, for Adam is the 'blood of God', and she needs to atone for shedding God's blood. Menahem Kister, in his study of AdRN B, points out that the other, usual form of the *midrash* with the blood, *ḥallah*, and light 'of the world' appears much earlier in the Palestinian Talmud and Midrash Genesis Rabbah (fourth–fifth century).[40] He cites S. Schechter's suggestion that this alternative version attributing them to God, not the world, may just be a mistake in the text, a nonfunctional doubling. However, he does not appear to be totally sure of this. He calls the expression the 'blood of God' 'amazing' but does not go any further. And indeed, this expression *is* amazing, more like a Shakespearian oath than a Jewish midrash. Is it relevant here that, as we saw earlier, the name *Adam* was derived from *dam*, blood, by R Joshua b Karḥa in the Midrash PRE? This late *midrash* has two chapters and many other aspects in common with AdRN.[41] Blood is, of course, taboo in Jewish tradition.[42]

Christian Body and Blood

Here I move sideways once more to note Christian tradition. In the synoptic gospel accounts, at the Last Supper, Jesus is described as taking bread and wine and proclaiming them as his body and his blood.[43] Jesus was presented by Christian theology as the redemptive corrective to the fallen Adam, while his crucifixion was seen as the last sacrifice. He too was seen as Divine.

Fishbane on Myth Revisited

I cited Michael Fishbane above, analyzing rabbinic myth as developing along two tracks, where the first has thematic continuity with mythic topics from earlier Near Eastern traditions. So should we site Adam as 'the blood of God' as an echo of the *Atrahasis*, the ancient Mesopotamian creation myth where a man of clay is animated by blood (and flesh) from a slaughtered god? Does this carry on to later Christian hints of Divine blood (and body)? The material is certainly suggestive, although the time scale is enormously long. Here I would note also that Jesus, the second Adam in Christian thought, also called himself 'the light of the world'.[44]

I turn now to Fishbane's second track of thematic innovation, the development of new mythic themes. Here I would site created man as the dough-offering or bread of the world – or of God (albeit with an echo of sacrifice contained in the dough-offering). The rabbinical *midrashim* expounding on the creation of Adam began around the fourth century in Palestine. These and later *midrashim* replace the primeval clay with dough, with God portrayed as foreshadowing a woman's function in working the dough and separating the *ḥallah*. Perhaps God as a woman baker is almost as amazing an image as God's blood. The fourth century was a time of the meteoric ascent of Christianity in the Roman Empire, which was affecting the very landscape of *provincia Palaestina* with its new and triumphant churches built in Jerusalem and Bethlehem, Jewish towns now taken over by Christianity.[45] In these churches, Christians were reading of the Fall of Mankind and the crucifixion of Jesus, which they saw as the last sacrifice. There too they would eat the wafers and drink the wine that they believed turned into the body and blood of Jesus. In contrast, Jews in their study-houses were taking the account of the creation of mankind and instead of concentrating on the 'Fall' (or the last sacrifice of the crucifixion) were metaphorizing Adam as the dough-offering, which would thus give women the opportunity to mend our fractured world.

About the Author

Susan Weingarten is a food historian and archaeologist living in Jerusalem. She is the author of *Haroset: A Taste of Jewish History* and *Ancient Jewish Food in Its Geographical and Cultural Contexts: What's Cooking in the Talmuds?*

Acknowledgements

I am grateful to Reba Hames and especially Gilla Rosen for discussing this paper.

Notes

1 My account here is knowingly unbalanced, concentrating on the creation of man. In fact, the text of Genesis and its rabbinic interpretation give a more important and complex role to Eve than I can accommodate here.

2 Isaiah 64. 7.

3 Jeremiah 18. 1–6.

4 'Like clay in the potter(*yotzer*)'s hands / Expanded or contracted at will / So are we in your hand / Guardian of lovingkindness / Look to the covenant, and disregard our inclination' (trans. by J. Sacks, *The Koren Yom Kippur Maḥzor* (Jerusalem, 2019), p. 151; see also Chava Alberstein, 'The Dove of Love', YouTube <https://www.youtube.com/watch?v=lXAP_eMtapg&t=65s> [accessed 26 February 2026]).

5 Midrash Genesis Rabbah 4. 7: *shamayim*, heaven, is a compound of *esh*, fire, and *mayim*, water; the verb is *p-q-kh*, to mix and possibly to knead. Different *midrashim* use different verbs for this. In BT Hagigah 12a, God beats them up (*t-r-f-n*) together to make the firmament; Midrash Genesis Rabbah 10. 3 notes God twining (*p-t-kh*) coils of snow and fire together to make the firmament.

6 For a brief explanation of *midrashim* and other talmudic sources, see Susan Weingarten, 'Nuts for the Children: The Evidence of the Talmudic Literature', in *Nurture: Proceedings of the Oxford Symposium on Food and Cookery 2003*, ed. by R. Hosking (Bristol, 2004), pp. 264–72.

7 Pirqei de Rabbi Eliezer (PRE) (Higger), Horev, ch 11, usually dated to the eighth century.

8 RV version.

9 BT Sanhedrin 38a–b.

10 PRE 11. This late *midrash* has been seen by Jeffrey Rubenstein as having 'an increased mythic quality' ('From Mythic Motifs to Sustained Myth: The Revision of Rabbinic Traditions in Medieval Midrashim', *Harvard Theological Review*, 89 (1996), pp. 131–59).

11 Genesis Rabbah 14.8 and parallels. Note the imagery of the embryo here. Rubenstein notes the myth of the Temple as cosmic centre in this *midrash*.

12 W. G. Lambert and A. R. Millard, *Atrahasis* (Oxford University Press, 1969), 1.208–50; similarly in the Gilgamesh epic 1.34, goddess Aruru created Enkidu from clay. See also Tikva Frymer Kemsky, 'Creation Myths: Atrahasis, An Introduction', in *Studies in Bible and Feminist Criticism* (Philadelphia, 2006), pp. 5–18.

13 Hesiod, *Theogony*, 507–45; Hesiod, *Works and Days*, 42–83.

14 Ovid, *Metamorphoses*, 1, 82 and see also Juvenal 14.35.

15 M. Fishbane, *Biblical Myth and Rabbinic Mythmaking* (Oxford University Press, 2003), p. 191; see also Rubenstein, 'From Mythic Motifs.'

16 Psalm 139. 16.

17 Avot xi 7.

18 Sanhedrin 38b.

19 AdRN B 8.

20 PT Niddah 50d.

21 BT Sanhedrin 65b.

22 BT Sanhedrin 38a–b.

23 The earth for Adam's buttocks is said to come from an obscure place in Babylonia, Akra d'Agma. We may suspect rabbinical satire here.

24 An alternative version has the *golem* animated by the word *emet*, truth, written on his forehead. When the initial letter *aleph* (e) is removed, the remainder spells *met*, dead, and the *golem* collapses.

25 Mary Shelley's *Frankenstein* is said to have been influenced by *golem* legends, but this has been questioned (see e.g. Ed Simon, 'Is Frankenstein's Monster the Golem's Son?', *Tablet Magazine*, 31 October 2018 <https://www.tabletmag.com/sections/arts-letters/articles/is-frankensteins-monster-the-golems-son> [accessed 26 February 2026]). On Tolkien's Gollum, see Woody Wendling, 'The Riddle of Gollum: Was Tolkien Inspired by Old Norse Gold, the Jewish Golem, and the Christian Gospel?', in *Inklings Forever: Published Colloquium Proceedings 1997–2016*, vol. 6 (2008), article 23 <https://pillars.taylor.edu/inklings_forever/vol6/iss1/23> [accessed 26 February 2026].

26 Athenaeus, *Deipnosophistae*, 3.112c8–d5.

27 Hesiod, *Works and Days*, 82.

28 There is a helpful table in *The Fathers according to Rabbi Nathan* (Abot deRabbi Natan) version B, trans. and commentary by A. J. Saldarini (Leiden, 1975), pp. 303–05 (based on the text ed. by S. Schechter (Vienna, 1887)).

29 *Avot deRabbi Nathan* (AdRN) exists in two versions, A and B, and was written after the fifth century CE and before the ninth century (Menahem Kister, *Studies in Avot de-Rabbi Nathan: Text, Redaction and Interpretation* (Jerusalem, 1998) in Hebrew and Saldarini, p. 220). Version B is considered the better text, but the *golem* is found only in version A, as well as other *midrashim*.

30 AdRN B 8 and see L. Ginzberg, *Legends of the Jews*, vol. 1, 59–60, and vol. 5, 79 n. 22, which writes of microcosmos and macroanthropos.

31 BT Hagigah 12a.

32 PRE (Higger), Horev ch 20 cites Adam relating to 'My body which His two hands worked or kneaded (*g-v-l*)'. The root *g-v-l* seems to relate to the whole process of mixing water with flour to make dough as well as the kneading of the dough. See on this J. C. Greenfield, 'The Root "gbl" in Mishnaic Hebrew and the Hymnic Literature from Qumran', *Revue de Qumran*, 2.6 (1960), pp. 155–62. Greenfield notes kneading dough among other meanings (pp. 156, 160).

33 Genesis Rabbah ed. by Theodor Albeck, Genesis 14, 2, 7 and cf. PT (Venice edn) Shabbat ii 5b.

34 Ex Rabbah Mishpatim 30, 13.

35 Numbers 15. 19–21: 'when [. . .] you eat of the bread of the land, you shall present a donation to the Lord. The first yield of your kneading troughs, a round loaf, you shall present in donation. Like the donation from the threshing floor, thus you shall present it. From the first yield of your kneading troughs you shall give to the Lord, a donation for your generations.' See also Mishnah Ḥallah.

36 PT Shabbat 5b; Midrash Genesis Rabbah 17; AdRN B 9 and 42; Tanhuma Noah 1; Tanhuma Metzora 17; Tanhuma Behalotkha 4; Yalqut Shimoni Genesis 23; Yalqut Shimoni Metzora 571; Yalqut Shimoni Judges; Yalqut Shimoni Proverbs 959.
37 In the words of Milton, *Paradise Lost*, 1, ll. 1–3.
38 For convenience I am relating to the three commandments in a different order from the text of AdRN B, which has blood, *ḥallah*, and light in that order.
39 Proverbs 20. 27 (RV For the spirit of man is the candle of the Lord).
40 Kister, *Studies*.
41 There appears to be a connection between AdRN B and PRE, as some chapters of the former are found at the beginning of the latter (Kister, p. 15; see also K. E. Keim Pirqei deRabbi Eliezer, *Structure, Coherence, Intertextuality* (Leiden, 2017)).
42 Deuteronomy 12. 23.
43 Luke 22. 19–20; Matthew 26. 26–28; Mark 14. 22–24; I Corinthians 11. 23–25.
44 John 8. 12; and see also Matthew 5. 14.
45 Jerome *ep* 120. 8.

32
Nourishing Atmosphere, Flow, and Floating Feelings in a Kyoto *Kaiseki* Kitchen
Chef Muroi and the Relationality of Food

Kate Willink

Introduction: The Elements as Culinary Foundation and Sacred Geography

At the traditional Japanese restaurant Muroi, nestled in a beautifully renovated old *kominka* farmhouse in Kyoto's Takagamine neighbourhood, Chef Shigeru Muroi has chosen this location 'to be close to the vegetables and to be inspired by nature', creating what becomes a compelling manifestation of how elemental forces animate both cuisine and cultural practice. When we consider the elements in the context of food and cookery, our attention naturally turns to those fundamental forces that shape not only what we eat but how we understand our relationship to the world through food. At Muroi, the classical elements of water, earth, and atmosphere are not merely ingredients or cooking mediums but vital participants in what I term the 'relationality of food' – how we conceive of ourselves as relational beings within food networks, how we relate to others through food, and how we relate across different levels of food movements, from the local to the global.

This area was once 'a remote farming village and artist's colony during the Edo period' and remains significant as the birthplace of the Rinpa school of art – 'an indigenous school of Japanese aesthetics that celebrated nature's beauty through the refined elegance of its compositions, nature's power through the use of vibrant colours and bold designs, and nature's harmony through a strong sense of rhythm and pattern'.[1] The Rinpa aesthetic tradition emerged as artists sought to develop a distinctly Japanese artistic philosophy that honoured nature's elemental forces, continuing to influence artists today across a range of forms – painting, ceramics, textiles, lacquerware, garden design, and in Muroi's case, cuisine.[2]

Theoretical Framework: Elemental Discernment and the Gardener of the Cosmos

My understanding of relationality builds on previous scholarship examining the Slow Food movement, where I have explored 'the concept of relationality as how we conceive of ourselves as relational beings within food networks, how we relate to others through food and how we relate across different levels of food movements'.[3] This theory of relationality operates on different registers 'from the aesthetic and personal, to the social and political' and is fundamentally concerned with what I call 'the participatory self' rather than the 'objectified self'. At Muroi, this participatory relationality manifests through a form of sensing that attunes to elemental forces as active participants in culinary practice rather than as passive materials to be manipulated.

This distinction between the objectified self and the participatory self operates as a foundational ontological framework for understanding food relationships. The objectified self, what some philosophers call the self-contained subject, conceives of oneself as an autonomous entity that only relates through nondefining relationships, existing in a system that assumes hierarchy and separation. In this paradigm, food becomes merely instrumental – something that serves to meet the physical energy needs of autonomous entities, where what matters is primarily nutritional content rather than relational meaning. As Deane Curtin argues, within the objectified self framework, 'We do not see that what we will count as food is a *choice*. Our true relationship to food is obscured'. In contrast, the participatory self argues that we gradually become people through relationship and openness to others. From this alternative ontology, relations with others are defining, not optional. Curtin argues, 'We are defined by our relations to the food we eat. To account for our openness to food requires a relational understanding of self. We are what we eat'.[4] This relational conception extends beyond the merely symbolic to encompass what I term the 'affective and energetic dimensions of food' – how food serves as a platform for conviviality and connection rather than merely as fuel for individual bodies.

Central to this analysis is what I call 'elemental discernment' – a form of sensing that emerges through relational engagement with fundamental forces rather than through imposing external frameworks upon natural materials. Chef Muroi embodies what David Cooper calls a 'gardener of the cosmos' – one who practices elemental discernment through 'respectful attention that emerges through sustained engagement with elemental forces and follows what Japanese tradition calls nature's request'.[5] Like the traditional Japanese gardener who must 'listen to what [materials] request, and modify or arrange them in order to follow their requests', Chef Muroi develops the sensory and spiritual capacity to recognize the inherent wisdom and agency of his elemental ingredients.[6]

This practice aligns with what Japanese aesthetic philosophy calls 'following nature's request', which requires practitioners to be 'good listener[s] of what they "request", and modify or arrange them in order to "follow their requests"'.[7] Such an approach fundamentally challenges Western notions of human dominance over nature, instead positioning humans as collaborative partners who must develop the sensory and spiritual capacity to recognize the inherent wisdom and agency of elemental forces.

Elemental discernment operates through the concept of 'interpretive discernment' – the awareness and mindfulness of live, sensory, and affective data and the process of sensing and interpreting affective registers.[8] This concept functions as both a methodological orientation and an analytical heuristic, particularly valuable for understanding how elemental relationships operate in culinary contexts.

Relationality, understood through the lens of affect theory, illuminates what I call 'the feltness of our relations' with both human and more-than-human worlds. Affect – the capacity to affect and be affected – operates as a force that arises through relationality and exchange, through contacts between myriad singularities and their assemblage. As Ame Kanngieser argues, affect constitutes 'a physical energy that unites a variety of human subjects through and with nonhuman objects, helping constitute social relations separate from and prior to our ability to decipher meaning'.[9] From this affective perspective of relationality, the microbes in Chef Muroi's 400-year-old soil and the energetic life force of their microbiome become just as vital as the *kyo-yasai* vegetables they nourish, the mountain water that flows through the landscape, and the evening's cuisine that emerges from these elemental relationships. This understanding of affect as relational reveals how food operates not merely as symbolic communication but as energetic and material exchange that circulates through bodies and environments, creating what might be understood as 'affective assemblages' where human skill, elemental forces, and culinary materials enter into co-creative relationships.

Central to this analysis is the concept of 'vitality affects', which Daniel Stern describes as 'dynamic forms of feeling that underlie all experience' and 'include forms such as movement, force, time, and directionality, and work to provide a choreography of sensing and making sense of the affective and the culinary'.[10] These vitality affects emerge through what Kathleen Stewart describes as 'atmospheric attunements' – those 'charged atmospheres of everyday life' that create the affective background of experience.[11] At Muroi, such atmospheric attunements manifest through Chef Muroi's embodied choreography as he moves through his cooking practice, creating what becomes a lived symphony of elemental engagement that diners witness and participate in through their own sensory attunement to his movements.

Vitality affects operate through what Stern describes as 'dynamic forms of feeling that underlie all experience', providing what he calls 'the experience of vitality in one's own movements and those of others'. These affects possess their own flow patterns based on time and intensity – what Stern categorizes as 'exploding, pulsing, or fading' – creating a choreography of sensing that does not give primacy to the cognitive.[12] At Muroi, such vitality affects manifest through multiple registers simultaneously: the rising steam from dashi pots creates ascending movements that carry aromatic molecules through space; the chef's knife work demonstrates force and directionality as blade meets ingredient; the temporal unfolding of the *kaiseki* progression marks time through seasonal ingredients and cooking techniques. These vitality forms work additively, combining to imbue each moment of the dining experience with what Stern calls 'the quality of lived experience' that, although ubiquitous, often escapes our conscious attention even as it profoundly shapes our embodied engagement with food and atmosphere.[13]

Sacred Waters and Daily Pilgrimage: Shinto Purification and Elemental Ritual

Perhaps no element demonstrates this philosophy of 'following the request' more clearly than water at Muroi: 'Shigeru begins each day by driving up Mt. Takagamine to collect water from the spring known as Funamizu at Kiyomi Pass. He uses this pure, mineral-rich water to make the rice and dashi stock that are the foundations of his cooking'.[14] The quality of water at Muroi carries particular significance within Japanese culinary tradition, where Japan's mountainous topography naturally produces soft water that enables optimal extraction of nutrients, tastes, and aromas compared to hard water.[15] Chef Muroi's mountain spring water enables what Japanese cuisine recognizes as optimal extraction of flavour components, particularly crucial for creating his dashi – 'the essence of Washoku' – where 'soft Japanese water has a complex taste but is also light and delicious'. This daily practice creates what might be understood as a 'hydrological flow state' – where Chef Muroi's movements align with water's natural tendency to flow from mountain to valley, from spring to kitchen, following paths of least resistance while carrying essential minerals and life force through the landscape and into the evening's cuisine (Figure 1).[16]

This daily water gathering exemplifies what might be understood as a contemporary form of *misogi* – the Shinto water purification ritual. In Shinto tradition, water serves as 'a medium of spiritual purification connection to the divine', where 'Japanese people have long believed that rivers, seas, and other bodies of water have guardians, who are *kami* (God) and other beings'.[17] Chef Muroi's practice connects to Kyoto's millennium-long relationship with water-based food networks, where 'Kyoto was the capital for a span of 1,000 years, so the flow of food supply was plenty and well developed, brought in from different parts of the country' through elaborate water transportation systems.[18] His daily mountain pilgrimage thus becomes a contemporary instantiation of these ancient water-based supply networks, connecting individual practice to historical foodways.

Chef Muroi's practice transforms this ancient tradition into what I call 'culinary *misogi*' – a daily ritual that not only purifies the water but creates what Stewart describes as 'atmospheric attunements' through 'the labour of attending to a space opening out of the charged rhythms of an ordinary'.[19] Through this daily enactment of culinary *misogi*, water becomes not merely

Figure 1. Traditional copper kettle over wood fire, embodying the hydrological flow state where water moves from mountain spring through fire to evening's cuisine. All photographs by the author.

an ingredient but a relational medium that carries the chef's embodied connection to sacred landscape into every bowl, transforming cooking into a form of spiritual practice.

The chef's use of water extends beyond cooking to encompass spiritual and aesthetic dimensions. He sprays water on top of each lacquer soup bowl before serving and on top of red and yellow maple leaves before plating food (Figure 2). These gestures invoke what Shinto philosophy recognizes as water's capacity to create 'nature mindfulness', where 'realising the value of water leads to mindful use and conservation'.[20] In this ritual sprinkling, water functions as what might be understood as a spiritual conductor, carrying the sacred essence of mountain springs into the intimate space of the dining encounter.

Figure 2. Lacquered soup bowl with golden autumn leaves sprayed with sacred mountain water, demonstrating how water functions as a spiritual conductor carrying sacred essence into the dining encounter.

Earth, Soil, and the 400-Year Terroir: Following the Request of Stressed Vegetables

The nutrient-rich, well-drained soil that nourishes and grows *kyo-yasai* – the heirloom Kyoto vegetables – represents another vital dimension of Chef Muroi's elemental practice: 'Higuchi Farm has been raising vegetables at Takagamine for over 400 years and has had a leading role in cultivating *kyo-yasai*'.[21] This extraordinary temporal depth creates what might be understood as an 'elemental archive' where centuries of agricultural practice have literally shaped the soil's capacity to nourish specific vegetable varieties.

Every morning, Chef Muroi engages in what becomes a form of elemental discernment by walking across the road to select vegetables directly from Higuchi Farm.[22] But his selection process embodies the Japanese aesthetic principle of 'following the request'. Chef Muroi seeks vegetables at their seasonal peak that have developed intensified flavours through environmental stress – those with crooked growth, awkward development, or unexpected shoots that reveal their concentrated life force and will become featured elements in the evening's multicourse presentation (Figure 3).[23] This practice of selecting

Figure 3. Seasonal course featuring kyo-yasai *and autumn maple leaf, demonstrating how stressed vegetables reveal their concentrated life force through elemental disclosure.*

for elemental disclosure rather than conventional appearance exemplifies how elemental discernment requires remaining open to what the materials themselves will reveal rather than imposing predetermined aesthetic or culinary expectations upon them.

The philosophy of 'following nature's request' that governs traditional Japanese garden-making finds perfect expression in Chef Muroi's kitchen, where, like the gardener who must 'distinguish between the essential features of the particular material and its adventitious, inessential, and irrelevant aspects', he practices a form of culinary pruning – selecting stressed vegetables for their intensified life force, allowing water's essential qualities to shape his dashi, and creating atmospheric conditions that enhance rather than mask the elemental character of his ingredients.[24]

Staging Elemental Atmospheres: A Choreography of Purification and Comfort

Each evening before opening, Chef Muroi performs what might be understood as a comprehensive elemental purification of the dining space. He washes the entryway with water, creating a ritual threshold that connects the external world with the internal sanctuary of the dining room (Figure 4).

Upon entering the restaurant's intimate eight-seat space, diners encounter what can only be described as a carefully orchestrated symphony of elemental comfort that creates what Sasha Engelmann calls 'floating feelings' – emotional states that emerge through atmospheric encounter and create forms of social-affective buoyancy.[25] The vestibule requires the removal of shoes, immediately shifting the body's relationship to earth and ground. Heated stones, warmed by the restaurant's fire element, await beneath each seat, radiating a deep somatic comfort that connects diners' bodies directly to earth's warming presence.

The atmospheric staging at Muroi operates through what might be understood as 'affective choreography' – a deliberate yet emergent orchestration of elemental forces that creates conditions for relational attunement among diners, chef, ingredients, and atmosphere. This choreography draws on principles of 'affective attunement', which refers to how individuals relationally orient themselves to spaces, encounters, and other selves through somatic interplay. Teresa Brennan argues that such attunement requires 'living attention' – an energetic and material form of presence that enlivens others without depleting one's own energy, creating what she calls 'connective ability'.[26] Chef Muroi's atmospheric staging embodies this principle: the heated stones beneath each seat do not merely provide warmth but create somatic conditions for diners to settle into receptivity; the rising steam and aromatic smoke do not simply signal cooking but establish olfactory pathways for anticipation and connection; the open kitchen arrangement does not merely display technique but creates visual and energetic flows between chef and guests. Through such careful calibration of elemental forces, the atmosphere becomes what Stewart describes as 'charged' – dense with potential for affective transmission and relational becoming.[27]

The open kitchen itself becomes a living element of the atmosphere, with billows of steam rising from large dashi pots as they simmer throughout the evening and wisps of aromatic smoke emanating from fish and meat grilling over open flames (Figure 5). This constant presence of steam and smoke creates

Figure 4. Entrance to Muroi showing the ritual threshold where diners transition from urban rhythms to seasonal temporalities.

Figure 5. Steam rising from dashi pots in the open kitchen, creating the atmospheric staging that heightens the mysterious quality of the dining experience.

a sensory veil that heightens the mysterious and ephemeral quality of the dining experience. Following Engelmann's insight that floating involves both physical buoyancy and social-affective cohesion, the atmospheric staging at Muroi suspends diners in a shared elemental environment where individual boundaries dissolve into collective attunement to Chef Muroi's elemental choreography. This carefully curated environment mirrors what Mihaly Csikszentmihalyi identifies as optimal conditions for flow states – clear goals, immediate feedback, and the balance of challenge and skill – creating what might be understood as 'atmospheric flow' where the boundaries between self and environment, like the boundaries between mountain water and soil, become permeable and mutually nourishing.[28] This collective receptivity embodies the Japanese aesthetic principle of following rather than controlling nature's expression, creating conditions where all participants – chef, ingredients, diners, atmosphere – remain open to elemental becoming. Throughout the dining space, seasonal floral arrangements created by Chef Muroi himself punctuate the restaurant's elemental staging, serving as seasonal calendars that connect diners viscerally to the temporal rhythms that govern both the *kyo-yasai* selection and the *kaiseki* progression (Figure 6).

Figure 6. Seasonal floral arrangement by Chef Muroi, serving as a temporal calendar connecting diners to the rhythms governing both vegetable selection and kaiseki *progression.*

The Performative Choreography of Elemental Cooking and the Art of Cutting

Central to Chef Muroi's practice is the Japanese aesthetic principle of *kire* (cutting), which extends far beyond mere technique to embody a fundamental philosophy of respectful separation. Drawing from the Rinzai School of Zen Buddhism, *kire* represents 'cutting off the root of life' to reveal true nature – whether severing flowers from their roots in ikebana to show their absolute rootlessness or cutting ingredients to release their essential character. Each precise cut of Chef Muroi's knife becomes an act of liberation, freeing the ingredient's hidden potential through what might be understood as 'compassionate separation'.[29]

With the fluid precision of a master calligrapher painting characters in air, Chef Muroi's blade moves through a piece of silvery fish or golden persimmon with such liquid grace that he seems to dance through the cooking process – a culinary ballerina whose parsimonious movements waste neither motion nor meaning, each cut releasing the ingredient's deepest secrets through the most economical of gestures (Figure 7). This embodied choreography exemplifies what Stern describes as vitality affects in action, where the chef's movements create 'dynamic forms of feeling' that transmit the temporal qualities of force, rhythm, and directionality directly to witnessing diners through the kinaesthetic poetry of his knife work.[30] Chef Muroi's practice embodies what Csikszentmihalyi describes as 'flow' – the psychological state where action and awareness merge, self-consciousness disappears, and the activity becomes autotelic. Like the mountain water that flows effortlessly from spring to kitchen, Chef Muroi's movements achieve what Csikszentmihalyi calls the 'paradox of control' – complete mastery that appears utterly natural, where technique dissolves into pure expression of elemental responsiveness.[31]

Figure 7. Chef Muroi demonstrating elemental discernment through knife work that exemplifies Csikszentmihalyi's flow state, where action and awareness merge in pure elemental responsiveness.

This embodied choreography exemplifies what performance scholars call 'interpretive discernment in action' – the capacity to sense and respond to affective registers in real time through cultivated bodily attention. Chef Muroi's movements demonstrate what I have elsewhere called 'affective attunement' in performative practice, where practitioners develop the ability to register imperceptible-yet-felt qualities that guide their engagement with materials. His knife work reveals a form of embodied knowledge that emerges through sustained relational engagement with ingredients: the blade's angle responds to the fish's grain, the cutting rhythm attunes to the persimmon's firmness, the hand's pressure adjusts to the vegetable's density. Such attunement requires what Brennan describes as 'discernment' – an awareness of live sensory and affective data coupled with the capacity to interpret these registers through an educated sensing body.[32] This form of knowing cannot be reduced to technique or procedure; it emerges only through the kind of patient, receptive engagement that characterizes elemental discernment, where the practitioner learns to follow what materials themselves request rather than imposing predetermined outcomes upon them.

The multicourse *kaiseki* experience at Muroi functions as elemental performance, where each course reveals different aspects of the relationships between elements, ingredients, and dining experience (Figure 8). As Chef Muroi prepares each course, he engages in

Figure 8. Fire element in Chef Muroi's open kitchen, where flames and smoke create sensory veils that transform cooking into elemental performance.

animated conversation with diners, seamlessly transitioning between Japanese and English, creating an intimate dialogue that bridges linguistic and cultural boundaries.

The counter arrangement where diners observe the cooking process creates conditions for interpretive discernment, where 'immediate sensory experiences unfold within larger processual frameworks'.[33] Diners witness the transformation of mountain water and stressed vegetables into culinary compositions, experiencing what Stewart describes as when 'the sense of something happening becomes tactile'.[34] This tactile witnessing transforms diners from passive observers into active participants in the elemental choreography, where the boundaries between chef, ingredient, and guest dissolve into a shared moment of becoming. As Engelmann observes, such atmospheric encounters create 'floating feelings' where bodies become collectively suspended in emotional states that blur individual boundaries, creating what she calls the 'social-affective cohesion and endurance of a collective'.[35] At Muroi, this collective emerges as diners experience temporal suspension. The ordinary rhythms of urban life give way to the seasonal temporalities governing Chef Muroi's elemental practice, in which each course arrives as an unpredictable revelation of what mountain water, stressed vegetables, and elemental forces might become when guided by respectful attention to their inherent requests. This shared experience exemplifies what Csikszentmihalyi describes as 'collective flow' – moments when groups achieve synchronized attention and mutual responsiveness, much like the way water molecules move together in a stream, with each individual element contributing to the larger current while maintaining its essential properties.[36]

Vitality Affects and the Rinpa Aesthetic: Nature's Power through Respectful Enhancement

At Muroi, vitality affects operate through Rinpa relationality – the creation of harmony through what Zen philosophy describes as a 'responsive rapport between and among various objects and phenomena' where 'harmony [. . .] is given rise to spontaneously by the co-arising of various phenomena and coexisting of various objects, each of which is simply asserting its own "thusness"'.[37] This aesthetic philosophy transforms the dining experience into what might be understood as a living mandala where each element – chef, ingredient, diner, atmosphere – contributes its essential nature to create emergent patterns of relational beauty (Figure 9). The result exemplifies what Engelmann calls the dual nature of floating, where physical suspension in the atmosphere becomes inseparable from social-affective suspension in shared emotional states. At Muroi, this floating manifests as wonder at elemental encounters that 'keep bodies and spaces open to the surprise of others' – where diners remain receptive to the unexpected gifts that emerge when Chef Muroi follows nature's request, allowing stressed vegetables to reveal their hidden intensities and mountain water to express its essential character through the medium of respectful culinary attention (Figure 10).[38] These moments of elemental disclosure cannot be predetermined or controlled but only discovered through the practice of respectful engagement that characterizes elemental discernment.

This aesthetic approach creates elemental becoming, where water, earth, atmosphere, and human participants enter into responsive relationships that generate new forms of relational possibility. The result demonstrates what Yuriko Saitō calls 'an improvement upon bare, untouched nature; however, it is an improvement according to nature's characteristics rather than in spite of or irrespective of them'.[39] In this way, Chef Muroi's practice embodies a form of culinary collaboration with elemental forces, where human skill serves not to dominate nature but to reveal its inherent wisdom and beauty.

Figure 9. Kaiseki *course demonstrating Rinpa aesthetic principles, where each element asserts its 'thusness' while contributing to emergent relational beauty.*

Figure 10. Kaiseki *course exemplifying Rinpa aesthetic principles, where autumn maple leaves, seasonal ingredients, and natural elements create 'responsive rapport between and among various objects and phenomena', each asserting its own 'thusness'.*

Conclusion: Following Nature's Request in Contemporary Culinary Practice

Chef Muroi's practice at his Kyoto restaurant reveals how attention to elemental relationships can create profound forms of relationality that connect individual dining experiences to broader ecological, cultural, and spiritual systems. His daily engagement with sacred mountain water, 400-year-old agricultural terroir, and carefully staged atmospheric conditions demonstrates elemental discernment – a form of interpretive discernment that emerges through relational engagement with fundamental forces.

This elemental approach to food, grounded in the Japanese aesthetic principle of 'following nature's request', offers insights that extend beyond individual restaurants or cooking practices. Drawing on Shinto water philosophy, atmospheric staging theory, and the Rinpa aesthetic tradition, Chef Muroi's practice demonstrates possibilities for what Saitō calls 'respectful enhancement' – approaches that honour the thusness of elemental materials while creating conditions for their optimal expression.

The significance of Chef Muroi's practice extends beyond its immediate aesthetic and culinary achievements to illuminate what might be understood as 'relational food networks' operating across multiple scales and temporalities. His daily pilgrimage to mountain springs connects individual cooking practice to sacred geography and water-based food supply networks that have sustained Kyoto for over a millennium. His relationship with Higuchi Farm's 400-year-old agricultural terroir situates contemporary cuisine within deep histories of soil cultivation and vegetable development. His evening preparation of *kaiseki* courses creates intimate encounters where diners become participants in these layered networks of elemental and cultural relationship. Such practices exemplify what I have elsewhere described as the 'participatory self' paradigm, where food serves not merely as fuel for autonomous individuals but as a platform for conviviality, connection, and what might be called 'affective nourishment' – sustenance that operates through energetic and relational exchange rather than purely nutritional transactions. This understanding challenges dominant Western food paradigms that reduce eating to instrumental consumption, instead revealing food as part of what Cecilia Condit calls 'four dimensional webs' that 'reconstitute matter and ideation as humans, discourse, and other beings within a dynamic field of many forces'.[40]

Through the lens of vitality affects and atmospheric attunements, we can understand how Chef Muroi's practice creates elemental intelligence – forms of relational knowing that emerge through sustained engagement with water, soil, atmosphere, and seasonal rhythms. This intelligence operates through networks of relational sensing that create an interpretive architecture for elemental engagement.

As we face contemporary challenges around climate change, food security, and cultural sustainability, practices like those at Muroi suggest possibilities for more relationally engaged approaches to food that honour what Stewart calls 'the feltness of our relations' with the more-than-human world.[41] Such approaches recognize food as a relational and cultural nexus, made material in a menu, a recipe, or a simple cooking utensil.

My hope is that this exploration of elemental discernment at Muroi becomes literally and metaphorically an 'ideal platform' for elaborating the importance and significance of the elements for cooking, gathering, and sustainable food politics within culinary epistemologies, foodways, cooking practices, and conviviality writ large. Through Chef Muroi's practice, we glimpse possibilities for culinary approaches that honour elemental wisdom while creating conditions for human flourishing – a model that suggests how food might serve as a bridge between ancient wisdom and contemporary ecological consciousness. In an era when the elements themselves face unprecedented disruption through climate change and environmental degradation, such practices offer vital blueprints for reimagining our fundamental relationships with the elemental forces that sustain us.

About the Author

Kate Willink is a Professor of Communication Studies at the University of Denver. Her work focuses on intercultural communication, with particular attention to how food and education function as relational nexuses in cultural life; over the past decade, she has examined the evolution of the Slow Food movement in contemporary Italy.

Notes

1 Terri Schiller, 'Muroi', *Oishi So Japan*, 2018 <https://oishisojapan.com> [accessed 15 November 2024].
2 Schiller.
3 Kate Willink, 'Food Culture, Relationality and the Slow Food Movement', *Journal of European Popular Culture*, 10.1 (2019), pp. 61–73 (p. 65).
4 Deane Curtin, 'Food/Body/Person', in *Cooking, Eating, Thinking: Transformative Philosophies of Food*, ed. by Deane Curtin and Lisa Heldke (Indiana University Press, 1992), pp. 3–22 (p. 11).
5 David E. Cooper, 'The Good Gardener? The Way of the Garden in East Asian Tradition', in *The Good Gardener?: Nature, Humanity and the Garden*, ed. by Annette Giesecke and Naomi Jacobs (Artifice Books on Architecture, 2015), pp. 38–48 (p. 38).
6 Cooper, p. 42.
7 Yuriko Saitō, 'Japanese Gardens: The Art of Improving Nature', in *Japanese Environmental Philosophy*, ed. by J. Baird Callicott and James McRae (Oxford University Press, 2017), pp. 145–58 (p. 151).

8 Salam T. Shukri and Kate G. Willink, 'Interpretive Discernment and Performative Listening: Feeling Our Way toward a Performative Understanding of Interviewing', *Departures in Critical Qualitative Research*, 9.3 (2020), pp. 48–77 (p. 52).
9 Qtd. in Angeli Alifuoco, '"Alive" Performance: Toward an Immersive Activist Philosophy', *Performance Philosophy*, 3.1 (2017), pp. 126–45 (p. 143).
10 Daniel N. Stern, *Forms of Vitality: Exploring Dynamic Experience in Psychology, the Arts, Psychotherapy, and Development* (Oxford University Press, 2010), p. 8; Kate Willink and Darrin Hicks, 'Teaching about Food Culture and Everyday Life in Japan', Biennial Perugia Food Conference (Perugia, Italy, June 2024), p. 3.
11 Kathleen Stewart, 'Atmospheric Attunements', *Environment and Planning D: Society and Space*, 29.3 (2011), pp. 445–53 (p. 445).
12 Stern, *Forms*, p. 8.
13 Stern, p. 5.
14 Schiller, 'Muroi'.
15 NIZ, 'You Think It's All Sushi in Japan?', 2019 <https://niz.com> [accessed 15 November 2024].
16 Reny Wiyatasari, '*Nihonnomizubunka*: Harmony between Water and Humans in Japan', *E3S Web of Conferences*, 317, 04014 (2021), doi:10.1051/e3sconf/202131704014.
17 Huei-Hua Chiang, Maumita Basu, and Yuki Sano, 'Water and Shintoism: Exploring the Link between Spirituality and Water Values in Japan', in *Indigenous and Local Water Knowledge, Values and Practices*, ed. by Maumita Basu and R. DasGupta (Springer Nature, 2023), pp. 67–78 (p. 71).
18 'The Art of Kyoto Cuisine Crafted with Kyoto Vegetables', *Kyoyasai Kyoto* <https://kyoyasai.kyoto/eng/2017/09/16900> [accessed 15 March 2026].
19 Stewart, 'Atmospheric Attunements', p. 448.
20 Chiang, Basu, and Sano, 'Water and Shintoism', p. 73.
21 Schiller, 'Muroi.'
22 Schiller.
23 Schiller.
24 Saitō, 'Japanese Gardens', p. 155.
25 Sasha Engelmann, 'Floating Feelings: Emotion in the Affective-Meteorological Atmosphere', *Emotion, Space and Society*, 40 (2021), 100803 (p. 1).
26 Teresa Brennan, *The Transmission of Affect* (Cornell University Press, 2004), p. 41.
27 Stewart, 'Atmospheric Attunements', p. 445.
28 Mihaly Csikszentmihalyi, *Flow: The Psychology of Optimal Experience* (Harper & Row, 1990), pp. 49–67.
29 Graham Parkes and Adam Loughnane, 'Japanese Aesthetics', in *Stanford Encyclopedia of Philosophy*, 6 December 2023, section 9 <https://plato.stanford.edu/entries/japanese-aesthetics/> [accessed 15 March 2026].
30 Stern, *Forms*, p. 8.
31 Csikszentmihalyi, *Flow*, p. 53.
32 Brennan, *Transmission of Affect*, p. 138.
33 Shukri and Willink, 'Interpretive Discernment', p. 55.
34 Stewart, 'Atmospheric Attunements', p. 449.
35 Engelmann, 'Floating Feelings' p. 1.
36 Csikszentmihalyi, *Flow*, p. 157.
37 Saitō, 'Japanese Gardens', p. 153.
38 Engelmann, 'Floating Feelings' p. 5.
39 Saitō, 'Japanese Gardens', p. 152.
40 Celeste M. Condit, 'Communication as Relationality', in *Communication as . . . : Perspectives on Theory*, ed. by Gregory J. Shepherd, Jeffrey St. John, and Ted Striphas (SAGE Publications, 2006), p. 3.
41 Stewart, 'Atmospheric Attunements', p. 451.

33
Cosmic Cooking
Changing Dietetic Patterns in Medieval Islam

Limor Yungman

In the ninth century, Ḥunayn ibn Isḥāq translated Galen's book *On the Properties of Foodstuffs* (*De alimentorum facultatibus*) from Greek into Syriac and Arabic (*Kitāb al-Aghdhiya*) as part of a larger movement of collection and translation of Greek texts of science and philosophy.[1] By doing so, he not only conserved an important body of knowledge that would have otherwise been lost or forgotten but also enabled a scholarly and scientific continuum from the shores of the northeastern Mediterranean to the heart of the Middle East in its political centre in Abbasid Baghdad. By translating these medicinal and philosophical texts, Ḥunayn and other prominent scholars revived and continued a tradition of thought and philosophy that was later developed and tested by other physicians and scientists.

Arab and Persian physicians pragmatically sought to balance their patients' health and solve medical issues by following guidelines that disassembled food into its components, each significant in the Galenic theory, while considering the patient's condition, temperament, age, and sex.[2] Their engagement with food likewise aligns with the breadth of their intellectual lives: many were not only physicians but also philosophers, astronomers, and poets. Echoing Aristotelian thought, Avicenna (d. 1037), for example, emphasized the importance of eating in moderation and avoiding excess as rooted in philosophical ethics. At the heart of these physicians' approach lay the idea that food and nutrition formed part of a larger story – that of human existence within the cosmos.

In the same city and at the same time, Baghdad witnessed a revival of old culinary traditions from Mesopotamia, borrowing from former and neighbouring empires, such as the Sassanian, to become the Abbasid gastronomic culture. These two parallel developments founded the Arab-Islamic medico-culinary, or dietetic, tradition.[3] Food, as apparent from the medical-dietetic sources, was not just sustenance or a source of material pleasure; it was part of a complex cosmological system that considered every aspect of human experience to maintain balanced health: eating, drinking, sleeping, and exercising. Rooted in the Hippocratic-Galenic tradition, this system viewed dietary practices as integral to sustaining physical and spiritual harmony, such as the regulation of humoural balance through food choices. These practices reveal prevailing cultural and scientific conceptions of the human body and its relationship to the natural world, including the role of seasons, geography, and other factors in maintaining health. This paper traces how Arabic cookery books, across the centuries, embodied a comprehensive worldview, entwining culinary expertise, health management, and reflections on the cosmic order.

Universal Cooking

This concept connecting the gastronomic world to a wider cosmological worldview that surpasses dietetics in its medicinal viewpoint was expressed as early as the tenth century in *Kitāb al-Ṭabīkh*, attributed to Ibn Sayyār al-Warrāq, the oldest extant Arabic cookbook from Iraq. Far more than a simple recipe collection, the text brings theory into practice, combining practical instructions with insights into health, morality, and the natural world. Through a complicated set of dietary principles, foods are categorized according to the humoural theory. It is unique among Arabic culinary texts because it is the only known example organized in multiple parts, giving cardinal attention to the humoural qualities of foodstuffs based on the cold/warm and dry/wet principles at the heart of Galenic theory.

Kitāb al-Ṭabīkh presents an intricate, layered structure that expresses a holistic cosmological vision. The first layer lies in the organization of the book itself, which unfolds in a circular progression: it begins with dietary theory, moves through cooking practices and recipes, and concludes with postmeal guidance on drinks, hygiene, table manners, and rest. This structure, in five parts, mirrors a conception of the meal as a microcosm of life processes and cosmic rhythms – integrating preparation, consumption, and bodily restoration into a unified cycle.

The first part consists of seventeen chapters dedicated to the humoural qualities of foodstuffs, organized by food category – spices, nuts and seeds, meat, dairy products, vegetables, grains – in a way that echoes Galen's *On the Properties of Foodstuffs*, which is similarly arranged into three sections: cereals and legumes, vegetables and fruits, and meat and fish.[4] The second part offers a variety of recipes for hundreds of different dishes – cold and warm, sour, sweet, and full of spices and flavours. Each chapter is dedicated to one type of dish and its variations.

The third part is a combination of the first and second parts – namely, cooked dishes for the sick, detailed in terms of which medical conditions they target. The main disorders are divided into five categories: imbalance in humours (excess in yellow bile, excess in hot properties); inflammatory and infectious conditions (fever, pleurisy, sore throat, cold); digestive and gastrointestinal issues (diarrhoea, upset stomach, stomachache, gastric ailments, colic, liver and abdomen ailments, jaundice, ulcers); respiratory conditions (cough, chest pain, asthma, wheezing); and joint and musculoskeletal issues. The fourth section is made up of medicinal theory and recipes for posteating activities: beverages, medicinal drinks, perfumed cleansing powders (*ushnān*), and preparation of toothpicks. The fifth and last part describes appropriate guest behaviour at dignitaries' tables and the benefits of taking a nap after eating. *Kitāb al-Ṭabīkh* is thus structured harmoniously.

From this structural layer, the book extends outward into further concentric spheres: the classification of cooking vessels, the matching of foods with bodily humours, and the application of cooking techniques, each reinforcing a vision of order and balance rooted in humoural and elemental thought. A second layer of cosmological theory in cooking emerges in al-Warrāq's second chapter, where he classifies the advantages of cooking vessels in great detail.[5] The emphasis on matching dish and vessel according to material properties conveys a belief that natural substances interact with food in ways that affect its balance, taste, and effect on the body. Al-Warrāq's list goes beyond practical advice; it conveys a vision in which food transformation is shaped not only by intrinsic properties but by the tools involved in its preparation:

> Soapstone pots are the best for meat dishes. The best *harāyis* (porridges) are made in tin-lined copper pots. Earthenware pots and tin-lined pots are suitable for making *jūdhābāt* (baked puddings topped with roasted meat), and the best of *tannūriyyāt* (dishes baked in the *tannūr* oven) are cooked in soapstone pots. Condensed *ḥalwā* (puddings) are to be cooked in brass cauldrons, and beverages in soapstone cauldrons. The most suitable frying pans for *zalābiya* (fried fritters) are brass ones and for frying fish, iron frying pans. Soapstone frying pans are to be used for *narjisiyyāt* (meat dish with sunny-side-up eggs) and *ʿujaj mabsūṭāt* (flat omelettes). Rounded soapstone pots are the best for stir-fried dishes.[6]

This passage reveals the first step in cooking: choosing the appropriate vessel, its material and form, to match the dish and ensure its successful transformation. Each material – soapstone, tin, copper, brass, or iron – interacts differently with ingredients, shaping their final qualities in accordance with humoural theory. This embodies a cosmological logic in which every aspect of the cooking process contributes to aligning the body with the natural world. The kitchen becomes a microcosm, where precision and harmony guide the preparation of food. Al-Warrāq then explains why certain materials are optimal for cooking specific dishes:

> Soapstone pots are used for cooking meat and soups because they are slow to heat, smooth from the inside, and easy to clean. Besides, they taste good and whatever flavours they absorb, they give back, unaltered and untransformed. [. . .] *Harāyis* are cooked in tin-lined copper pots because such foods are usually beaten vigorously and stirred for a long time until they thicken. Stone pots will never stand all this beating as they might break in the process.[7]

This passage adds a new dimension to the logic of choosing vessels by foregrounding their mechanical and material properties. According to al-Warrāq, the pot becomes an active participant in the cooking process, faithfully preserving taste and texture and enhancing the dish through its stable interaction with ingredients. Soapstone pots are preferred for cooking meat dishes and soups due to the material's qualities: high thermal retention, even heat distribution, and flavour neutrality, which contribute to both the cooking process and the resulting taste. In contrast, tin-lined copper is recommended for *harāyis*, thick puddings that require vigorous beating and extended stirring; such physical demands would risk breaking a more fragile stone pot. These explanations show how practical experience and embodied knowledge converge in the cook's selection of materials.

In his sixth chapter, 'Beneficial Foods for the Young and the Elderly', al-Warrāq adds a personalized layer, listing the best foods for people according to their dominant humour: blood, yellow bile, black bile, and phlegm. This chapter elaborates on the best meat, flavourings, fruits, vegetables, composed dishes, and sweets for each humour. A person with blood type, for example, should eat dishes cooked in sour agents such as vinegar, pomegranate juice, verjuice, or sumac, while people with dominant phlegm should dip their foods into mustard.

The last consideration in tailoring food is found in Chapter 10, classifying cooked dishes according to their techniques (liquid, fried, roasted) and the conditions they best suit. Fried fatty meat, for instance, was considered 'hard and slow to digest but has great nourishing properties. In any case, it is heavier on the digestion than meat cut up and grilled on the fire. Frying it in olive oil would make it relatively lighter'.[8] Al-Warrāq draws humoural information from books of ancient philosophers and physicians' manuals, regrettably without exact titles or authors, though he later mentions Galen's now-lost *Book of Foods in Common Use* (*Kitāb al-Aghdhiya al-Musatʿmala*).

These chapters, unfolding like nested circles, demonstrate great thought and care for every component in a meal. Each layer addresses a different aspect of nourishment – from the ingredients themselves to the tools and vessels used, to the bodily constitution of the eater and the methods of preparation – yet all are unified by a common vision of maintaining cosmic balance through food. This creates a complex system requiring in-depth knowledge of the intricacies of the theory and the systems governing nature, each evoking a deeper connection between food, body, and cosmos. As the sources show, the humoural thought could go even deeper and cover more aspects of human life.

Dietetic Matrix

Among the many medieval physicians who addressed dietary theory, Ibn Buṭlān's *Taqwīm al-Ṣiḥḥa* (known in Latin as *Tacuini sanitatis*) stands out for its systematic organization of dietary advice in tabular form.[9] Alongside Ibn Jazla's *Taqwīm al-Abdān*, this work reflects a medical tradition concerned with the relationship between food, health, and cosmological order yet is distinguished by its structured, accessible presentation.[10] A comparison with Ibn Sayyār al-Warrāq's *Kitāb al-Ṭabīkh* further highlights shared categories, differences in classification, and the enduring integration of gastronomic and humoural thought.

Ibn Buṭlān's *Taqwīm al-Ṣiḥḥa* is structured in detailed tables according to these categories: fruits, cereals (grains), breads, fresh herbs, spices and condiments, vinegar and pickles (effectively vegetables), sauces and condiments (*muṭayyibāt*), cold dishes (*bawārid*), braised poultry dishes (*muṭajjanāt*), dairy products and eggs, meats (divided into livestock, game meat, poultry, fish, organ meats, and innards), cooked dishes (*maṭbūkh*), wrapped sandwiches (*bazmāward*), fried dishes, kebab and *kardanāj*, roasted meat and puddings (*jawādhīb*), and sweeteners and sweetmeats.[11] Each entry evaluates a dish or ingredient based on its nature, level, usage, potential harm, and remedy for its harmful effects and recommends the appropriate consumer age, season, and location for optimal health. Alongside practical reasons, such as facilitating access to medical information and avoiding philosophical digressions in works intended for a general readership, the tabular form itself conveys a cosmological worldview. Modelled on astronomical tables, it merges dietary advice with a structured and systematic view of the natural order.[12]

A direct comparison between Ibn Buṭlān's tables and the recipes found in al-Warrāq's *Kitāb al-Ṭabīkh* reveals notable correspondences. Two particularly relevant tables from *Taqwīm al-Ṣiḥḥa* list cooked dishes and their varieties (*al-maṭbūkh wa-alwānuhu*),[13] comparable to those in *Kitāb al-Ṭabīkh*. Ten dishes – *sikbāj* (beef stew in vinegar), *isfidhbāj* (white stew), *zīrbāj* (chicken stew in vinegar), *maṣliyya* (stew cooked with yogurt whey), *liftiyyā* (turnip stew), *bahaṭṭa* (rice pudding), *ḥiṣrimiyya* (stew based on sour juice of unripe grapes), *rībāsiyya* (rhubarb stew), *rummāniyya* (pomegranate stew), and *summāqiyya* (sumac stew) – are common to both.

Certain dishes, such as *summāqiyya*, *ḥiṣrimiyya*, and *rummāniyya*, are classified as having similar effects in the two texts. Both authors note this by simply stating, 'In this respect, *rībāsiyya*, *rummāniyya* and *summāqiyya* are similar to it (*ḥiṣrimiyya*)'.[14] However, *Kitāb al-Ṭabīkh* provides further detail, explaining that their similarity lies in the use of acid agents – verjuice, sumac, pomegranate, and rhubarb – which produce a sour taste.

Both Ibn Sayyār al-Warrāq in Chapter 10 and Ibn Buṭlān place dishes at the centre of their dietary discussions using similar vocabulary, though sometimes diverging in classification. This inconsistency highlights the dynamic and interpretive nature of humoural theory, which allowed for flexibility based on context, temperament, and culinary traditions. Differences also appear in their classifications. Two prominent dishes of Persian heritage, *sikbāj* and *zīrbāj*, are classified differently: Ibn Buṭlān categorizes both as having a cold nature, while al-Warrāq classifies them as moderately hot. Moreover, while al-Warrāq considers *zīrbāj* best suited for individuals with dominant yellow and black bile, Ibn Buṭlān associates it with the production of balanced blood.

The two authors also differ in their organizational approach. Ibn Buṭlān focuses closely on the qualities of each dish according to specific parameters: the eater's age, temperament (hot/cold), the season in which the dish should be consumed, and the geographic location and climate (northern/southern, cold/warm) where the food is eaten. His more intricate system, balancing multiple variables for each dish, demonstrates a heightened sensitivity to the complex interaction between individual constitution, environment, and food – a logic that became even more pronounced during certain phases in the production and copying of manuscript cookery books. In contrast, in Chapter 6 of his work, al-Warrāq organizes his material around humoural types, but not in a tabular form, and does not systematically provide information about the adaptability of a dish to specific seasons, climates, or even age, despite the chapter's title referring explicitly to foods for the young and the elderly.

Changing Focus

The structure of al-Warrāq's *Kitāb al-Ṭabīkh* distinguishes it from the nine other surviving Arabic cookbooks. It is the only one to include a dedicated section on the humoural qualities of foodstuffs and to organize culinary knowledge systematically around philosophical principles in a circular structure covering all issues related to food: its storage, preparation, cooking, cooking vessels, and consumption. This circular arrangement mirrors a cosmological vision in which the acts of preparing and consuming food were seen as part of maintaining the natural balance of the body and, by extension, harmony within the larger cosmic order.

One possible explanation for why it is the only cookbook with this type of structure lies in the changing priorities of culinary manuals over time. Earlier texts like *Kitāb al-Ṭabīkh* may have been designed to embody medico-dietetic philosophies, serving as both culinary guides and theoretical treatises. This can be seen in the *adab*-like structure of *Kitāb al-Ṭabīkh*, which aims not only to preserve and transmit knowledge and remind the cook of culinary practices but also to educate in related realms such as health and social etiquette.

In contrast to *Kitāb al-Ṭabīkh*, later cookbooks appear to have prioritized practical utility over theoretical discourse, aligning with shifts in societal needs and literary trends. This shift is especially evident in the compilers' decision to omit theoretical information about the use of dishes for specific conditions or humours, focusing instead on pragmatic culinary instructions. Dietetic information became diffuse, embedded within recipes or adjunct to them, rather than systematically organized. This shows a long, uninterrupted tradition of considering dietetics. The fourteenth-century Egyptian cookbook *Kanz al-Fawāʾid* mentions the humours primarily in counterfeit dishes (*muzawwarāt*)[15] for the sick in Chapter 8 on 'Counterfeit Dishes for the Nourishment of the Sick' and in digestive syrups and confections, sauces, and condiments.[16] The shift occurred on several levels. First, the structure of the cookery book changed: the section containing dietetic information became adjunct to the recipe rather than presented separately, and the information was organized in a nonsystematic manner. Second, the shift is detectable in the types of dishes and drinks that receive dietetic mention or are related to dietetics. Instead of defining which type of dish benefits which type of humour, as in *Kitāb al-Ṭabīkh*, *Kanz al-Fawāʾid* uses the category of *muzawwarāt* to address the correction of various ailments, such as fever, excess yellow bile, nausea, and heartburn. The dietetic advice is oriented more toward curing disease than preserving health, marking a shift from preventive to curative applications.

By the fifteenth century, the Syrian cookery book *Kitāb al-Ṭibākha* (or *Ṭabbākha*) contains no references to dietetic theory.[17] Its alphabetized, succinct recipes suggest a different mindset on the part of its author, Ibn ʿAbd al-Hādī (Ibn Mibrad) – perhaps a personal collection or aide-mémoire. This form attests to changes in authorship, readership, and manuscript production: from a work composed for the ruling elites to a product shaped by and for an urban class.

Despite textual evolution, the manuscript tradition shows that knowledge of humoural theory persisted in marginalia, as seen in a thirteenth-century manuscript, also titled *Kitāb al-Ṭabīkh*, by Muḥammad ibn al-Ḥasan al-Baghdādī, where dietetic notes were added by another hand, referencing Ibn Jazla's *Minhāj al-Bayān*. For example, for the recipe for *ḥiṣrimiyya*, the marginal note adds the following information: 'suppresses yellow bile, cools [the body] and is constipating if made from fresh unripe grapes'.[18] The same dish is considered by both al-Warrāq and Ibn Buṭlān as cold (and by Ibn Buṭlān also as dry), thus aligning with Ibn Jazla's ideas.

Holistic Approach

Changes in the ways dietetic knowledge was incorporated into culinary literature attest to broader cultural and intellectual transformations, such as the preoccupations of the author-compilers and their identities, social status, and objectives in compiling their cookbooks. Even when theoretical material disappeared from later texts, the underlying holistic view of food – linking body, health, and cosmos – persisted.

All this shows that Arabic cookbooks were more than culinary manuals; they functioned as comprehensive repositories of medico-dietetic, cultural, and societal knowledge. At the same time, the way these books were produced, organized, and used reveals changing understandings of dietetic knowledge itself – not only in how it was transmitted but in how it was experienced, interpreted, and applied by different readers over time. This tradition of dietetic and medical knowledge influenced how food was conceived, prepared, and consumed across medieval Arabic culture.

About the Author

Limor Yungman is a postdoctoral fellow at the ERC project Medieval Placebo Effect at Tel Aviv University. Widely published in medieval Arabic foodways, she is working on a book exploring the origins of Arab cuisine.

Notes

1 John Wilkins and Siam Bhayro, 'The Greek and Syriac Traditions of Galen, *De Alimentorum Facultatibus*', *Galenos*, 7 (2014), pp. 95–113 (p. 96). These texts consisted only of prose works considered part of the scientific realm – including medicine and, by extension, cooking.

2 David Waines, 'Dietetics in Medieval Islamic Culture', *Medical History*, 43.2 (1999), pp. 228–40 (p. 230), doi:/10.1017/S0025727300065108.

3 Waines, p. 232.

4 Claude Galen, *On the Properties of Foodstuffs*, trans. with introduction and notes by Owen Powell (Cambridge University Press, 2003); Claude Galien, *Tome V, Sur les facultés des aliments*, trans. by John Wilkins (Les Belles lettres, 2013).

5 Charles Perry treated the subject of the humours of cooking utensils in his article 'The Humours of the Pot', in *The Cooking Pot: Proceedings of the Oxford Symposium on Food & Cookery 1988*, ed. by Tom Jaine (Prospect Books, 1988), pp. 123–24.

6 Ibn Sayyār al-Warrāq, *Kitāb al-Ṭabīkh*, MS Signum 3035 A. 3548, Nuruosmaniye Library, Istanbul, fol. 11r; *Annals of the Caliphs' Kitchens*, trans. by Nawal Nasrallah (Brill, 2007), pp. 84–85. The opening of Chapter 2 is missing in the manuscripts used by Öhrnberg and Mroueh for their edition. Ibn Sayyār al-Warrāq, *Kitāb al-Ṭabīkh*, ed. by Kaj Öhrnberg and Sahban Mroueh, *Studia Orientalia*, 60 (Finnish Oriental Society, 1987).

7 Al-Warrāq, MS Signum 3035 A. 3548, fol. 11v; Öhrnberg and Mroueh, p. 11; Nasrallah, p. 85.

8 Al-Warrāq, MS Signum 3035 A. 3548, fol. 20v; Öhrnberg and Mroueh, p. 26; Nasrallah, p. 108.

9 Dror Weil, 'Translating Medical Experience in Tables: The Case of Eleventh-Century Arabic Taqwīm Works', in *Premodern Experience of the Natural World in Translation* (Routledge, 2022), pp. 232–33.

10 Weil, pp. 232–33.

11 Ibn Buṭlān, *Taqwīm al-Ṣiḥḥa*, MS Or 1347, British Library, fol. 22v–25r; al-Mukhtār ibn al-Hasan ibn ʿAbdūn ibn Saʿdūn Ibn Buṭlān, *Le Taqwīm al-Ṣiḥḥa (Tacuini sanitatis) d'Ibn Buṭlān: un traité médical du XIe siècle*, ed. by Hosam Elkhadem (Peeters, 1990), pp. 92–95 (Arabic text), 188–93 (French translation).

12 Astronomical treatises use tables of star positions to track times of prayer, for calendrical use, and for navigation and astrological calculations.

13 Ibn Buṭlān, *Taqwīm al-Ṣiḥḥa*, MS Or 1347, British Library, fol. 22v–25r.
14 Al-Warrāq, *Kitāb al-Ṭabīkh*, MS Signum 3035 A. 3548, fol. 21v; Öhrnberg and Mroueh, p. 26; Nasrallah, p. 109.
15 On these types of dishes, see Manuela Marín and David Waines, '*MUZAWWAR*: Counterfeit Fare for Fasts and Fevers', *Der Islam*, 69.2 (1992), pp. 289–301.
16 *Kanz al-fawā'id fī Tanwī' al-mawā'id*, ed. by Manuela Marín and David Waines (Dār al-Nashr; Franz Steiner, 1993); Nawal Nasrallah, English translation, with an introduction and glossary, *Treasure Trove of Benefits and Variety at the Table: A Fourteenth-Century Egyptian Cookbook* (Brill, 2018). Al-Warrāq also includes chapters for *muzawwarāt* for the sick (no. 105). Al-Warrāq, *Kitāb al-Ṭabīkh*, MS Signum 3035 A. 3548, fols. 150v–152v; Öhrnberg and Mroueh, pp. 281–84; Nasrallah, pp. 433–37.
17 Jamāl al-Dīn Yūsuf ibn Ḥasan Ibn 'Abd al-Hādī (Ibn al-Mibrad), 'Kitāb al-Ṭibākha', ed. by Ḥabīb Zayyāt, *al-Mashriq*, 35 (1937), pp. 370–76. English translation: Charles Perry, 'Kitāb al-Ṭibākha: A Fifteenth-Century Cookbook', *PPC*, 21 (1985), pp. 17–22; reprint in Maxime Rodinson, A. J. Arberry, and Charles Perry, *Medieval Arab Cookery* (Prospect Books, 2001), pp. 467–75.
18 Muḥammad ibn al-Ḥasan al-Karīm al-Baghdādī, *Kitāb al-Ṭabīkh*, MS Ayasofya 3710, Süleymaniye Library, Istanbul, fol. 8r. Charles Perry's translation does not include these dietetic pieces of information (Charles Perry, 'A Baghdad Cookery Book', *PPC*, 79 (2005), pp. 1–127).

34
Wok Hei in the Making
Changing Tools, Skills, and Tastes in the Guangzhou Kitchen

Zien Zhi

Elusive *Wok Hei*

'Do you want soy sauce chicken or white-cut chicken?' my grandmother asked over the phone. 'And how about steamed fish, beef and kale stir-fry, and *Luohan Zhai*?' Then she added with a chuckle, 'You must have missed stir-fries with *wok hei* – something you just cannot find abroad'.[1]

It was a week before my flight to Guangzhou. My grandmother, ever thoughtful, promised to prepare all the dishes I loved as a child for my first meal back home. Though thunderstorms delayed my arrival by three hours, I knew she would wait. When I finally stepped into my grandmother's apartment around 8 p.m., four dishes sat neatly under a mesh food umbrella, alongside a big pot of soup. My grandmother, half-asleep in front of the TV, woke up, gently scolding me for not calling earlier. While bustling to reheat the food, she sighed, 'These dishes have been sitting for too long. They lost their *wok hei* [C. *mou saai wok hei*]'.[2]

I understood her frustration. For her, stir-fries must be relished immediately, while still steaming hot, to fully savour the fleeting aroma and the almost-burning sensation on the tongue – the essence of *wok hei*. Reheated dishes, no matter how delicious to me, could never recapture that magic of the stir-fry. As I listened to her mild complaints and tasted the slightly faded flavours of reheated dishes, I realized how deeply *wok hei* mattered and how little I understood it.

This moment marked the beginning of my journey into *wok hei*. Between April and June 2024, I conducted exploratory fieldwork on Cantonese cooking. Leveraging family ties and personal networks, I reached a diverse group of informants – retirees, housewives, students, chefs, and office workers ranging in age from their 20s to 80s – all deeply rooted in Guangzhou. I was grateful for their generosity in sharing memories of their favourite dishes, kitchens, and cooking techniques and for inviting me over for dinners. I documented their approaches to achieving *wok hei* through photos and videos and also experimented with how to wield a wok, ignite the right flame, and chase that elusive breath of the wok.[3]

One theme emerged repeatedly during my fieldwork: *wok hei*. It defies easy translation. Though stir-frying is widespread across Chinese cuisines, *wok hei* – the smoky, rich flavour and fleeting energy born from a well-seasoned wok, fierce heat, simple seasoning, and quick, skilful movement of the cooks – seems uniquely Cantonese. Non-Cantonese friends rarely recognized the term, and even among locals, explanations were often tinged with impatience: 'Only Cantonese understand'. For many, *wok hei* was not something you explained – it was something you tasted and knew. Yet *wok hei* is more than a flavour or a cooking technique. It is a potent symbol of local pride, a marker of cultural belonging, and a sensory ideal deeply embedded in the everyday lives of local people. This local sensibility is simultaneously taken for granted and fiercely guarded as a part of what it means to be Cantonese. Thus, relying solely on informants proved difficult, since it was too deeply ingrained to be articulated.

I returned to Guangzhou for another two-month stay from August to September 2024. This second visit led me to a broader question: Under what historical, social, and economic conditions did *wok hei* become so significant? To complement my ethnographic approach, I turned to archival research, focusing particularly on *Yangcheng Evening News*, one of Guangzhou's most influential newspapers. Its food columns published from 1996 to 2016 offered a valuable window into more elaborate discussions around *wok hei*.[4]

Combining fieldwork and media analysis, this research explores *wok hei*. While it is often understood as an adjective, such as 'delicious', and a marker of Cantonese identity, this research aims to capture it as an ever-evolving practice – one that embodies intricate physical and technical skills, aesthetic and sensory appreciation, and dynamic adaptation to ongoing social, material, and technological changes.

This research hopes to contribute to broader discussions in material culture, sensory studies, and cooking by foregrounding the role of changing kitchen environments in shaping how cooks achieve and experience *wok hei*.

Material culture studies have long examined how people relate to objects in domestic life. Daniel Miller, for instance, argues that the arrangement of household items reflects and reinforces identities such as gender, age, and class.[5] Meanwhile, homemaking, in his view, is a reciprocal process: while people shape their homes, the homes also shape their inhabitants. However, Miller's work tends to emphasize the symbolic and sentimental dimensions of domestic space and items, giving less attention to the functional value of kitchen objects in daily practice.

Later studies, such as those by Elizabeth Shove, highlight the coevolution of social practices and material objects.[6] Specifically, they argue that technologies not only shape users by introducing specific scripts but also transform users' behaviours and competency, as well as the norms and meanings of a certain practice. However, the discussion rarely extends to material objects' influence on sensory experiences.

Recent ethnographic work on cooking adds a pinch of 'salt' by drawing attention to the sensory and bodily dimensions of culinary practice.[7] These studies vividly illustrate how cooks develop their sensory techniques through interactions with the local social, material, and ecological world and consider how such skills become embedded in everyday routines. They offer important insights into how sensory knowledge is acquired and enacted. However, less attention has been paid to how these embodied skills are maintained – or recalibrated – when those sociomaterial conditions begin to shift.

This study builds on and extends these literatures by asking, How do Cantonese cooks continue to produce and recognize 'good taste', especially *wok hei*, in the face of rapid technological and spatial transformations in the kitchen? Rather than viewing taste as static or embodied once and for all, it is considered a practice that is constantly adjusted through sensory engagement, technical improvisation, and negotiation with changing tools and environments.

The following sections trace *wok hei* across three generations of Cantonese kitchens. The second section begins in the countryside kitchens and urban communal cooking spaces before the 1970s, where *wok hei* emerged under material constraints – imperfect woks, unpredictable flames from wood- and coal-burning stoves, and limited ingredients. These limitations, however, also gave rise to remarkable resilience and resourcefulness in home cooking.

The third section shifts the focus from domestic kitchens to the bustling *dai pai dong* (open-air food stalls) during their heyday in the 1990s and 2000s. In this period, *wok hei* was transformed beyond the smoky aroma and flavour of daily cooking into a captivating sensory culinary performance and became an important part of the vibrant culinary culture of Guangzhou.

The fourth section returns to the home, exploring how the transition to modernized, well-equipped kitchens reshaped *wok hei*-related practices. Despite the convenience and efficiency brought by new appliances, home cooks continued to value *wok hei* as essential to local taste and home-cooked meals – demonstrating how the commitment to the local sensibility endures and adapts within changing domestic kitchens.

Wok Hei Before the 1970s

Wood, Coal, and Lard

While living at my grandmother's home in Guangzhou for two weeks, I realized it was impossible to neglect the discussion of *wok hei*, as it surfaced at nearly every meal. Its frequency made it seem timeless, yet it raised a question: When did *wok hei* become central to Cantonese daily life? Seeking clues, I asked everyone about their earliest memories of *wok hei* and searched *Yangcheng Evening News* archives dating back to 1982.

Although I could not pinpoint an exact origin, an article from 2013 introduced a restaurant founded in 1927 called Jinlianji, which had been stir-frying Fujian noodles over coal fires with lard for more than 80 years. While its appearance was said to be unappealing to locals' sharp senses, its smoky *wok hei* flavour, complemented by fragrant lard, evoked nostalgia in many informants.[8]

While this was merely a restaurant review – possibly commercially influenced – it revealed key associations between *wok hei* with coal fire and lard. It also suggests that the commitment toward *wok hei* at least can be traced back to a time when coal was used as the main source of cooking. Notably, many of the people I interviewed, now in their 40s or 50s, tended to almost instinctively associate *wok hei* with *dai pai dong* – a food business that gained prominence in Guangzhou in the 1990s. Earlier generations, instead, had a different perception of *wok hei*. For those who grew up cooking with coal or eating food prepared over a coal fire, before gas stoves became widely available in home kitchens, the most authentic *wok hei* was characterized by the smoky flavour uniquely imparted by this traditional cooking method.

Pao Wok *or Not*

Conversations about *wok hei* often led my grandmother to reminisce about her early married life in Qingyuan during the 1960s, where she cooked for her patrilocal family. The hearth stove was a two-arm-long, one-arm-wide, and waist-high stage holding a massive cast-iron wok. At that time, they relied on wood collected from streets and fields, with occasional coal purchases.

'No *pao wok*'! she exclaimed.[9] In her rural kitchen, the heavy, fixed wok made lifting impossible. Instead, she used a spatula, flicking small portions of food into air to mimic the *pao wok* effect. She also pushed food to the wok's sloped edges to aid evaporation, adding pieces of wood to feed the fire. For her, this method produced the real *wok hei*.

After moving to Guangzhou in the mid-1970s, my grandmother's family lived in a small room on Gaodi Street, sharing an eight-square-metre kitchen with three other households. My aunt, only 8 or 9 years old, managed much of the family's cooking. She described juggling two stoves – one coal-burning and one wood-burning – keeping a small flame constantly alive for daily use and reserving intense heat for stir-fries. Although urban kitchens shifted from wood to coal after 1957 forest conservation policies, neither fuel was easily affordable.[10] Liao, a 55-year-old informant, recalled how her family burned discarded herbal medicine from hospitals to conserve coal, creating weak, smoky flames barely sufficient for boiling water.

With smaller family units came smaller woks – typically eighteen or twenty inches – better suited to confined kitchens and reduced meal sizes. Charcoal stoves, in contrast to the stable hearths in the past, produced what locals called 'living fire' (C. *sang fo*): a flickering, unpredictable flame that, in my aunt's words, 'jumped' frequently. This required new techniques – tilting the wok to 'catch' the changing fire, lifting it quickly when the flame surged, and performing *pao wok* to avoid scorching. However, cooking with worn-out pots and patched woks added another challenge. Many woks had cracks reinforced with wire, making tossing a matter not just of flavour but of careful control – to keep ingredients from slipping through the gaps.

Times of Shortage

Along with mastering stoves, managing groceries was a big challenge. Food scarcity dominated the memories of informants who grew up during the 1960s and 1970s. Urban households relied on government-issued food coupons, which restricted access to pork, fish, grain, and oil until reforms in the late 1980s.[11]

My aunt recalled lining up before dawn to buy fatty pork for the New Year feast. Fatty cuts were prized because they had richer flavour and families could render them into precious lard for cooking. Lard was both essential and a rare treat. One informant, Dong (58 years old), still vividly remembered the simple luxury of a bowl of rice mixed with lard and soy sauce after surgery in his childhood. Today, lard is often replaced by oils like peanut, sunflower, or olive oil, largely for health reasons. Yet many older Cantonese lament that there was something missing in the stir-fries made with types of oil other than lard.

Daily diets mostly featured vegetables stir-fried with small amounts of meat and lard. Informal food markets flourished alongside official rationing. Some families kept chickens at home for eggs, while others relied on relatives working for railway companies to procure food directly from rural areas without coupons needed. Despite these networks, diets remained limited most of the year, except during major festivals. Cooks adapted by slicing ingredients thinly to stretch meals and speed cooking; Eugene Newton Anderson and M. L. Anderson note that stir-frying, among cooking methods, maximizes flavour and portion size with minimal ingredients and fuel.[12]

Thus, the wok, the thin-sliced ingredients, the modest amount of fat, the dynamic flame, and the skilful motion of tossing all coalesced into the sensory phenomenon of *wok hei*. The prevalence of stir-frying, the emergence of *pao wok* techniques, and the meticulous control of heat were not incidental developments – they were strategies forged in response to fuel shortages, crowded living conditions, and limited food supplies during times of material scarcity.

Yet *wok hei* was never merely a product of necessity or pragmatism. It embodied craftsmanship and a spirit of making the most of what was available. Achieving *wok hei* demanded precision, skill, and sensory attention: ingredients had to be sliced uniformly, heat and timing judged carefully, and subtle seasoning used that did not overpower the dish. It was a practice of ingenuity as much as economy.

Taken together, it is safe to say that *wok hei* is not exclusive to *dai pai dong* and restaurants, which have a much shorter history compared to the local sensibility of *wok hei*. More importantly, the practice of creating *wok hei* – shaped by cooking techniques, ingredients, and cultural meanings – has gradually evolved in response to changing social and material conditions over time. This will be further illustrated by examining the context of Guangzhou in the late 1990s and early 2000s, a period marked by dramatic shifts in people's residence, cooking, and eating.

Elevated *Wok Hei* from the 1980s and the 2000s

Dai Pai Dong

In my 'excavating' project of *wok hei* in the food columns of *Yangcheng Evening News*, I found that the term rarely appeared in headlines until 1999. From that point onward, however, mentions of *wok hei* surged. Across fifty-five articles between 1999 and 2016, *wok hei* was frequently discussed and even a recurring theme in food writings, often headlining reviews and recommendations of newly opened or trendy venues in town. This surge coincided with the rise of *dai pai dong* – open-air food stalls characterized by their noisy, untidy environments; affordable dishes; and semiopen kitchens where flames and sizzling sounds became part of the dining experience.

Originating in the post-World War II period in Hong Kong, *dai pai dong* gained popularity in Guangzhou in the late 1990s and 2000s. Its success was bolstered not only by the influence of Hong Kong but also by broader economic liberalization following Deng Xiaoping's 1978 reforms, which dismantled the previous restrictions on private business and created new entrepreneurial opportunities.

In food reviews, *wok hei* was no longer simply a fleeting descriptor or synonym for 'fragrant' or 'tasty'. It was elevated into a signature in the Cantonese culinary landscape. A new phrase emerged, *huoqi xiaochao* (stir-fries with *wok hei*), suggesting that *wok hei* was not just a quality but a genre in itself. A food writer celebrated the technique and drama of cooking, offering detailed, almost cinematic descriptions of chefs at *dai pai dong*:

> The shirtless chef heats a large iron wok and pours a large spoonful of peanut oil into it, making a loud, broken roar. He tosses in rice noodles and lifts and *pao wok* with one arm nonstop. The noodles are two inches high over the wok, while the flame kisses the rim. He then drops in sprouts, onions, and beef and drizzles soy sauce from the edge of the wok.[13]

Wok hei was now inseparable from visual and auditory performance. The movement of the chef, the rhythm of the spatula, the sound of the flame – all became part of the appeal. Another article praised the mastery of flame:

> A strong flame is blue and makes crackling, popping sounds, while a weak one is yellow. If it takes five minutes to cook a dish that should be done within three minutes, the food is overcooked and loses its aroma.[14]

In fact, technological advancements in heat devices further deepened the divide between domestic and professional cooking. While both homes and restaurants once relied on coal or firewood, the introduction of gas stoves created new benchmarks for heat intensity and control. Commercial burners, capable of reaching 15,000 to 30,000 BTUs, far surpassed the 8000 to 14,000 BTUs of domestic models. This power not only enabled faster cooking but also allowed for more dramatic wok movements and flame effects – culinary theatrics that, as vividly captured in the food columns, became a hallmark of *dai pai dong* but remained difficult, if not impossible, to reproduce in the average home kitchen.

In addition, the semiopen design of *dai pai dong* kitchens allowed diners to appreciate the cooking process. Unlike formal restaurants that emphasized decor, *dai pai dong* offered spectacle: the clang between metal wok and specula, the sudden bursts of flame, sizzling oil, and the lively choreography of chefs in motion. This performative cooking style drew eaters not only for flavour but for the thrill of watching *wok hei* made visible, audible, and aromatic. At *dai pai dong*, *wok hei* was elevated from a sensory quality to a full multisensory experience and an embodiment of Cantonese culinary pride.

The Best Formula for Stir-Fry King

Most stir-fried dishes have no fixed formula. Cooks often made the most of what they had on hand – a culinary tradition shaped by necessity during periods of scarcity. Yet many of these once-humble home dishes have since been reimagined by *dai pai dong* chefs into complex and refined creations.

Take Stir-Fry King (C. *siu caau wong*), for instance. Originating in Shunde, this dish was born in the domestic kitchen as a thrifty, flexible stir-fry that made creative use of seasonal and preserved ingredients. It typically featured vegetables and freshwater seafood such as small shrimp, clam meat, and squid – often dried to concentrate their umami and extend shelf life.[15]

With urban food scarcity fading in the postrationing era, *dai pai dong* and restaurants sought to elevate the Stir-Fry King into a more delicate and standard dish.[16] Ingredient selection became more prescriptive: overly moist vegetables were said to make the dish soggy and diminish its flavour, while overly dry ones risked burning. According to these standards, chives emerged as the quintessential ingredient for their balanced moisture and aromatic quality, and therefore, chives became a staple in nearly every restaurant version.[17]

Chef Deng, for example, shared his signature Stir-Fry King recipe in the *Yangcheng Evening News*, featuring a broader variety of ingredients than the typical home version: dried silver fish, dried shrimp, chives, bean sprouts, celery, and cashew nuts. For him, the well-balanced texture of tenderness, crispness, and crunchiness – achieved through both precise heat management and thoughtful composition – was the key to a successful stir-fry.[18] In this sense, *wok hei* became something not only to taste but also to calibrate – to measure, test, and perfect.

During the 2000s, *wok hei* has been both elevated and standardized – transformed from a flexible, improvisational practice rooted in everyday constraints into an iconic feature of Cantonese cuisine. No longer just the invisible 'breath of the wok', it became a full-blown culinary performance – defined by open kitchens; high-powered burners; choreographed, theatrical routines performed with exaggerated flair; and a fixed formula of dishes. In this process, *wok hei* was not only celebrated but also codified and refined into a reproducible standard by the dining business.

Yet *wok hei* is still important in domestic cooking. The next section turns to the domestic sphere to explore how home cooks pursue *wok hei* – not through spectacle but through tools, embodied techniques, memory, and adaptation.

Enduring *Wok Hei* from the 1990s to the 2020s

A Precise Control of Heat

As Terence Conran suggests, it is the kitchen that serves the most compelling example, more than any other room within a home, of the great socioeconomic changes occurring outside the domestic sphere.[19] Building on Conran's observation, it is equally important to examine how broader changes and the new, modern kitchen inform cooking practice and taste. This section examines how *wok hei* has been reconfigured in Guangzhou's urban kitchens from the late 1990s through the 2010s, particularly in relation to evolving cooking tools, domestic expectations, and contrasting ideals between commercial and home cooking.

The turn of the century marked significant improvements in economic and residential conditions in Guangzhou. Following housing reforms in the 1980s and the rise of private homeownership in the 1990s, more urban residents moved into self-contained flats with dedicated kitchens. By 2006, the average urban kitchen in Guangzhou exceeded eight square metres.[20] Modelled after the Frankfurt kitchen, urban kitchens not only were bigger but also had better efficiency, workflow, and appliance integration.[21]

Alongside spatial transformation, Guangzhou's energy supply also changed. By the 1990s, gas penetration reached over 40.9 percent, enabling more precise control of heat.[22]

This transition in cooking fuel was evident not only in the *Yangcheng Evening News*'s advertisements about gas cooktops and gas safety devices (to prevent dangerous accidents caused by gas leaks) but also in cooking instructions for home cooks. In the late 1990s, many recipes included precise requirements for heat, with high heat, medium heat, medium-high heat, and low heat corresponding to different ingredients. For instance, a recipe for *Huangpu dan* (*Huangpu* omelette) emphasized the importance of gentle heat and patience in making this simple home-style dish:

> Heat the pan first, add lard, and move it around the wok to coat the bottom of the wok evenly. Then, it is important to turn down [*zhuan di*] the heat and slowly pour in the egg mixture, while pushing the liquid toward the large slope of the wok with a spatula until the egg liquid thickens and sets.[23]

Similarly, low temperature and careful attention were also the key to infuse a smokiness and *wok hei* with the sauce.[24] Vegetable stir-fries, on the contrary, were said to require high heat and a generous amount of oil in order to bring out the flavour of the aromatics.[25]

In practice, however, managing heat remained challenging. It was more than just switching the knob. Mei, an experienced home cook in her 50s, explained that controlling flame required listening as much as looking.

She put a wok over high heat first while continuously moving it to make the heat distribute evenly. She then reduced the heat to a moderate level for a few seconds and raised it back to high heat again before turning off the heat and immediately adding ginger slices and garlic. I could hear a series of crispy, brittle, sizzling sounds like tiny bubbles that kept bursting. It was the right sound for Mei. Otherwise, the sizzling would drag on and be loud because the water did not evaporate fast enough when the ingredients hit the wok, she explained.

Through decades of cooking, including with coal stoves, Mei developed an intuitive sense of timing and temperature. Gas stoves, though more responsive, raised the expectations of cooks' mastery of timing and heat. Experienced cooks like Mei were asked to blend traditional skills honed on coal stoves with modern precision. This allowed her to perfectly time the addition of aromatics, main ingredients, and sauces to create a rich flavour infused with smoky goodness.

Wok Hei *in a Gadget-Filled Kitchen*

While a variety of modern kitchen tools have transformed the tempo and efficiency of cooking, they have not replaced the central role of stir-frying or diminished the pursuit of *wok hei*. In fact, as kitchens have become more advanced, expectations around timing, sequencing, and heat control have only grown – not just for perfecting a single dish but for coordinating multiple dishes to be served together at their best.

Compared to my grandparents' generation, when meals typically featured one stir-fried and one steamed dish – made possible by juggling a coal- and a wood-burning stove – today's home-cooked meals often include three or more dishes, plus a slow-simmered soup. While gastronomes may visit *dai pai dong* solely to savour its famous stir-fries, home meals typically include a variety of dishes that complement one another in flavour and cooking style. Stir-fried dishes are tempered by milder steamed, boiled, or cold dishes to offset the internal 'heat' (*re qi*) associated with intense, oily cooking.[26] *Wok hei*, then, is no longer just about flash and flame; at home, it involves coordination and wholesomeness.

Those evolving expectations around cooking and meal preparation were a result of technological advancement at home. Televisions, fans, washing machines, and refrigerators were the first to be adopted into the homes of my informants. Microwaves followed, though their initial reception was uneasy. Between 1994 and 1996, several articles in the *Yangcheng Evening News* discussed why microwaves were unpopular in the Guangzhou market, despite Guangdong province being the major production site. The underlying reason was Cantonese people's strong attachment to *wok hei*.

Microwaves faced many criticisms. For example, publications said 'food cooked by microwaves [was] incompatible with that cooked in "living fire" [C. *sang fo*], in terms of the colour, the aroma and flavour', and they cited the microwave as a less appealing cooking tool, which '[made] food with zero *wok hei*'.[27] In other words, these sentiments underscored the significance of *wok hei* in home-cooked meals, where the presence of *wok hei* was clearly valued more highly than convenience or efficiency.

A modest shift in public perception began in 1997, when a series of microwave recipes appeared in the food column sponsored by Guangdong-based appliance manufacturer Galanz. By highlighting microwaves' functions of baking, stewing, reheating, and steaming and offering recipes of Western-style cakes and hamburgers and non-Cantonese dishes, microwaves were repositioned not as rivals but as helpers to the main cooking task.

Indeed, today's urban kitchens are more spacious, well lit, and better equipped, while the design of the workflow still revolves around the stove. Other appliances also play supporting roles in the culinary choreography of a multidish meal: rice cookers quietly ensure perfectly timed rice. Electric kettles rapidly boil water for blanching vegetables (before being stir-fried) or making soup. Electric steamers, often combined with microwaves and ovens, allow for hands-free preparation in steaming fishes or reheating broths. Powerful range hoods, now a standard feature, extract smoke and oil vapours produced by stir-frying. Sliding glass doors, added to narrow U-shaped or I-shaped semiopen kitchens, help isolate cooking smells from the rest of the home.

These tools enhance sequencing, precision, speed, and cleanliness – but they do not diminish the significance of stir-frying or the pursuit of *wok hei*. Instead, they have been integrated into everyday cooking routines and allow home cooks to focus on making stir-fries. With such coordination, multiple dishes can arrive hot and perfectly timed to the dining table. In this sense, *wok hei* remains the centre of the kitchen and the meal. Modern appliances support timing and multitasking, but the gas stove and the wok still define the taste, rhythm, and spirit of a Cantonese home-cooked meal.

Ephemeral *Wok Hei*

One informant, Fang, told me about a trending topic on social media: '*wok hei* essence' (*huoqi xiangjing*), a flavour additive claiming to replicate the aroma of stir-fries. 'It's ridiculous and funny', Fang said with a mix of humour and unease. Though he doubted it could truly mimic real *wok hei*, the very idea unsettled him. 'If it works, what else are we eating that's not real?' he asked.

His frustration echoed a central question in my research. *Wok hei* – so dependent on freshness, timing, intense heat, the cooks' embodied skills, and the immediacy of serving – is inherently unstable, unpredictable, and difficult to reproduce. Paradoxically, its magic lies precisely in its fragility. The fact that it cannot be guaranteed every time has made its presence more cherished, even nostalgic – intensifying people's emotional attachment to it.

This elusive quality is rooted in the resourcefulness of earlier generations of Cantonese cooks, who, amid scarcity, made the most of what they had through improvisation and embodied know-how.

However, as Guangzhou's kitchens and food systems have undergone decades of sociomaterial transformation, the practice of *wok hei* has also evolved. Notably, it has at times drifted toward recipe-ization and standardization. The rise of prescriptive stir-fry formulas, high-efficiency and more precise kitchen tools, and highly choreographed cooking routines reflects a collective attempt to make *wok hei* more repeatable, more accessible, and even more 'master-able'.

Yet some aspects of *wok hei* resist this transformation. Essential steps – bringing the wok to the right temperature, catching the flame, executing a well-timed *pao wok*, adjusting seasoning – remain irreducibly sensory, embodied, and intuitive. These are not techniques that can be fully codified or outsourced to machines.

It is, therefore, reasonable to conclude that *wok hei* has persisted and permeated not in spite of but because of modern transformation. Its meanings and practices continue to be redefined in response to changing kitchens, tools, and values. Nevertheless, one thing seems to hold constant: *wok hei* ultimately defies being fully measured, replicated, or disembodied.

About the Author

Zien Zhi is a master's student in cultural anthropology at the University of Tokyo, focusing on sensory anthropology and Cantonese food culture.

Notes

1 Some sources refer to it as *wok hay*, but both *wok hei* and *wok hay* refer to the same concept. This paper uses *wok hei*, which is more common in English-language food writing.

2 Cantonese terms and sayings mentioned in this paper are indicated by 'C.', followed by their Cantonese pronunciation. All translations are done by me with the assistance of online Cantonese dictionaries, unless otherwise cited. Other Chinese terms without the 'C.' indication are in Mandarin Chinese.

3 A cooking vessel with a big slope and a rounded bottom.

4 My turn to local media was inspired by Jacob Klein's study of gastronomic writing in Guangzhou's daily press (*Changing Tastes in Guangzhou: Restaurant Writings in the Late 1990s* (Routledge, 2006)), which highlighted *Yangcheng Evening News* as the first to dedicate a regular section to food in the mid-1990s. Launched in 1957 and widely circulated by the 1990s, the newspaper offered a rich site for tracing public discourse on food and taste. This research focuses on its food columns from 1996 to 2016, when such writing flourished before gradually declining.

5 See Daniel Miller, 'Appropriating the State on the Council Estate', *Man*, 23 (1988), pp. 353–72; *Home Possessions: Material Culture behind Closed Doors*, ed. by Daniel Miller (Routledge, 2001).

6 See Elizabeth Shove and others, *The Design of Everyday Life* (Berg, 2007).

7 For examples, see David E. Sutton, *Secrets from the Greek Kitchen: Cooking, Skill, and Everyday Life on an Aegean Island* (University of California Press, 2014); and Katharina Graf, *Food and Families in the Making: Knowledge Reproduction and Political Economy of Cooking in Morocco* (Berghahn, 2024).

8 Yu Jian and Chen Shaokang, 'Zhuyou, Tanhuo: Chao Chu 80 Nian Lishi de Fujian Mian' ('Lard and Charcoal: Stir-Frying Fujian Noodles with 80 Years of History'), *Yangcheng Evening News*, 19 October 2013, B13; see also Liang Xuhua, 'Guang KL, Tiao Tiao Dalu You Zhenwei' ('Walking through Kuala Lumpur, Every Street Has Its Own Delicacy'), *Yangcheng Evening News*, 9 March 2013, B14; Liang Xuhua, 'Zhoufenmianfan: Didiao Shenjia, Neihan Shizu' ('Congee, Noodles and Rice: Humble in Value, Rich in Flavour'), *Yangcheng Evening News*, 8 March 2014, B13.

9 *Pao wok* is a cooking technique involving jerking the wok in a small circular motion to toss the food, and this technique is widely seen in *dai pai dong*.

10 Guangzhou Shizhi Editorial Committee, *Guangzhou Shizhi 1979–2000* (*Guangzhou Yearbook 1979–2000*), vol. 5 (part 2) (Guangzhou Publishing House, 1998).

11 Guangzhou Shizhi Editorial Committee, *Guangzhou Shizhi 1979–2000* (*Guangzhou Yearbook 1979–2000*), vol. 6 (Guangzhou Publishing House, 1996).

12 Eugene Newton Anderson and M. L. Anderson, 'Modern China: South', in *Food in Chinese Culture: Anthropological and Historical Perspective*, ed. by Kwang-chih Chang (Yale University Press, 1977), pp. 317–82.

13 Liang Xuhua, 'Zhoufenmianfan.'

14 Liu Xingtong, 'Jiangchang Yuecai: Jiulou de 'Xunwei' Zhuzhang' ('Home-style Cantonese Cuisine: The Search for Flavour in Jiulou'), *Yangcheng Evening News*, 9 January 2011, B13.

15 Wang Qi, 'Xipin Guangzhou de Fengcheng Ziwei' ('Savouring the Flavors of Fengcheng in Guangzhou'), *Yangcheng Evening News*, 30 June 2007, B15.

16 Before the 1980s, food in Guangzhou was strictly controlled under the rationing policy, as discussed. However, the abolition of the food rationing policy in 1985 gradually led to a more open market for meat, vegetables, condiments, and other food items and gave urban consumers access to better and more diverse food options (Guangzhou Shizhi Editorial Committee, *Guangzhou Shizhi*, vol. 6, pp. 257–58, 263–64, 284).

17 Wang Min, 'Xiaochaowang Huoqi Zu' ('Stir-Fry King Full of *Wok Hei*'), *Yangcheng Evening News*, 3 July 2010, B9.

18 Wang Min, 'Xiaochaowang Xiangqibiren' ('Stir-Fry King with an Irresistible Aroma'), *Yangcheng Evening News*, 23 February 2016, B8.

19 Terence Conran, *The Kitchen Book* (Crown, 1983), p. 1.

20 Zhou Yanmin and Lin Jingyi, 'Woguo Chengshi Zhuzhai Chufang de Yanjin Licheng Yu Weilai fazhan qushi' ('Urban Residential Kitchen Evolution and Future Trends'), *Zhuangshi* (*Arts & Design*), 11 (2010), pp. 20–25.

21 Ran Tao and others, 'Exploring the New Kitchen Layout in China by Analyzing the Cooking Route and Cuisine Culture', *Journal of the Science of Design*, 3.1 (2019), pp. 45–52.

22 Guangzhou Shizhi Editorial Committee, *Guangzhou Shizhi*, vol. 5, pp. 747–48.

23 Fan Yanxi, 'Jiandan de Mingcai' ('Simple but Famous Dish'), *Yangcheng Evening News*, 31 December 1996, p. 28.

24 Zhu Zhu, 'Hunjiang Chu, Baiwei Sheng' ('From a Mix of Sauces, a Hundred Flavors Arise'), *Yangcheng Evening News*, 2 January 1999, p. 20.

25 Hu Xueming, 'Meishi Xinxiang' ('Gourmet Mailbox'), *Yangcheng Evening News*, 30 January 1999, p. 24.

26 A typical Cantonese home meal does not rely on stir-fries alone. While at least one high-heat, wok-fired dish is often present, it is balanced by a range of milder offerings – steamed, boiled, or cold-cut dishes – that help temper the internal 'heat' (*re qi*) produced by stir-frying. This combination reflects not only nutritional and sensory balance but also a cultural emphasis on harmony in everyday dining.

27 Wan Changyan, Yin Yingxia, and Li Yuan, 'Guangdong Shi Dianre Qiju de Zhuyao Shengchandi, Weibolu Weihe Zai Guangzhou Shou Lengluo?' ('Guangdong Is a Major Producer of Electric Heating Appliances, but Why Are Microwaves Unpopular in Guangzhou?'), *Yangcheng Evening News*, 24 June 1994, p. 8; Tang Xinzi, 'Bianlu Suixiang' ('Thoughts on the Hotpot'), *Yangcheng Evening News*, 12 December 1996, p. 18.